Thierry of Chartres

The Commentary on the *De arithmetica* of Boethius

Unlike the commentaries on Plato's *Timaeus* and on Boethius's *Consolatio philosophiae*, medieval exegesis of Boethius's *De arithmetica* has seldom been subjected to comprehensive and systematic enquiry. Inhabiting the shifting boundary between philosophy and history of science, the *De arithmetica* itself has been neglected by most medievalists. Yet, from the Carolingian renaissance onward, when the scholarly curriculum came to be based on the seven liberal arts, Boethius's work soon became a canonical text for the study of arithmetic. Indeed, the growing interest in it during the twelfth century is attested by the large number of commentaries surviving in manuscript.

The commentary on the *De arithmetica* preserved in Stuttgart, Württembergische Landesbibliothek, Cod. math. 4° 33 and edited here for the first time can be securely attributed to Thierry of Chartres. It belongs to a phase when the Chartrian master's interests were mainly directed toward the liberal arts. We can also discern in Thierry's commentary on the *De arithmetica* themes and problems developed in his *Tractatus de sex dierum operibus* and more elaborately in his commentaries on Boethius's *Opuscula sacra*. Indeed, the discovery of this commentary on the *De arithmetica* might legitimately be said to clarify not only the more intractable passages in Thierry's theological writings but also to illuminate his philosophical project as a whole. At the heart of that vision is a developing trend in twelfth-century philosophy that places number and proportion at the heart of the physical cosmos. In this profoundly 'mathematical Platonism,' all things are based on number and follow the rule of number; or, to quote Thierry himself, "creatio numerorum, rerum est creatio."

STUDIES AND TEXTS 191

Thierry of Chartres

The Commentary on the *De arithmetica* of Boethius

Edited with an Introduction by
IRENE CAIAZZO

PONTIFICAL INSTITUTE OF MEDIAEVAL STUDIES

Library and Archives Canada Cataloguing in Publication

Thierry, de Chartres, approximately 1100–approximately 1150
[Commentum super Arithmeticam Boethii]
The commentary on the De arithmetica of Boethius / Thierry of Chartres ; edited with an introduction by Irene Caiazzo.

(Studies and texts ; 191)
The text of the commentary on Boethius's De arithmetica edited here is based on the sole surviving manuscript, Stuttgart, Württembergische Landesbibliothek, Cod. math. 4° 33, fols. 1ra–34ra.
Original Latin title of commentary: Commentum super Arithmeticam Boethii.
Includes bibliographical references and index.
Issued in print and electronic formats.
Text in Latin; introduction and notes in English.
ISBN 978-0-88844-191-1 (bound). – ISBN 978-1-77110-361-9 (pdf)

1. Boethius, –524. De institutione arithmetica. 2. Boethius, –524 – Manuscripts. 3. Arithmetic – Early works to 1800. 4. Science, Medieval. 5. Philosophy, Medieval. 6. Education, Medieval – France – Chartres. 7. Ecole de Chartres. 8. Boethius, –524 – Influence. 9. Thierry, de Chartres, approximately 1100–approximately 1150. 10. Württembergische Landesbibliothek. Manuscript. Cod. math. 4° 33, fols. 1ra–34ra. I. Caiazzo, Irene, editor II. Pontifical Institute of Mediaeval Studies, issuing body III. Württembergische Landesbibliothek. Manuscript. Cod. math. 4° 33, fols. 1ra–34ra IV. Title. V. Title: De arithmetica of Boethius. VI. Series: Studies and texts (Pontifical Institute of Mediaeval Studies) ; 191

PA6231.A73T55 2015 513 C2014-908503-6 C2014-908504-4

59 Queen's Park Crescent East
Toronto, Ontario M5S 2C4
Canada
www.pims.ca

MANUFACTURED IN CANADA

"Ad pedes eius prima artis huius rudimenta accepi."

Paulo Lucentini, amatissimo magistro meo
(Macerata, 1937–Florence, 2011)

Contents

Preface

The works of Thierry of Chartres, that "most learned of explorers in the liberal arts" in the twelfth century, have long been bound up with the scholarly life of the Pontifical Institute of Mediaeval Studies in Toronto. His principal texts, the commentaries on Boethius's *Opuscula sacra,* edited by Nikolaus Häring, and on Cicero's *De inventione* and the *Ad Herennium,* edited by Karin Fredborg, were published in the Institute's signature series of Studies and Texts. Edouard Jeauneau has taught there for many years, and his pioneering studies of Thierry's *Heptateuchon* and on 'mathematics and theology' have proved important milestones in the historiography of twelfth-century thought. Several years ago, I began studying medieval commentaries on Boethius's *De arithmetica,* which had hitherto never been subjected to comprehensive and systematic enquiry in the way the commentaries on Plato's *Timaeus* and on Boethius's *Consolatio Philosophiae,* by contrast, were. Boethius's work inhabits the shifting boundary between philosophy and history of science and, as a consequence, has been neglected by most medievalists. Yet, from the Carolingian renaissance onward, the scholarly *curriculum* was based on the seven liberal arts, and Boethius's *De arithmetica* soon became a canonical text for the study of arithmetic, and the growing interest in the work during the twelfth century is attested by the large number of surviving commentaries. The commentary on the *De arithmetica* preserved in Stuttgart, Württembergische Landesbibliothek, Cod. math. 4° 33 and edited here for the first time can be attributed with certainty to Thierry of Chartres. It belongs to a phase when the Chartrian master's interests were mainly directed toward the liberal arts; however, we can also distinguish in it themes and problems that were abundantly developed in his commentaries on Boethius's *Opuscula sacra.* Indeed, thanks in large measure to the discovery of the commentary on the *De arithmetica,* some of the more intractable or enigmatic pages in the theological writings, as well as from the *Tractatus de sex dierum operibus,* have become more intelligible, and Thierry's whole philosophical project has gained greater clarity. I hope this volume will also further understanding of a developing trend in twelfth-century philosophy, that puts number and proportion at the heart of the physical cosmos. In this profoundly 'mathematical Platonism,' everything is based on number and follows the rule of number; to quote Thierry himself: "creatio numerorum, rerum est creatio."

A French first draft of this book was part of the dossier submitted to the Ecole pratique des Hautes Etudes, Paris in December 2012 in fulfillment of the requirements of the "Habilitation à diriger des recherches." Professor Danielle Jacquart was my mentor and I would especially like to thank her for her penetrating critique and also her suggestions. She has always encouraged me to explore new avenues, although she was a little surprised when I decided to send my manuscript to Toronto.

But in Toronto it has also found its proper home. I am grateful to the Pontifical Institute of Mediaeval Studies for the warm welcome it has shown me, in particular to Edouard Jeauneau, who has supported and guided my work with unfailing kindness, and to Fred Unwalla, Editor in chief, who has helped me in every way. Many thanks as well to the three anonymous referees who appraised my manuscript with great skill and patience; they made significant corrections and suggested numerous ways in which I might improve the text. I am also indebted to a long list of colleagues and friends. Barbara Obrist, who read the last draft of my French manuscript with the critical eye that is the mark of true friendship, helped me to find a reliable translator, Rachel Tapley from the University of Wisconsin. William J. Courtenay, Michael Chase, Fred Unwalla, and his editorial staff helped further refine the introduction and the text. Dominique Poirel generously agreed to review the final draft of the Latin edition, which has benefited considerably from his rigorous scrutiny. I must not forget here Charles Burnett, David Juste, Claudia Rabel, Sabine Rommevaux, and Patricia Stirnemann for their many thoughtful observations and suggestions. Finally, I would like to express appreciation to the Laboratoire pour l'étude des monothéismes, my *équipe de recherche* at the Centre national de la recherche scientifique, which graciously financed the English translation; my warm thanks to Olivier Boulnois, who was its director until March 2014, and also to Félicia Yuste, who resolved several administrative dilemmas.

My parents, my husband Xavier, and my son Emile, have supported my work with patience and love; my gratitude to them is boundless and, ultimately, inexpressible in words.

Paris, December 2014

Abbreviations

AHDLMA	*Archives d'histoire doctrinale et littéraire du Moyen Age*
CCCM	Corpus Christianorum, Continuatio Mediaevalis. Turnhout: Brepols, 1971– .
CCL	Corpus Christianorum. Series Latina.Turnhout: Brepols, 1953– .
CSEL	Corpus scriptorum ecclesiasticorum latinorum. Vienna: Holder– Pichler – Tempsky, 1866– .
MGH	Monumenta Germaniae Historica.
SS	Scriptores. Hannover: Hahn, 1826.
SS rer.Germ.	Scriptores rerum Germanicarum in usum scholarum separatim editi. Hannover: Hahn, 1871– .
PL	Patrologiae cursus completus. Series Latina, ed. J.P. Migne. 221 vols. Paris, 1844–1864.

Thierry of Chartres

The Commentary on the *De arithmetica* of Boethius

Introduction

Thierry of Chartres and the School of Chartres

Every cathedral school should have its own Alexandre Clerval! It is thanks to the meticulous scholarly studies of this priest in the diocese of Chartres (who was born in 1859 and died in 1918) and who then became titular canon – as well as to the works of Reginald Lane Poole and Barthélémy Hauréau – that the school attached to the cathedral of Chartres became so important in medieval historiography and had so much written about it throughout the twentieth century.[1]

There are multiple reasons for this success. First, the cathedral itself is still standing and some parts of it date back to the twelfth century. Its royal portal with the representation of the seven liberal arts is a monument, both literally and figuratively, to the grandeur of the cathedral school and, above all, is visible and tangible testimony to the scholarly teaching of the Middle Ages. Edouard Jeauneau and others have devoted many pages to this portal and the representation of the seven liberal arts, and it is thus unnecessary to repeat their observations here.[2] Another reason for the renown of the school of Chartres lies in the indisputable talent of the masters once associated with it. Some historians have tried to diminish the significance of the school of Chartres, but the fact remains that Fulbert of Chartres, Thierry of Chartres, William of Conches, Gilbert of Poitiers, and John of Salisbury (to cite just a few) were remarkable thinkers and writers.

Still, there has been no shortage of serious attempts to dismantle or at least weaken the school's prestige. Richard Southern notably called into

1 See Alexandre Clerval, *Les écoles de Chartres au Moyen Age du Ve au XVIe siècle* (Chartres, 1895); Reginald Lane Poole, *Illustrations of the History of Medieval Thought and Learning*, 2nd ed. (London, 1920); Reginald Lane Poole, "The Masters of the Schools at Paris and Chartres in John of Salisbury's Time," *English Historical Review* 35 (1920), 321–342; and Barthélemy Hauréau, *Histoire de la philosophie scolastique*, 2 vols. (Paris, 1872).

2 Edouard Jeauneau, *L'âge d'or des écoles de Chartres* (Chartres, 1995); trans. Claude Paul Desmarais as *Rethinking the School of Chartres* (Toronto, 2009). On royal portal, see Adolf Katzenellenbogen, *The Sculptural Programs of Chartres Cathedral: Christ, Mary, Ecclesia* (Baltimore, 1959); Robert Branner, *Chartres Cathedral* (London, 1969); Willibald Sauerländer, *Das Königsportal in Chartres: Heilsgeschichte und Lebenswirklichkeit* (Frankfurt, 1984).

question the importance of the school of Chartres in the twelfth century in a book published in 1970, arguing that very few surviving documents speak of the actual presence of all these great thinkers and their teaching in Chartres, a small provincial city compared to Paris in the first half of the twelfth century. Asserting, with reason, that the school's renown in this era before the birth of universities stemmed from a few exceptional figures, Southern questioned the presence of masters such as William of Conches and Thierry of Chartres, whose teaching he situated in Paris rather than in Chartres.[3]

There was no lack of responses: Nikolaus Häring took stock of new textual discoveries relating to the school of Chartres unknown to Southern. Roberto Giacone patiently studied cartularies and registers for evidence of the masters, of the *magistri* – recognizing that it was only in the 1160s that the *licentia docendi* was instituted by Pope Alexander III, and that before this date the term *magister* designated anyone who had finished a cursus of studies in the liberal arts, without necessarily implying a teaching post.[4]

Most modern scholars who have sought to determine the impact of the school of Chartres in the twelfth century have based their arguments on John of Salisbury's account in the *Metalogicon* of the studies he completed in France during his youth. His statements, which are not without ambiguities of their own, have been pulled in one direction or another by historians seeking either to emphasize or to weaken the role played by the cathedral school of Chartres. For example, Katharine Keats-Rohan argues that John of Salisbury would have spent time with and listened to the greatest masters of his time in Paris, between 1136 and 1147.[5] In 1138, he "moved" to follow William of Conches, with whom he stayed for three

3 Among the works of Richard W. Southern, see: *Medieval Humanism and Other Essays* (New York, 1970; 2nd ed., Oxford, 1984); *Platonism, Scholastic Method and the School of Chartres* (Reading, 1979); "The Schools of Paris and the School of Chartres," in *Renaissance and Renewal in the Twelfth Century*, ed. Robert L. Benson and Giles Constable (Cambridge, MA, 1982; repr. Toronto, 1991), pp. 113–137.

4 Peter Dronke, "New Approaches to the School of Chartres," *Anuario de estudios medievales* 6 (1969), 117–140 [despite the date of publication, it is indeed a response to Richard Southern]; Nikolaus M. Häring, "Chartres and Paris Revisited," in *Essays in Honour of A.C. Pegis*, ed. J. Reginald O'Donnell (Toronto, 1974), pp. 268–329; Roberto Giacone, "Masters, Books and Library at Chartres According to the Cartularies of Notre-Dame and Saint-Père," *Vivarium* 12 (1974), 30–51.

5 Katharine S.B. Keats-Rohan, "John of Salisbury and Education in Twelfth Century Paris from the Account of his *Metalogicon*," *History of Universities* 6 (1986–1987), 1–45; "The Chronology of John of Salisbury's Studies in France: A Reading of *Metalogicon* II.10," *Studi Medievali*, 3rd ser., 28 (1987), 193–203. See also David Bloch's recent overview, *John of Salisbury on Aristotelian Science* (Turnhout, 2012), pp. 1–25.

years, until 1141. Agreeing with Southern, Keats-Rohan contends that John would have moved outside of Paris, that is to say, merely outside the *île de la cité*.[6] After carefully reconstructing the chronology, Olga Weijers concluded that it was indeed in Chartres that John of Salisbury was taught by William of Conches.[7]

Beyond questions of "geography," which are extremely difficult to answer definitively, we must recognize that Southern's critiques – he returned to the question of the school of Chartres several times[8] – have also proved beneficial. His scepticism has compelled us to reflect on what is specifically "Chartrian" and, consequently, to be circumspect in using the term. It is true that certain scholars use "Chartrian" too broadly, applying it to any author in the twelfth century who writes in elegant Latin and demonstrates a good knowledge of the classics and a predilection (even a moderate one) for Plato's *Timaeus*.

It seems reasonable to define as Chartrian an author who passed physically, either as a master or as a student, through the cathedral school of Chartres, as well as an author who openly proclaims an affiliation or an intellectual link with a master from Chartres. From this point of view, the cases of William of Conches and Thierry of Chartres deserve to be subjected to a new examination in light of recent research and previously unknown documents. Indeed, and fortunately, the *Metalogicon* is not the only source available to us for reconstructing their careers – I will come back to John of Salisbury and Thierry of Chartres at a later point. To Southern's question about the place where these masters would have taught, I would reply, with Edouard Jeauneau, that as far as William of Conches is concerned, there are no arguments for situating his teaching in Paris rather than Chartres.[9] In fact, when we take into account especially William's hitherto unpublished *Glosses*

6 John of Salisbury, *Metalogicon* 2.10 (CCCM 98: 71): "… me ad grammaticum de Conchis transtuli, ipsumque triennio docentem audivi." There has been so much written about "me transtuli" and where John of Salisbury would have listened to William of Conches. Richard Southern rightly remarks that Paris was quite small in the first half of the twelfth century: what is now the Left Bank was considered to be outside of Paris.

7 Olga Weijers, "The Chronology of John of Salisbury's Studies in France (*Metalogicon*, II.10)," in *The World of John of Salisbury*, ed. Michael Wilks (Oxford, 1984), pp. 109–116.

8 See *Foundations* and *The Heroic Age*, the two volumes comprising Richard W. Southern's *Scholastic Humanism and the Unification of Europe*, 2 vols. (Oxford, 1995–2001); cf. John Marenbon's review of *Foundations* in *International Journal of the Classical Tradition* 6 (2000), 569–577.

9 Jeauneau, *Rethinking the School of Chartres*, pp. 24–26.

on Priscian, the burden of evidence clearly points to Chartres.[10] I would add that even if John of Salisbury had attended William of Conches's lectures in Paris during the years 1138 to 1141, this does not mean that William did not teach at Chartres before that time. We know that masters moved around easily in the first half of the twelfth century; we find the same William at the court of the Duke of Normandy in 1147. Moreover, in the case of Thierry of Chartres, the links with Chartres and its cathedral school seem very solid, as will be shown in the pages that follow.

Little is known about the life of Thierry of Chartres, and the few details we do possess are often vague. The collective labours of the great historians of the school of Chartres, such as Alexandre Clerval, Peter Dronke, Nikolaus Häring, Edouard Jeauneau, André Vernet, and Karin M. Fredborg, on whose work I rely in the exposition below, have allowed us to sketch Thierry's biography and intellectual trajectory. We will attempt to keep to the principal facts, the authenticity of which has been proven or is generally accepted.

Born at the end of the eleventh century, somewhere in Brittany, as he clearly notes in his *Commentary on Cicero's "De inventione"* – *Ecce Theodoricus Brito, homo barbarae nationis*[11] – Thierry's reputation is impressive among his contemporaries. He had many students who speak of him in terms full of praise. Clarembald of Arras, to whom we will return in greater detail later, states that Thierry is *totius Europae philosophorum praecipuus.*[12] Hermann of Carinthia dedicates his translation of Ptolemy's *Planisphaerium*, completed

10 Edouard Jeauneau, "Deux rédactions des gloses de Guillaume de Conches sur Priscien," *Recherches de théologie ancienne et médiévale* 27 (1960), 212–247, at 231–232, repr. in Edouard Jeauneau, *Lectio philosophorum: Recherches sur l'école de Chartres* (Amsterdam, 1973), pp. 354–355; Karin M. Fredborg, "Some Notes on the Grammar of William of Conches," *Cahiers de l'Institut du Moyen Age grec et latin de Copenhague* 37 (1981), 23–24; Jeauneau, *Rethinking the School of Chartres*, pp. 44–45. Several references to the city of Chartres are found in the first redaction of William of Conches's *Glosae super Priscianum*; they have since been reunited and commented on in William of Conches, *Glosae super Platonem*, ed. Edouard Jeauneau (Turnhout, 2006), pp. xxi–xxiv.

11 *The Latin Rhetorical Commentaries by Thierry of Chartres*, ed. Karin M. Fredborg (Toronto, 1988), p. 107, l. 16. Thierry of Chartres writes his "self-portrait"; see ibid., pp. 107–108; English translation in Peter Dronke, "Thierry of Chartres," in *A History of Twelfth-Century Western Philosophy*, ed. Peter Dronke (Cambridge, 1988), pp. 358–385, at 363.

12 Clarembald of Arras, *Epistola ad Dominam* 3, ed. Nikolaus M. Häring, *Life and Works of Clarembald of Arras* (Toronto, 1965), p. 226.

in 1143, to Thierry,[13] and Bernard Silvestris dedicates his *Cosmographia* in 1147–1148.[14] Vernet had called attention to a rubric in a manuscript of the *Cosmographia* which clearly indicates the addressee of the work was the chancellor and archdeacon Thierry of Chartres: "Ad virum litteratissimum et philosophantium amantissimum magistrum Therricum Carnotensis ecclesiae cancellarium et archidiaconum."[15] This rubric agrees with what is reported in the obituary published in the *Cartulary* of the church of Notre-Dame de Chartres, cited later, and in the verse epitaph discovered by Vernet, in which the last couplet reads:

> Doctorem, protholevitam simul et logothetem
> Hunc habuit Cartis vix habitua parem.[16]

Chartres and Brittany also appear in another *Epitaphium Theoderici,* previously published by Wilhelm Wattenbach, after a manuscript from the second half of the twelfth century:

> Emicuit nato Britannia, risit alumpno
> Gallia, Carnotis corpore, quaeque suo.[17]

According to Vernet, the rest of *Epitaphium Theoderici* contains ordinary information and statements that could apply to many people. This does not mean that these verses are not true. The epitaph belongs to a collection of poems all or mostly written apparently by the same scribe; among the historical people cited, we find the counts of Flanders and the Norman kings of England. The last date mentioned is the death of the archbishop Samson of Liège, which occurred in 1161. In any case, however reliable the facts mentioned in this collection of poems may be, it is certain that Thierry was the archdeacon of Dreux from 1136[18] onward and, from 1141 onward, chancellor of Chartres, when he succeeded Gilbert of Poitiers, who was named bishop

13 Hermann of Carinthia, *De essentiis*, ed. Charles Burnett (Leiden, 1982), Appendix II, pp. 348–349.
14 Bernard Silvestris, *Cosmographia*, ed. Peter Dronke (Leiden, 1978), p. 96.
15 André Vernet, "Une épitaphe inédite de Thierry de Chartres," in *Recueil de travaux offert à M. Clovis Brunel*, 2 vols. (Paris, 1955), 2: 660–670, at 663 n2.
16 Ibid., p. 670.
17 Wilhelm Wattenbach, "Beschreibung einer Handschrift mittelalterlicher Gedichte (Berl. Cod. theol. oct. 94)," *Sitzungsberichte der königlich preussischen Akademie der Wissenschaften zu Berlin* 8 (1895), 123–156, at 149.
18 Lucien Merlet and René Merlet, *Dignitaires de l'Eglise Notre-Dame de Chartres: Listes chronologiques* (Chartres, 1900), p. 194.

of Poitiers. Thierry also left his private library to the cathedral of Chartres; in comprising more than 45 volumes, among them the *Heptateuchon*, his compilation on the seven liberal arts, as well as volumes on Roman law, the collection was impressive for its time:

> (Nonis novembris) Obiit magister Teodoricus, cancellarius et archidiaconus alme Marie, qui dedit huic ecclesie Bibliothecam septem liberalium artium, et de legibus Romanis librum Institutionum Justiniani, librum Novellarum constitutionum ejusdem et librum Digestorum, et preter hec quadraginta quinque volumina librorum.[19]

Unfortunately, we do not have a complete list of the books bequeathed.

Since Vernet's discovery of the epitaph, historians agree that Thierry retired to a monastery at the end of his life. In one of his last contributions, Vernet repeats the hypothesis that was previously put forth that the monastery in question was none other than the Cistercian abbey of Vaux-de-Cernay in the valley of Chevreuse between Paris and Chartres, a place that is still peaceful today and possessed of a rare beauty.[20] As for the date of Thierry's death, it is generally accepted that he died between 1153 and 1155.

Thierry of Chartres should undoubtedly be identified as the master Thierry who attended the council of Soissons in 1121 and supported the accused, Peter Abelard.[21] Thierry was also present at the council of Reims in 1148, where Gilbert of Poitiers was condemned.[22] Finally, he was with Gerland of Besançon and Albero, archbishop of Trier (from 1131 to 1152), at the Diet of Frankfurt held on 15 August 1149, according to the testimony of Balderich, Albero's biographer.[23] This trip to Germany is particularly impor-

19 Eugène de Lépinois and Lucien Merlet, *Cartulaire de Notre-Dame de Chartres*, 3 vols. (Chartres, 1862–1865), 3: 206.

20 André Vernet, "Thierry de Chartres, moine cistercien?" in *Vie spéculative, vie méditative et travail manuel à Chartres au XIIe siècle* (Chartres, 1998), pp. 11–17. Cf. Vernet, "Une épitaphe inédite," pp. 668–669.

21 Peter Abelard, *Historia calamitatum mearum*, ed. Jacques Monfrin (Paris, 1967), p. 88: "Quo audito Terricus quidam, scolaris magister, irridendo subintulit illud Athanasii: 'Et tamen non tres omnipotentes, sed unus omnipotens.'"

22 Jean Leclercq, "Textes sur saint Bernard et Gilbert de la Porrée," *Mediaeval Studies* 14 (1952), 107–128, at 109.

23 Balderich, "Gesta Alberonis archiepiscopi" 32 (MGH SS 8: 257): "Magistrum quoque Jarlandum Bisintinum et magistrum Teodericum Carnotensem, duos fama et gloria doctores nostri temporis excellentissimos, secum in sua ducens navali cammenata, in illorum disputatione et collatione valde delectatus est, et a curia domum reversus, decentibus donis largiter honoratos ad propria letos

tant, as it could account for the existence of his commentary on the *De arithmetica*, published here, in a manuscript written somewhere in southern Germany around the middle or in the second half of the twelfth century. I have already referred to some of Thierry's students. Others should also be mentioned. Adalbert II of Mainz (bishop from 1131 to 1142), according to the testimony of Anselm of Havelsberg, studied the arts of the *trivium* with Thierry in Paris[24] at the beginning of the 1130s.[25] Petrus Helias was also Thierry's student for a long time and, in any case, before 1145, according to William of Tyre who himself studied in Paris in the years 1145–1155.[26]

John of Salisbury and Alexander Neckam also mention Thierry's teaching. Despite the fact that John of Salisbury's testimony in the *Metalogicon* 2.10 is very ambiguous, proceeding as it does through an association of ideas and events without taking into account strict chronological progression, it seems as though he attended Thierry's lectures in rhetoric, most likely in the 1130s. John explains that after taking his leave of William of Conches, he studied rhetoric again – which he had already previously studied to no avail with master Thierry, and which he finally understands, thanks to

remisit." See also *A Warrior Bishop of the Twelfth Century: The Deeds of Albero of Trier by Balderich*, trans. Brian A. Pavlac (Toronto, 2008), pp. 70–71.

24 Anselm of Havelsberg, "Vita Adalberti II Moguntini," ed. Philipp Jaffé, in *Bibliotheca Rerum Germanicarum*, 6 vols. (Berlin, 1864–1873), 3: 565–603, at 589–592.

25 Fredborg, *The Latin Rhetorical Commentaries*, p. 10.

26 William of Tyre, *Chronicon* 19.12 (CCM 63/63A: 880): "Fuerunt autem nobis hoc medio tempore, quo in partibus transmarinis nostram in disciplinis transegimus adolescentiam, in paupertate voluntaria literarum studiis etatis nostros dedicavimus dies, in liberalibus artibus doctores precipui, viri venerabiles et pia recordatione digni, scientiarum vasa, thesauri disciplinarum, magister Bernardus Brito, qui postea fuit in patria unde ortus fuerat episcopus Cornualensis, magister Petrus Helie, natione Pictavensis, magister Ivo, genere et natione Carnotensis. Hii omnes magistri Theodorici, senioris viri litteratissimi, per multa tempora auditores fuerunt; horum tamen novissimus, magister Ivo, magistri Gilleberti Porrea Pictavensis episcopi, quem post magistrum Theodoricum audierat, doctrinam profitebatur. Hos alternatim, secundum quod eorum negocia presentes eos nobis permittebant vel absentes, annis audivimus circiter decem. Audivimus et alios etsi non assidue, tamen sepius et maxime disputationis gratia, viros eximios et omni laude prosequendos magistrum Albericum de Monte, magistrum Robertum de Meleuduno, magistrum Mainerum, magistrum Robertum Amiclas, magistrum Adam de Parvo Ponte, qui videbantur quasi maiora luminaria. In theologia autem virum in ea scientia singularem, cuius opera que exstant prudentum chorus cum veneratione amplectitur et colit cum reverentia, virum sana doctrina per omnia commendabilem, magistrum videlicet Petrum Lonbardum, qui postea fuit Parisiensis episcopus, annis sex continuis diligenter audivimus. Magistrum quoque Mauricium, qui ei postmodum in eodem episcopatu successit, sepius audivimus."

the teaching of Petrus Helias.[27] In another chapter of the *Metalogicon*, he remembers that Thierry was making fun of the *Topics* of Drogo of Troyes, "although he taught them on occasion."[28] Alexander Neckam also makes reference to Thierry's teaching on logic. Notably, in his *Corrogationes Promethei*, he reports that the Chartrian master did not understand a passage of Aristotle's *Sophistici Elenchi* because he was reading "plunulas" instead of "lunulas." It was Adam of Petit Pont who correctly understood the text in question:

> In geometria dicuntur lunule quedam portiuncule circuli, et in hac significatione Aristoteles in libro Elenchorum ubi de quadratura circuli loquitur, ubi Therricus deceptus legit "plunulas" (*SE* 171b15 and 172a3), antequam liber iste venisset in manus magistri Adae Parvipontis.[29]

It is certainly inexcusable for a master trained in the arts of the *quadrivium* not to recognize the geometric term "lunula," although in Thierry's defense it must be said that the text of the *Sophistici Elenchi* in his *Heptateuchon* seems to be corrupted or, at least, unclear. Chartres, Bibliothèque

27 John of Salisbury, *Metalogicon* 2.10 (CCCM 98: 72): "Relegi quoque rethoricam, quam prius cum quibusdam aliis a magistro Theodorico tenuiter auditis paululum intelligebam. Sed eam postmodum a Petro Helia plenius accepi." For the interpretation of this passage, in addition to the contributions cited above in notes 5 and 7, see Karin M. Fredborg, "The Grammar and Rhetoric Offered to John of Salisbury," in *The Classics in the Medieval and Renaissance Classroom: The Role of Ancient Texts in the Arts Curriculum as Revealed by Surviving Manuscripts and Early Printed Books*, ed. Juanita Feros Ruys, John O. Ward, and Melanie Heyworth (Turnhout, 2013), pp. 103–130, at 107–108.

28 John of Salisbury, *Metalogicon* 4.24 (CCCM 98: 162): "Magister Theodoricus, ut memini, topica non Aristotilis sed Trecassini Drogonis irridebat. Eadem tamen quandoque docuit."

29 Paul Meyer, "Notice sur les *Corrogationes Promethei* d'Alexandre Neckam," *Notices et extraits des manuscrits de la Bibliothèque nationale et autres bibliothèques* 35 (1897), 641–682, at 677. Alexander Neckam's testimony is only mentioned, not cited, in Richard W. Hunt, *The Schools and the Cloister: The Life and Writings of Alexander Nequam (1157–1217)*, ed. Margaret Gibson (Oxford, 1984), p. 5. The passage of *Corrogationes* is however cited in Enzo Maccagnolo, *Rerum universitas: Saggio sulla filosofia di Teodorico di Chartres* (Florence, 1976), p. 17 n22. See Karin M. Fredborg, "The Commentary of Thierry of Chartres on Cicero's *De inventione*," *Cahiers de l'Institut du Moyen Age grec et latin de Copenhague* 7 (1971), 1–36, at 35; Fredborg, *The Latin Rhetorical Commentaries*, p. 2. In the margins of Adam of Petit Pont's *Ars disserendi*, we find references to "Theodoricus"; see *Twelfth Century Logic: Texts and Studies*, ed. Lorenzo Minio-Paluello, 2 vols. (Rome, 1956–1958), 1: xix, xxiii. See also Gian Carlo Garfagnini, "*Ratio disserendi* e *ratiocinandi via*: Il *Metalogicon* di Giovanni di Salisbury," *Studi medievali*, 3rd ser., 12 (1971), 915–954, at 939.

municipale, MS 498, which contains part of the work, was destroyed during the Second World War. At line 19 on fol. 5va and line 4 on fol. 5vb we can indeed read "plunulas"; but in the photograph on which we must now rely (the microfilm being illegible at these points) the bar through the stem of the *p*, which would allow us to read "per lunulas" in both cases, is not very neat, and seems to suggest a correction. In his *Suppletio defectuum*, Alexander Neckam names a "Terricus monachus" among the logicians of his time:

> En, Porretanus, Albricus, Petrus Alardi,
> Terricus monachus, Gualo sophista potens,
> Temporibus micuere suis quasi lumina terre
> Et nostro fulgens tempore sidus Adam.[30]

Historians identify the "Terricus monachus" as Thierry of Chartres. This seems to fit with the epitaph discovered by Vernet, which also praises Thierry as a logician.[31] Moreover, in his commentary on Martianus Capella's *De nuptiis Philologiae et Mercurii*, Alexander Neckam adduces the opinions of *magister Theodoricus*, possibly our Thierry of Chartres, no fewer than three times.[32]

These testimonies, differing and scattered as they are, however, allow us to draw a portrait in broad strokes of Thierry the master. Much of Thierry's own work, as well as its authorship, dating and place of composition,

30 Alexander Neckam, *Suppletio defectuum* 2.1581–1585 (CCCM 221: 182).The editor inserted, without good reason, a comma between "Terricus" and "monachus," and he identifies the monk as Robert of Melun (see his commentary's notes, ibid., pp. 306–307). This correction is not justified and it is better to refer to the text as it is cited in Hunt, *The Schools and the Cloister*, pp. 5–6.

31 Vernet, "Une épitaphe inédite," p. 666.

32 Alexander Neckam, *Commentum super Martianum* 1, ed. Christopher J. McDonough (Florence, 2006), p. 61: "Terram autem constat matrem esse deorum, unde et simulacrum eius cum clave pingitur. Nam terra tempore verno aperitur, clauditur hyemali; vel secundum magistrum Theodoricum mater deorum, id est planetarum, Cibile, id est terra, dicitur. Singulis enim diebus cum in orizonte apparent, de terra oriri dicuntur;" p. 62: "Vestam ei deam ignis adherere fingit Martianus, quia in visceribus terre ignis invenitur; videmus enim ex silicibus ignem excudi. Hoc Remigio videtur. Magister autem Theodoricus dicit ignem illum non de silice provenire, set aera inter duos lapides clausum nimio impetu concurrentes in igneam naturam versum esse;" and 2, p. 242–243: "Lacteus circulus est syderei caloris infusio, ... Magister autem Theodoricus altissimam partem celi eam esse et ideo videri nobis stellas sine intervallo coniunctas et multum lumen reddere." On these quotations, cf. Edouard Jeauneau, "Note sur l'Ecole de Chartres," *Studi Medievali*, 3rd ser., 5 (1964), 821–865, repr. in Edouard Jeauneau, *Lectio philosophorum*, pp. 5–49, at 14–17.

remains obscure. Let me therefore start by surveying what we know with some certainty.

We might usefully begin with Thierry's *Heptateuchon* or *Biblioteca septem artium liberalium,* to give the work its full title. The manuscript was copied and illuminated in Chartres in 1140s, according to Patricia Stirnemann,[33] perhaps around the time of his arrival there in 1141. Thierry bequeathed his *Heptateuchon* to the cathedral chapter before leaving teaching and active life around or after 1150. It is Alexandre Clerval who found this collection of texts on the seven liberal arts in Chartres, Bibliothèque municipale, MS 497 and MS 498. As mentioned above, both manuscripts were destroyed during the Second World War, in 1944. Luckily, they had been microfilmed a few years earlier, and the microfilms can be consulted in several research libraries around the world. As for the original authorship of the *Heptateuchon,* there is no doubt; Thierry is explicitly named in the title that precedes the prologue:

> Incipit prologus Theoderici in *Eptatheucon.* <V>olumen septem artium liberalium, quod Greci *Eptatheucon* vocant, Marcus quidem Varro primus apud Latinos disposuit, post quem Plinius, deinde Marcianus. Sed illi sua. Nos autem non nostra sed precipuorum super his artibus inventa doctorum quasi in unum corpus voluminis apta modulatione coaptavimus, et trivium quadruvio ad generose nationis phylosophorum propaginem quasi maritali federe copulavimus.[34]

As pointed out by Edouard Jeauneau, who has published the prologue twice,[35] the title refers to the Heptateuch, the first seven books of the Bible (Genesis, Exodus, Leviticus, Numbers, Deuteronomy, Joshua, Judges). Thierry of Chartres's *Heptateuchon* will thus be a kind of "Bible" of the seven liberal arts. The two manuscripts of the *Heptateuchon* originally comprise around 350 folios each. Of MS 497, there are about 130 fragments left,

33 Patricia Stirnemann, "Où ont été fabriqués les livres de la Glose ordinaire dans la première moitié du XIIe siècle?" in *Le XIIe siècle: Mutations et renouveau en France dans la première moitié du XIIe siècle,* ed. Françoise Gasparri (Paris, 1994), pp. 257–301.

34 Chartres, Bibliothèque municipale, MS 497, fol. 2r.

35 Edouard Jeauneau, "Le *Prologus in Eptatheucon* de Thierry de Chartres," *Mediaeval Studies* 16 (1954), 171–175, repr. in Edouard Jeauneau, *Lectio philosophorum,* pp. 87–91; "Note sur l'Ecole de Chartres." On Thierry's *Prologus,* see also Rita Copeland, "Thierry of Chartres and the Causes of Rhetoric: From the *Heptateuchon* to Teaching the *Ars Rhetorica,*" in *The Classics in the Medieval and Renaissance Classroom,* ed. Juanita Feros Ruys, John O. Ward, and Melanie Heyworth, pp. 81–102. Cf. also the studies cited in n37 below.

corresponding to folios 21–29 and 32–82, containing Priscian's *Institutiones grammaticae*. As for MS 498, we have 140 fragments, corresponding to folios 180–245, which contained the *Preceptum Canonis Ptolomei* and the astronomical tables of al-Khwārizmī.[36] The several descriptions made of the contents of the *Heptateuchon* are all easily accessible, and it is unnecessary to repeat them here in detail.[37] However, it may be useful to reproduce the more detailed account found in the catalogue of the chapter library of Chartres produced at the end of the sixteenth century.[38]

> 81. Eptateucon seu volumen septem artium liberalium. – In hoc volumine continentur libri plures Donati, Prisciani, prima et 2ª rhetorica Tullii, cum alia rhetorica, libri Porphyrii, liber praedicamentorum Aristotelis, libri Peri hermeneias, libri priorum et liber topicorum; post sequitur: Incipit prima editio Donati grammatici urbis Romae; in- fol., optimae notae.[39]

36 These fragments have been carefully restored. Unfortunately, they are still out of order, and I have been asked by the Institut de recherche et d'histoire des textes (IRHT), Centre national de la recherche scientifique (CNRS) Paris to put them together, as part of a project on the fragments of manuscripts from Chartres: see "A la recherche des manuscrits de Chartres: Etude et renaissance virtuelle d'un fonds sinistré" at http://www.manuscrits-de-chartres.fr.

37 Clerval, *Les écoles de Chartres*, pp. 222–223; Menso Folkerts, *"Boethius" Geometrie II: Ein mathematisches Lehrbuch des Mittelalters* (Wiesbaden, 1970), pp. 8–9; Gillian R. Evans, "The Uncompleted *Heptateuchon* of Thierry of Chartres," *History of Universities* 3 (1983), 1–13; Charles Burnett, "The Contents and Affiliation of the Scientific Manuscripts Written at, or Brought to, Chartres in the Time of John of Salisbury," in *The World of John of Salisbury*, ed. Michael Wilks, pp. 127–160; Colette Jeudy and Yves-François Riou, *Les manuscrits classiques latins des bibliothèques publiques de France*, Vol. 1: *Agen–Evreux* (Paris, 1989), pp. 450–455; Lucio Toneatto, *Codices artis mensoriae: I manoscritti degli antichi opuscoli latini d'agrimensura (V–XIX sec.)* 3 vols. (Spoleto, 1994–1995), 1: 447–457; Charles Burnett, "La réception des mathématiques, de l'astronomie et de l'astrologie arabes à Chartres," in *Aristote, l'école de Chartres et la cathédrale* (Chartres, 1997), pp. 101–107.

38 *Catalogue général des Manuscrits des Bibliothèques Publiques de France, Départements*: Vol. 11: *Chartres* (Paris, 1890), p. xv. Details of another catalogue are provided on p. xxi: "On trouve enfin un dernier catalogue de la bibliothèque du chapitre de Chartres, au XVIIe siècle, dans le *Catalogus catalogorum* (1675) de Charles Le Tonnelier (Bibliothèque de l'Arsenal, ms. 4630, fols. 224–234)." The content of the *Heptateuchon* is described on fol. 230v. This catalogue is still unpublished, but I was able to consult the file containing André Vernet's notes at the IRHT (CNRS, Paris). I thank Hanno Wijsman, and Claudia Rabel at the IRHT for their generosity and help.

39 *Catalogue général des Manuscrits, Départements*, Vol. 11: *Chartres*, p. xv n4: "En marge: *Pulpito 3° ad sinistram, gros.*"

82. In hoc volumine continentur libri elenchorum Aristotelis, logica, rhetorica, Boetii arithmetica et musica, mathematica Julii Firmici. Ibidem Martini junioris geometria, canones, tabulae et alia de astronomia; in- fol., optimae notae.[40]

Recall that in the prologue Thierry, following the example of Martianus Capella, claims to have effected a marriage of the *trivium* and the *quadrivium* and has harmoniously arranged the writings of different authors on the arts in a single volume. In the section on the *trivium*, it is not surprising to encounter such authors as Priscian, Donatus, and Cicero. What is more surprising is the presence of all the texts of Aristotle's *Organon*, the *Logica Vetus* as well as the *Logica Nova*, with the exception of the *Posterior Analytics*, the date of whose entry into the Latin West remains uncertain. Among these, for example, is the *Prior Analytics* in a version, known precisely as the *versio Carnotensis*, which adheres very close to Boethius's translation. This version was used by Peter Abelard but, as Lorenzo Minio-Paluello observed, it was reorganized and widely used by medieval writers. We leave to one side the question of how Thierry had access to a version that was "very close to the original Greek."[41] The *Heptateuchon* also contains the "purest" text of the *Sophistici Elenchi*, discussed above.[42] As for the text of the *Topics*, it is identical, save for a few minor errors, to the text contained in Oxford, Trinity College, MS 47, an English manuscript from the beginning of the twelfth century.[43] We now know that this manuscript has links with Chartres and with the teaching of Adelard of Bath.[44] We can assume that Thierry did his best to obtain the most reliable versions available at the time for his *Heptateuchon*. This abiding interest in logic is recognized in Thierry's epitaph:

40 Ibid., p. xv n5: "En marge: 3° *pulpito ad sinistram*."

41 On this version, see Lorenzo Minio-Paluello, "Note sull'Aristotele Medievale Latino VIII: I *Primi Analitici*: La redazione carnutense usata da Abelardo e la 'Vulgata' con scolii tradotti dal greco," in his *Opuscula: The Latin Aristotle* (Amsterdam, 1972), pp. 229–241; "Note sull'Aristotele Medievale Latino XI: Le due redazioni boeziane (carnutense e fiorentina) e la contaminazione vulgata (ambrosiana) dei *Primi Analitici*," in *Opuscula*, pp. 377–383.

42 Lorenzo Minio-Paluello, "The text of Aristotle's *Topics* and *Elenchi*: The Latin Tradition," in *Opuscula*, pp. 299–309, at 303, 308.

43 Ibid., pp. 300, 303.

44 On this manuscript, see Charles Burnett, "Adelard, Music and the *Quadrivium*," in *Adelard of Bath: An English Scientist and Arabist of the Early Twelfth Century*, ed. Charles Burnett (London, 1987), pp. 69–86. See also Charles Burnett's review of *Glosae super Platonem*, by Bernard of Chartres, ed. Paul E. Dutton, *The Journal of Medieval Latin* 3 (1993), 234–235.

Dissolvens Logice nodos penetravit ad illa
Que non adtigerant tempora nostra prius:
Primus Analeticos primusque resolvit Helencos,
E Gallis Graecas accumulavit opes.[45]

In the *Heptateuchon*, the *quadrivium* occupies slightly less space than the *trivium* (from fol. 86r of MS 498); we should note the presence of several "classical" texts, among them Boethius's *De institutione arithmetica* and *De institutione musica*, book 7 of Martianus Capella's *De nuptiis Philologiae et Mercurii*, the different geometries attributed to Boethius, the Roman *agrimensores*, Gerbert of Aurillac's scholia on the *De arithmetica* II, 1 (also likely known from the beginning of the eleventh century as *Saltus Gerberti*),[46] as well as some relatively little-known texts and other texts that were entirely new in the first half of the twelfth century: al-Khwārizmī's *Tables*, translated by Adelard of Bath and, perhaps, revised by Hermann of Carinthia; the *Preceptum Canonis Ptolomei*, a Latin adaptation completed in Rome in the sixth century;[47] and a portion of Euclid's *Elements* (books 7–9 and 14.8–15.5), without proofs, referred to as the "Adelard II" version, in a translation attributed to Robert of Chester,[48] of which, according to its editors, MS 498 is the oldest surviving witness. But a note dating from the sixteenth century and written on the verso of fol. 1 indicates there were originally even more texts:

In hoc volumine continentur:
liber elencorum Aristotilis
logica

45 Vernet, "Une épitaphe inédite," p. 670.

46 We do not have the attribution to Gerbert of Aurillac. See fol. 114ra: "Explicatio cuiusdam dicti Boecii circa finem primi capituli secundi libri arithmeticae."

47 The oldest preserved manuscript of the *Preceptum* was originally from Fleury and it dates from the eleventh century. See David Pingree, "The *Preceptum Canonis Ptolomei*," in *Rencontres de cultures dans la philosophie médiévale: Traductions et traducteurs de l'antiquité tardive au XIVe siècle*, ed. Jacqueline Hamesse and Marta Fattori (Louvain-la-Neuve and Cassino, 1990), pp. 355–375. See also *Preceptum Canonis Ptolomei*, ed. David Pingree, Corpus des Astronomes Byzantines 8 (Louvain-la-Neuve, 1997); Charles Burnett, "King Ptolemy and Alchandreus the Philosopher: The Earliest Texts on the Astrolabe and Arabic Astrology at Fleury, Micy and Chartres," *Annals of Science* 55 (1998), 329–368.

48 MS 498 has been studied in *Robert of Chester's (?) Redaction of Euclid's Elements, the So-Called Adelard II Version*, ed. Hubert L.L. Busard and Menso Folkerts, 2 vols. (Boston, 1992), 2: 39–40. Cf. Max Lejbowicz, "Le premier témoin scolaire des *Eléments* arabo-latins d'Euclide: Thierry de Chartres et l'*Heptateuchon*," *Revue d'histoire des sciences* 56 (2003), 347–368. On the *Elements*, see p. 31 below.

rethorica —
arismetica — Boecii
musica —
Matematica Iulii Firmici
Materni iunioris[49]
Geometria
Canones, tabule et alia de astronomia.

Some physical damage to the manuscript must have occurred. There is a large lacuna after fol. 140 and before fol. 141 in the current numbering. In fact, 13 quaternions are obviously missing, that is, a total of 104 folios. The text of Boethius's *De institutione musica,* which occupies fols. 125r–140v, is mutilated and stops abruptly in the middle of the chapter, at book 2, chapter 21. Then, at fol. 141ra–va, we have a fragment of Euclid's *Elements* (books 14.8–15.5), in the "Adelard II" version. The *Mathesis* of Firmicus Maternus (a relatively rare text for which we have no manuscript prior to the eleventh century) was still included in the manuscript at the end of the sixteenth century, since it is mentioned in the catalogue of the chapter library; indeed, it was still there in the seventeenth century, when Charles Le Tonnelier compiled the *Catalogus Catalogorum*:

Firmicus Mathematica Iulii Firmici
Matern. Iuni. Materni Iunioris Geometria[50]

When Alexandre Clerval discovered the *Heptateuchon,* MS 498 was already in the state that we know it today. The mutilation must therefore have occurred after 1675 (the year of composition of the *Catalogus Catalogorum*) and before 1840, the year in which Chasles and Rossard de Mianville published their catalogue of the manuscripts of the Chartres library.[51] Based

49 Jeudy and Riou, *Manuscrits classiques latins,* p. 450, write that another scribe added, in regard to the *Mathesis*: *Hanc suspectam credo.* On the microfilm at the IRHT (CNRS, Paris) I was not able to find the annotation in question. However, in the photos of MS 498 that Claudia Rabel IRHT (CNRS, Paris) sent to me (and for which I thank her here) I can see a note that is not very legible. I thank my friend David Juste for his suggestions concerning the *Mathesis.*

50 Paris, Bibliothèque de l'Arsenal, MS 4630, fol. 230v. See above, n38.

51 Victor E.P. Chasles and L.M.A. Rossard de Mianville, *Catalogue des manuscrits de la bibliothèque de la ville de Chartres* (Chartres, 1840), p. 36: "En effet, on ne rencontre dans ce Ms. aucune pièce de Julius Firmicus Maternus. Peut-être en existait-il dans la lacune de 104 feuillets, à peu près, que nous avons signalée." (In fact, in this MS there is no work by Julius Firmicus Maternus. Perhaps there was some in the lacuna of about 104 folios that we have indicated.)

on the note on fol. 1, and on subsequent descriptions of the manuscript, the missing leaves must have contained the rest of the *De institutione musica,* Firmicus Maternus's *Mathesis* and a larger portion of the *Elements* – most likely books 1–6 and 10–14.7.[52]

Two other works, whose attribution to Thierry has not been questioned, are the commentaries on Cicero's *De inventione* and the *Rhetorica ad Herennium,* already cited above and which will be discussed further below.[53] These texts are difficult to date in relation to Thierry's other works because they do not treat subjects that are similar enough to form the basis of comparison. Only a short chapter of the commentary on the *De inventione* on the goal of art is repeated, almost word for word, in the *Lectiones,* which can be dated securely to around 1148.[54] According to Karin M. Fredborg, the fact that Petrus Helias's *Summa* on the *De inventione,* a work that is significantly dependent on Thierry's commentary, could not have been composed after 1138, nevertheless allows the rhetorical commentaries to be dated with reasonable certainty to the 1130s.[55]

The *Tractatus de sex dierum operibus,* a commentary on the beginning chapters of *Genesis,* is transmitted anonymously in the manuscripts.[56] However, it is unquestionably the work of Thierry, as we learn from his

52 Books 7–9 of the *Elements* are on fols. 122r–124v. For suggestions concerning the books of Euclid's *Elements* found in MS 498, see Burnett, "The Contents and Affiliation of the Scientific Manuscripts," p. 142.

53 Cf. Karin M. Fredborg's contribution, cited above in n29.

54 Fredborg, *The Latin Rhetorical Commentaries,* p. 11.

55 Ibid., pp. 9–12; on Petrus Helias's *Summa,* see at pp. 11–12: "This was composed, at the latest, not long after 1138 as can be inferred from the fact that the papal schism of Pope Anacletus II (1130–1138) is cited as an example of the rhetorical issue, *constitutio negotialis.* Since Petrus Helias firmly understood that this particular issue could be used only for litigation concerning unprecedented legal matter, his use of Anacletus's ordinations would only be appropriate if he composed the work close to the end of the schism, that is around 1138. For already in 1139, the Lateran Council had provided the occasion to regulate the legality of Anacletus's ordinations and hence they could no longer serve as an example of litigation concerning unprecedented legal matters. Because of the dependence of Petrus Helias's rhetorical commentary on that of Thierry's, Thierry's commentaries must be earlier. They were probably composed in the 1130s, when, as we have seen, Thierry also taught rhetoric to Adalbert of Mainz." On dating, see Fredborg, "The Commentary of Thierry of Chartres,"; John Ward, "The Date of the Commentary on Cicero's *De inventione* by Thierry of Chartres (*ca.* 1095–1160?) and the Cornifician Attack on the Liberal Arts," *Viator* 3 (1972), 219–273.

56 The list of manuscripts is in Nikolaus M. Häring, *Commentaries on Boethius by Thierry of Chartres and His School* (Toronto, 1971), pp. 42–52.

disciple Clarembald of Arras, who addresses it, preceded by a preface, to a lady Matilda whom scholars identify as Empress Matilda, wife of Emperor Henry V and later of Geoffrey Plantagenet, count of Anjou and duke of Normandy.[57] The version prepared by Clarembald is transmitted in two manuscripts, Cambrai, Bibliothèque municipale, MS 339 and Paris, BnF, MS lat. 3584. The attribution can be further corroborated by an excerpt from the *Tractatus* discovered by Nikolaus Häring in a twelfth-century manuscript, Heiligenkreuz, Stiftsbibliothek, MS 153, which is explicitly attributed to Thierry: "Magister Tirricus super locum illum posuit deus firmamentum in medio aquarum" (fol. 110v).[58] It is undoubtedly significant that the Cistercian abbey of Heiligenkreuz was founded in 1133 by Leopold III, Margrave of Austria, at the urging of his son Otto, later Bishop of Freising, who completed his studies in Paris between 1126 and 1134, the year he became a monk at the Cistercian abbey of Morimond.[59] According to Häring, the *Tractatus* goes back to 1130–1140, a period in which Thierry was mainly interested in the liberal arts.[60] He adds that since Thierry identifies the world soul with the Holy Ghost, the *Tractatus* was necessarily written before the council of Sens, which was held in May 1140 (or 1141) and the condemnation of Abelard

57 This preface was already cited above in n12. Cf. Clarembald of Arras, *Epistola ad dominam* 3 (ed. Häring, pp. 225–226): "Direxi itaque Vestre Sublimitati libellum quem magister Theodoricus, meus doctor, de sex dierum operibus edidit quem Roma iam suis conmisit archivis."

58 Häring, *Commentaries on Boethius*, p. 46. Richard Southern has analyzed the testimony of the Heiligenkreuz manuscript and believes the *Tractatus de sex dierum operibus* is a sort of *reportatio* by Clarembald of Arras of the lectures of Thierry of Chartres on *Genesis*; see Southern, *Platonism*, pp. 32–34. I have found a new twelfth-century manuscript containing Thierry's *Tractatus*, which will be the subject of a forthcoming study.

59 The link between the manuscript of the *Tractatus* from Heiligenkreuz and Otto of Freising was suggested by Southern, *Platonism*, pp. 33–34.

60 Häring, *Commentaries on Boethius*, p. 47. Anneliese Stollenwerk is of the same view as Häring: see her *Der Genesiskommentar Thierrys von Chartres und die Thierry von Chartres zugeschriebenen Kommentare zu Boethius "De Trinitate"* (Cologne, 1971), pp. 21, 24. Peter Dronke, "Thierry of Chartres," p. 360, in contrast, dates the writing of the *Tractatus* after that of the commentaries on the *Opuscula sacra*, more specifically after 1150. Enzo Maccagnolo dates the writing of the *Tractatus* to just after the middle of the twelfth century; see his *Rerum universitas*, p. 213: "… e permette di concludere che il *Tractatus*, come i commenti al *De Trinitate*, sia stato scritto negli anni intorno alla seconda metà del XII secolo." Cf. also the insightful observations of Robert M. Ziomkowski, "Science, Theology, and Myth in Medieval Creationism: Cosmogony in the Twelfth Century" (PhD diss., Cornell University, 2000), pp. 200–215. On the *Tractatus*, see also Willemien Otten, "Nature and Scripture: Demise of a Medieval Analogy," *Harvard Theological Review* 88 (1995), 257–284.

that followed.[61] Since several authors writing after the council of Sens continue to identify, despite the condemnation, the world soul with the Holy Ghost Häring's argument is not entirely convincing; still, it is reasonable to date the writing of the *Tractatus* between 1130 and 1140 on the basis of arguments that will be laid out in further detail in the following section.

The attribution of the commentaries on Boethius's *Opuscula sacra* listed below to Thierry of Chartres has been the object of lively debate. Häring gathered all the commentaries from Thierry's pen as well as those bearing traces, more or less marked, of his teaching into a group and published them in a single volume. Readers unfamiliar with the problems of authorship that these texts pose might be puzzled by such a decision. Here is the complete list:

1. *Commentum super Boethii librum De Trinitate*, inc. "Librum hunc"
2. *Commentum super Boethii librum De hebdomadibus*
3. *Lectiones in Boethii librum De Trinitate*, inc. "Quae sit"
4. *Commentarium in Boethii librum Contra Eutychen et Nestorium* (Fragmentum Londinense)
5. *Glosa super Boethii librum De Trinitate*, inc. "Aggreditur propositum"
6. *Tractatus De Trinitate*
7. *Abbreviatio Monacensis (De Trinitate, De hebdomadibus, Contra Eutychen)*
8. *Commentarius Victorinus (De Trinitate)*, inc. "In titulo"
9. *Glosa Victorina (De Trinitate)*
10. *Tractatus de sex dierum operibus*

Scholars generally agree that the *Commentum*, the *Lectiones* and the *Glosa* on the *De Trinitate* (texts 1, 3, and 5 in the list) are indeed by Thierry – or at least they are the most "Theodorician" of the commentaries on the *Opuscula sacra* that have been preserved. It is perhaps worth recalling Louis J. Bataillon's judgement:

> It is well-established that the three commentaries come from the school of Chartres. That all three of them are from the same author seems doubtful to me. One of the three may be Thierry's, but it is not proven. It is something of a shame that these three texts, very interesting in and of themselves, have been published under a name with so little evidence;

61 On the council of Sens, see Pietro Zerbi, *"Philosophi" e "Logici": Un ventennio di incontri e scontri – Soissons, Sens, Cluny (1121–1141)* (Milan, 2002), and Constant J. Mews, "The Council of Sens (1141): Abelard, Bernard, and the Fear of Social Upheaval," *Speculum* 77 (2002), 342–382.

> it would have been far preferable to propose this attribution of authorship in the preface, for now it risks being too easily accepted without question. As for the attribution of the third commentary in the Oxford MS, Bodl. Lyell 49 to Petrus Helias, it also undoubtedly requires more clarification, but at least that attribution has a contemporary attestation; it can thus be temporarily retained as a hypothesis.[62]

Despite Bataillon's reservations, the arguments put forth for attributing these three commentaries to Thierry, or connecting them with his teaching, seem relatively solid. The *Commentum* shares many doctrines with the *Tractatus*, whose authorship is not at all in question. For his part, Clarembald of Arras states, in his own commentaries on Boethius's *Opuscula sacra*, that he followed Thierry's teaching (and that of Hugh of Saint Victor) on the *Opuscula*;[63] in fact, he adopts as his own several points of doctrine from the *Commentum*, the *Lectiones*, and the *Glosa*. Since Clarembald uses all three commentaries in his own writings, it seems difficult to deny attributing them to Thierry. Nevertheless, the authorship of the *Commentum* has been called into question on the basis of an attribution placed at the head of the text in a manuscript from the twelfth century originating from Admont and found today in Oxford, Bodleian Library, MS Lyell 49. It is folios 80 through 100 of the manuscript that most interests us. Folio 80r is blank; on its verso are found two figures speaking to each other and, at the bottom of the page, an *ex-libris* that reads: "Hic claustri sancti Blasii liber est ademunt."[64] The *Commentum* begins on fol. 81r, where it is preceded by the title: "Commentum Helye cuiusdam magistri Gallicani super Boetium de Trinitate." Häring argued that the scribe undoubtedly meant the French master Petrus Helias. However, Helias is known for his *summae* on grammar and rhetoric; no one has ever attributed readings (*lectiones*) of the *Opuscula sacra* to him.[65] This attribution seems without foundation. The *Commentum* ends on fol. 99v, where the commentary on the *De hebdomadibus* (text 2 in

62 Louis Jacques Bataillon, "Bulletin d'histoire des doctrines médiévales, IV: Le douzième siècle; V: Le début du treizième siècle," *Revue des sciences philosophiques et théologiques* 46 (1962), 499–542, at 508 (my translation).

63 Clarembald of Arras, *Tractatus super librum Boetii De Trinitate*, Epistola ad Odonem 3 (ed. Häring, p. 64).

64 We can distinguish an "o" written over the "u."

65 Häring, *Commentaries on Boethius*, pp. 20–21. On the *summae* of Petrus Helias, see Karin M. Fredborg, "The Dependence of Petrus Helias' *Summa super Priscianum* on William of Conches' *Glose super Priscianum*," *Cahiers de l'Institut du Moyen Age grec et latin de Copenhague* 11 (1973), 1–57, and "Petrus Helias's *Summa* on Cicero's *De inventione*," *Traditio* 64 (2009), 139–182.

the list), known only through this manuscript, begins without interruption: "Explicit Tractatus super Boetium. Incipit de Bono ad Iohannem." The commentary on the *De hebdomadibus* stops abruptly in the middle of fol. 100v with the words "nec participat entitate," and Abelard's *Theologia summi boni,* written in the same hand, begins on fol. 101r.[66] In addition, we might note that Tours, Bibliothèque municipale, MS 300, the best manuscript witness of the *Commentum* according to Häring, also contains the *Mathematicus* of Bernard Silvestris, who dedicated his *Cosmographia* to Thierry of Chartres.[67] As for the date of the composition of the *Commentum,* it is after the *Tractatus* and a *terminus ante quem* of 1148 is likely, for a number of related reasons. The *Commentum* contains the prophecy of the Spanish Sibyl, which was used for political ends in the years 1146–1147 in order to promote the second crusade to the Holy Land. Thierry, however, subjects it to a "Trinitarian" rather than a political interpretation (we will return to the Spanish Sibyl later, for the prophecy is cited in the commentary on the *De arithmetica* preserved in the Stuttgart manuscript). Häring also noted that the *Commentum* contains an allusion to the Trinitarian doctrine of Gilbert of Poitiers: "Et peccat et desipit qui quemadmodum homo dicitur ab humanitate sic deum dici estimat a deitate."[68] Taking into account the fact that allusion to the "heresies" of Gilbert of Poitiers are much more numerous and explicit in Thierry's *Lectiones*[69] and especially in the *Glosa,* it seems reasonable to date these two texts after the council of Reims, and thus after 1148. We might also recall that Häring observed that the allusion to Gilbert of Poitiers at *Glosa* 2.44 may have its actual source in the *capitula,* the list of errors attributed

66 Häring, *Commentaries on Boethius,* p. 27. See Nikolaus M. Häring, "A Third Manuscript of Peter Abelard's *Theologia 'Summi boni'* (MS Oxford, Bodl. Lyell 49, fols. 101–128)," *Mediaeval Studies* 18 (1956), 215–224, at 215–216.

67 Tours, Bibliothèque municipale, MS 300, fols. 72r–73v, *Mathematicus* of Bernardus Silvestris, inc. "Semper ut ex aliqua felices parte querantur / Leges humane conditionis habent"; fol. 73v: "Liber et explicitus ad mea vota meus"; fol. 74r, blank, followed by the remainder of the text which ends on fol. 75r: "quicquid vult mime (?) proferat cvis cro." The Tours manuscript was noticed by the first editor of the *Mathematicus*: see Barthélemy Hauréau, in *Notices et extraits des manuscrits de la Bibliothèque nationale et des autres bibliothèques,* 42 vols. (Paris, 1787–1933), 28.2 [1878]: 372; 29.2 [1880]: 343, 347. Cf. now Deirdre M. Stone, "Bernardus Silvestris, *Mathematicus:* Edition and Translation," *AHDLMA* 63 (1996), 209–283, at 221.

68 Thierry of Chartres, *Commentum super Boethii librum De Trinitate* 4.2 (ed. Häring, p. 95).

69 Thierry of Chartres, *Lectiones in Boethii librum De Trinitate* 2.56 (ed. Häring, p. 173): "Est autem in hoc loco cavendum a veneno quorundam imperitorum qui dicunt: deus est a deitate deus. Quod omnino hereticum est. A nullo enim deus et nullo participat. Immo esse omnium est."

to Gilbert by his adversaries, rather than in Gilbert's works themselves, thus further corroborating the view that the *Glosa* most certainly were written after 1148.[70]

To this corpus of works written by Thierry between 1130 and 1150, it is now possible to add a commentary on the *De arithmetica*, most likely composed in the 1140s.[71] Another, still unpublished commentary on book 8 (*De astrologia*) of Martianus Capella's *De nuptiis Philologiae et Mercurii*, previously studied by Edouard Jeauneau, could be the work of Thierry of Chartres or of one of his students.[72] Moreover, Karin Fredborg put forth the hypothesis that Thierry may have written a commentary on Priscian's *Institutiones grammaticae* – based on testimony preserved in a Vatican manuscript – that compares the opinions (*sententiae*) of the masters Alberic, William, and Thierry.[73] Finally, we must recall that Thierry himself, in his *Lectiones* on Boethius's *De Trinitate*, refers to subjects he could have discussed when teaching logic: "Sed non est concedendum: Socrates est hec res homo que est species: homo est hec res animal quod est genus. Sed in logica de hoc satis diximus."[74] We might be tempted to conclude that Thierry did indeed compose a "Logica"; however, at present, no text on logic has been properly attributed or linked to him.

Commentary and Commentator

At a colloquium held nearly sixty years ago, Raymond Klibansky, speaking of the works of Thierry of Chartres, declared that "to the writings generally recognized as his we can add his Commentary on Boethius'

70 Häring, *Commentaries on Boethius*, p. 33.

71 The discovery of this commentary is announced in: Irene Caiazzo, "Il rinvenimento del commento di Teodorico di Chartres al *De arithmetica* di Boezio," in *Adorare caelestia, gubernare terrena*, ed. Pasquale Arfé, Irene Caiazzo, and Antonella Sannino (Turnhout, 2011), pp. 183–203.

72 Jeauneau, "Note sur l'Ecole de Chartres," in *Lectio philosophorum*, pp. 17–23; Fredborg, "The Commentary of Thierry of Chartres on Cicero's *De inventione*," 2–4. This commentary will be briefly discussed on pp. 44–46 below.

73 Karin M. Fredborg, "*Tractatus Glosarum Prisciani* in MS Vat. lat. 1486," *Cahiers de l'Institut du Moyen Age grec et latin de Copenhague* 21 (1977), 21–44, at 43: "Aliter enim dicit magister Thedricus (*sic*). Significat substantiam et qualitatem i.e. suppositum de quo loquimur communiter vel proprie ita quod conveniat uni vel pluribus, et sic male disiuncta, ut nomen et appellatio; bene coniungit Priscianus (II.IV.18) et vocabulum i.e. nomina rerum incorporearum."

74 Thierry of Chartres, *Lectiones in Boethii librum De Trinitate* 4.44 (ed. Häring, p. 201). This reference had already been pointed out by Fredborg, "The Commentary of Thierry of Chartres on Cicero's *De inventione*," 3–4, 35–36.

De arithmetica."[75] Several years later, in the introduction to her edition of Thierry's commentaries on Cicero's *De inventione* and on the *Rhetorica ad Herennium*, Karin Fredborg identified the manuscript discovered by Klibansky as Bern, Burgerbibliothek, Cod. 633, fols. 19ra–27ra.[76] The commentary in question is fragmentary and written in a Carolingian script by a French hand from the second half of the twelfth century; it deals with only a small part of the *De arithmetica* and stops abruptly in the middle of an analysis of book I, chapter 1.[77] A cursory reading reveals that the author relied in part on other works by Thierry, such as the commentaries on the *Opuscula sacra* and undoubtedly a commentary on the *De arithmetica*, as well as on Hugh of Saint Victor's *Didascalicon*.[78]

Another commentary on the *De arithmetica*, noted by the historians of science Arno Borst and Menso Folkerts, has completely escaped the attention of historians of medieval philosophy.[79] It is preserved in a manuscript written near the middle or in the second half of the twelfth century by several scribes from south Germany in a Carolingian script characterized by very archaic features. The manuscript, originally from the Benedictine library of Komburg – where it was perhaps copied – is now part of the manuscript collection of Württembergische Landesbibliothek of Stuttgart, under the shelfmark Cod. math. 4° 33. This miscellany of 86 folios transmits some very interesting texts, for some of which it is the only surviving witness currently known. The fact that the collection of mathematical manuscripts at the Württembergische Landesbibliothek has never been the object of a printed catalogue is probably the reason why Cod. math. 4° 33 has not been adequately appreciated by scholars.[80] The commentary on the

75 Raymond Klibansky, "The School of Chartres," in *Twelfth-Century Europe and the Foundations of Modern Society*, ed. Marshall Clagett, Gaines Post, and Robert Reynolds (Madison, 1961), pp. 3–14, at 5.

76 Fredborg, *The Latin Rhetorical Commentaries*, p. 3.

77 A transcription of the Bern commentary can be found in Vera Rodrigues, "*Creatio numerorum*: Nature et rationalité chez Thierry de Chartres" (PhD diss., Ecole Pratique des Hautes Etudes, IVème section, 2006).

78 This commentary is also cited by Charles Burnett, "Scientific Speculations," in *Twelfth-Century Western Philosophy*, ed. Dronke, pp. 151–176, at 167.

79 Arno Borst, "Rithmimachie und Musiktheorie," in *Geschichte der Musiktheorie*, Vol. 3: *Rezeption des antiken Fachs im Mittelalter* (Darmstadt, 1990), pp. 253–286, at 256 n3; Menso Folkerts, "Die *Rithmachia* des Werinher von Tegernsee," in *Vestigia mathematica: Studies in Medieval and Early Modern Mathematics in Honour of H.L.L. Busard*, ed. Menso Folkerts and Jan Hogendijk (Amsterdam-Atlanta, 1993), pp. 107–142. Cf. Caiazzo, "Il rinvenimento del commento."

80 A handwritten catalogue from the nineteenth century is preserved at the Stuttgart library, where it must be consulted on-site. I thank Carmen Kämmerer, director

De arithmetica, which we must ascribe to the Chartrian master, is found on fols. 1ra–34ra, with the *incipit*: "Quoniam ars ista, ars scilicet arithmetica."

All the texts contained in the manuscript deal with the disciplines linked to the *quadrivium*: treatises on the astrolabe by Hermann of Reichenau and by Gerbert of Aurillac (attributed), the computus of Gerlandus, astronomical tables, and a calendar for the monastery of Komburg.[81] We also have three anonymous lemmatic commentaries on the texts of Hermann and Gerlandus and one text attributed to Gerbert, for which, in any case, the Stuttgart manuscript is the only known witness. These three commentaries and the commentary on the *De arithmetica* are preserved in the first codicological unit (fols. 1–40), which is very homogenous; they may represent the written record of a course given by a master in the first half of the twelfth century, before the assimilation of Arabic science freshly introduced to the West by way of translations. Several of these texts, it should be noted, are either dated or mention a date. First, a lemmatic commentary, apparently the only one identified, on Gerlandus's *Computus* – a text that was widespread in the Middle Ages and has only just now been published[82] – is found on fols. 34ra–37va (where it follows without interruption the commentary on the *De arithmetica*). It was most probably written in 1143 or 1150.[83] In fact, the anonymous author explains that in order to carry out the calculations, it is necessary to take a complete year, and that for the year 1143, it is necessary to keep the year 1142. If the author adopts Gerlandus's chronology – a fact of which we cannot be sure – the year 1143 would really indicate the year 1150 of the Dionysian era, since Gerlandus subtracted seven years from Dionysus Exiguus's chronology. We might also add that a good part of the lemmatic commentary in the Stuttgart manuscript is found in the form of glosses on Gerlandus's *Computus,* in the margins and between the lines of Troyes, Bibliothèque municipale, MS 1558. Originally from the abbey of Saint-Claude in the Jura,[84] this manuscript, written by a French hand in the twelfth century, was part of the private library of the Bouhier family, where

of the department of manuscripts, for having sent me some bibliographical information pertaining to the Cod. math. 4° 33.

81 Cf. the description of the manuscript, pp. 81–85 below.

82 *Der "Computus Gerlandi": Edition, Übersetzung und Erläuterungen,* ed. Alfred Lohr (Stuttgart, 2013).

83 For the commentaries on the *Computus,* and on the treatises on the astrolabe, see Irene Caiazzo, "Sur l'astrolabe et le comput au XIIe siècle: nouveaux textes inédits," *Sudhoffs Archiv: Zeitschrift für Wissenschaftsgeschichte* 96 (2012), 28–38.

84 Auguste Castan, "La bibliothèque de l'abbaye de Saint-Claude du Jura: Esquisse de son histoire," *Bibliothèque de l'Ecole des Chartes* 50 (1889), 301–354, at 346 n° 113.

it had the shelfmark E. 83 according to the last catalogue, compiled in 1721.[85] However, the Stuttgart manuscript is not a direct copy of the manuscript of Troyes, which has, in some places, more lacunae; we must therefore suppose the existence of at least one antigraph that, in any case, has not been preserved.[86] Other evidence internal to the Stuttgart manuscript is significant for its dating: a second codicological unit contains an excerpt and the table of fixed stars "according to the moderns" from the *Liber de astrolabio* of the astrologer Raymond of Marseilles, written a little before 1141;[87] a table of computus that starts with the year 1140; and an excerpt from the *Liber de astrolabio* of Ibn al-Saffār, in the translation of John of Seville (*floruit* 1133–1153). Finally, a short note cites, in all likelihood, Raymond of Marseilles's *Liber cursuum planetarum*, a work written in 1141 containing an adaptation of the tables of Toledo for the meridian of Marseilles. As well, the work of Raymond of Marseilles seems also to have been linked to Chartres in the twelfth century.[88]

To summarize, the Stuttgart manuscript brings together texts written in the 1140s, some of which are of French origin. It is thus very plausible that the commentary on the *De arithmetica* of the Stuttgart manuscript was also written in the same period.

There is further evidence linking the commentary on the *De arithmetica* to Thierry of Chartres. It is very likely that Thierry commented on or read Boethius's *De arithmetica* in front of an audience – according to the method of *lectio* used in schools of the twelfth century. His writings – the *Tractatus de sex dierum operibus*, and the commentaries on the *Opuscula sacra*, in particular – contain numerous citations from the *De arithmetica*, although

85 Albert Ronsin, Raymond Etaix, and André Vernet, *La Bibliothèque Bouhier: Histoire d'une collection formée du XVIe au XVIIIe siècle par une famille de magistrats bourguignons* (Dijon, 1971), pp. 228, 241.

86 Alfred Lohr, the editor of Gerlandus's *Computus*, arrives at the same conclusions and situates the Stuttgart and Troyes manuscripts in the same branch of the *stemma*.

87 On these various excerpts, see Irene Caiazzo and Charles Burnett, "New Manuscripts of *On the astrolabe* by Raymond of Marseilles," *Scriptorium* 65 (2011), 338–349.

88 Burnett, "The Contents and Affiliation of the Scientific Manuscripts," pp. 133–136, 140; Marie-Thérèse d'Alverny, "Abélard et l'astrologie," in *Pierre Abélard-Pierre le Vénérable: Les courants philosophiques, littéraires et artistiques en Occident au milieu du XIIe siècle*, ed. René Louis, Jean Jolivet, and Jean Châtillon (Paris, 1975), pp. 613–614.

Nikolaus Häring did not identify them comprehensively or systematically in his apparatus of sources. Moreover, several chapters of the *De arithmetica* have inspired some of the most original and significant ideas of the Chartrian master, such as his reflections on the relationship between unity, alterity, and plurality, and between equality and inequality; the identification of the Son of God with the first tetragon; and the doctrine of *complicatio* and *explicatio*. Unfortunately, there is no exhaustive examination of these questions and we must content ourselves with scattered indications in different studies, among which Edouard Jeauneau's contribution remains pioneering.[89] A rigorous analysis would require highlighting the presence of the *De arithmetica* in Thierry's known and published writings, while at the same time comparing them with the commentary preserved in the Stuttgart manuscript and published here for the first time. The issue here, however, is not to draw up a complete catalogue of citations from the *De arithmetica* in the work of Thierry of Chartres. A few key examples demonstrating the role Boethius's treatise played in the development of Thierry's philosophy will suffice.

Already in the commentary on the *Rhetorica ad Herennium*, from the 1130s, we find an explicit citation of the *De arithmetica*:

> Nota: prologus ille ante rem factus esse dicitur in quo agit de arte extrinsecus, scilicet de materia, de officio artis et talibus, ut in secundo *Arithmeticae* prologo: "Neque praeter rem, etc." Praeter rem dicitur illud prooemium in quo nihil dicitur de arte extrinsecus, sed in eo quaedam artis vel dictaminis commendatio vel auditoris benivolentiae captatio continetur. Ubi autem Tullius dicit se neque ante rem neque praeter

89 Edouard Jeauneau, "Mathématiques et Trinité chez Thierry de Chartres," in *Miscellanea Mediaevalia*, Vol 2: *Die Metaphysik im Mittelalter*, ed. Paul Wilpert (Berlin, 1963), pp. 289–295; Gillian Evans, "*Alteritas:* Sources of the Notion of 'Otherness' in Twelfth-Century Commentaries on Boethius' *Opuscula Sacra*," *Archivum Latinitatis Medii Aevi* 40 (1975–1976), 103–113; Klaus Riesenhuber, "Arithmetic and the Metaphysics of Unity in Thierry of Chartres: On the Philosophy of Nature and Theology in the Twelfth Century," in *Nature in Medieval Thought: Some Approaches East and West*, ed. Chumaru Koyama (Leiden, 2000), pp. 43–73; Bernard McGinn, "Does the Trinity Add Up? Transcendental Mathematics and Trinitarian Speculation in the Twelfth and Thirteenth Centuries," in *Praise No Less Than Charity: Studies in Honor of M. Chrysogonus Waddell, Monk of Gethsemani Abbey*, ed. E. Rozanne Elder (Kalamazoo, 2002), pp. 237–264; Vera Rodrigues, "Thierry de Chartres, lecteur du *De Trinitate* de Boèce," in *Boèce, ou La chaîne des savoirs*, ed. Galonnier, pp. 649–663.

rem locutum fuisse, notat se in libris superioribus nullum, igitur, prooemium, sed solos ad litteram descensus aut introitus fecisse.[90]

In the *Tractatus de sex dierum operibus* there is a specific reference to *De arithmetica* II, XXV–XXVII, chapters in which the subject is the construction of figures with equal sides, figures with one side longer than the other by one unit, figures whose sides differ by more than one unit, and figures in which all sides are of unequal lengths:

> Prior igitur generatio numerorum facit tantummodo tetragonos, vel cubos vel circulos vel speras que equalitatem dimensionum custodiunt. Posterior vero numerorum generatio facit parte altera longiores vel antelongiores vel ceteras figuras que ad inequalitatem laterum pertinent.[91]

Here, Thierry evokes some very difficult and perplexing chapters of the *De arithmetica*, showing his familiarity with the text and its content. But the doctrine itself of the creation of the square or tetragon from unity – explicitly associated with the Son of God in the commentary on the *De arithmetica* preserved in the Stuttgart manuscript and in the *Commentum super Boethii librum De Trinitate* – seems to have as its main source a passage from Boethius's *De arithmetica*, which is worth citing here:

> Hoc autem in hac est dispositione divinum quod omnes angulares numeri tetragoni sunt. Tetragonus autem dicitur, ut brevissime dicam quod post latius explicabitur, quem duo aequales numeri multiplicant, ut in hac quoque dispositione est. Unus enim semel unus est, et est potestate tetragonus.[92]

Unity as a potential square is also mentioned in another chapter of the *De arithmetica*, which seems to have inspired some passages of the *Tractatus de sex dierum operibus*:[93] "Recte igitur unitas propria virtute perfecta est, quod et prima est et incomposita et per se ipsam multiplicata sese ipsam conservat."[94]

90 Thierry of Chartres, *Commentarius super Rhetoricam ad Herennium* 4.1.1 (ed. Fredborg, p. 313).
91 Thierry of Chartres, *Tractatus de sex dierum operibus* 37 (ed. Häring, p. 571).
92 Boethius, *De arithmetica* I, XXVII, 7 (ed. Guillaumin, p. 56).
93 Thierry of Chartres, *Tractatus de sex dierum operibus* 38–39 (ed. Häring, p. 571); the text is quoted in n186 below.
94 Boethius, *De arithmetica* I, XX, 9 (ed. Guillaumin, p. 45).

The intersection, and mutual enrichment, of Augustine's *De doctrina christiana* with some passages of the *De arithmetica*, also informs Thierry's most original doctrines, to which I will return in more detail below. In what was considered the last and the most accomplished of the series of commentaries on Boethius's *Opuscula sacra*, the *Glosa in Boethii librum De Trinitate*, we find another explicit reference to the *De arithmetica*:

> Quod non videtur eo quod numerus prior est omni accidente. Numerus enim, ut habet *Arithmetice* prologus, principale exemplar extitit in mente conditoris. Quod si ex accidentibus profluit numerus, in substantias ergo preiacent numero accidentia. Quare exemplar conditoris precessisse videntur. Quod falsum est.[95]

This is a point on which Thierry insists in all his works: number is not prior to the creator. The commentary on the *De arithmetica* preserved in Stuttgart expresses the same point of view:

> Non tamen ideo quis dicat vel quid vel numerum priorem deo esse. Hoc enim aperte falsum est. Dico itaque quoniam vis numerorum et figurarum in principalibus causis est rerum, ex numerorum et figurarum causa habet esse.[96]

In several places in the commentary on the *De arithmetica*, the author states that everything is made according to number, after the model of number: "Ad exemplar numeri dicitur deus res creasse, quia omnia ex discretione, quae est ex numeris, habent esse."[97] This seems to echo the idea that creation and the creation of numbers are identical with one another, expressed by Thierry in the *Tractatus de sex dierum operibus* in a well-known and oft-cited passage: "Sed creatio numerorum, rerum est creatio."[98] As for

95 Thierry of Chartres, *Glosa super Boethii librum De Trinitate* 1.38 (ed. Häring, p. 267). Cf. also *Lectiones in Boethii librum De Trinitate* 1.51 (ed. Häring, p. 149–150), and the illuminating analysis by Vera Rodrigues, "Pluralité et particularisme ontologique chez Thierry de Chartres," in *Arts du langage et théologie aux confins des XIe-XIIe siècles: Textes, Maîtres, Débats*, ed. Irène Rosier-Catach (Turnhout, 2011), pp. 509–536, at pp. 526–528.

96 Thierry of Chartres, *Commentum super Arithmeticam Boethii* II, XXXI, 1 (p. 186 below).

97 Thierry of Chartres, *Commentum super Arithmeticam Boethii* I, I, 8 (p. 114 below). Cf. ibid., I, II, 1 (p. 118 below): "Est autem haec dignitas numeri: … ."

98 Thierry of Chartres, *Tractatus de sex dierum operibus* 36 (ed. Häring, p. 570).

the *Commentum super Boethii librum De Trinitate*, we have another specific and explicit reference to the *De arithmetica*:

> Res ergo in sua inmutabilitate, ut predictum est, conprehendere summa quidem est sapientia. Ideoque dictum est sapientiam esse comprehensionem veritatis rerum que sunt i.e. que in sui natura inmutabiles sunt. Unde et idem philosophus in secundo *Artis arismethice* prologo: *Sunt autem,* inquit, *hee forme qualitates quantitates,* et cetera. Videns autem Pythagoras paucos et fere nullos res in sua puritate – que summa "sophia" est – comprehendere posse, veritatis inquisitores non "sophos" appellare ausus est sed "philosophos" quasi sapientie studentes.[99]

In the *Lectiones in Boethii librum De Trinitate,* we also discern a "generic" yet important reference to arithmetic: "De hoc autem in arimethica melius dicitur quia in ea agendo de numero ista subtilitas docetur ostendendo quid sit numerus et cuius rei sit numerus et quid in rebus efficiat."[100] In fact, in the *Lectiones* the whole interpretation of *De Trinitate* III relies strongly on the "arithmetical triplex way of understanding the multiplication of the unity."[101]

But the imprint Boethius's treatise on arithmetic leaves on Thierry's work is also more fundamental. In his prologue to the *De arithmetica* Boethius had proclaimed, loudly and clearly, the unity of knowledge by which different disciplines are but parts of a homogenous and unified whole; it is this ideal of knowledge, above all, that Thierry advocates and to which the perfected library of the seven liberal arts of the *Heptateuchon* corresponds. Whoever wishes to attain perfection in philosophical speculation, Boethius declares following Pythagorean doctrine, must first learn and master the four disciplines of the *quadrivium*; of these, arithmetic constitutes the initial step, for it allows us to approach and understand the other three:

99 Thierry of Chartres, *Commentum super Boethii librum De Trinitate* 2.7 (ed. Häring, p. 70). Cf. Boethius, *De arithmetica* I, 1, 2. See also Clarembald of Arras, *Tractatus super librum Boetii De Trinitate* 2.14 (ed. Häring, p. 112): "Secundum huiusmodi ergo esse hic idem phylosophus in secundo *Arithmeticae* prologo sapientiam diffinivit dicens eam esse comprehensionem veritatis rerum quae sunt quaeque sortiuntur immutabile esse sui. Unde quoniam difficillimum est in huiusmodi veritate res comprehendere, facillimum erat huius comprehensionis compotes insolescere et arrogantes fieri. Et iccirco Pytagoras pertimuit eos cum primum *sophi* appellarentur *sophos* vocare. Sed vocabulo eorum paululum humiliato decrevit eos phylosophos appellandos esse i.e. sapientiae amatores."

100 Thierry of Chartres, *Lectiones in Boethii librum De Trinitate* 1.42 (ed. Häring, p. 146).

101 Thierry of Chartres, *Lectiones in Boethii librum De Trinitate* 3.5–7 (ed. Häring, p. 178): "Est autem, ut ex arismeticis habetur, triplex consideratio numeri i.e. repeticio unitatis. Una consideratio est numeri in se absolute et simpliciter ..."

music, geometry, and astronomy. The *De arithmetica* thus represents a fundamental part of the *Heptateuchon*; indeed, its prologue, which serves as an introduction to this collection of texts on the seven liberal arts, opens with a definition of philosophy taken from the *De arithmetica:* "Phylosophya autem est amor sapientie; sapientia vero est integra comprehensio veritatis eorum que sunt,"[102] a definition Thierry uses again in the *Commentum,* his first commentary on Boethius's *De Trinitate,* in two places: "Philosophia ergo est studium sapientie. "Philos" enim amor vel studium dicitur. "Sophia" vero sapientia interpretatur. Sapientia vero est conprehensio veritatis eorum que sunt i.e. inmutabilium."[103]

These definitions, it is true, are very common among the authors of the twelfth century; but it is interesting to highlight the distinction drawn between philosophy and wisdom – and the fact that former leads, in a way, to the second, which, for Thierry, is nothing other than the understanding of the forms that are in God. In his prologue to the *Heptateuchon* Thierry writes: "We have blended together *trivium* and *quadrivium* for the augmentation of the noble race of philosophers." A few lines later, he returns to the same idea: the *trivium* and *quadrivium* are intimately connected, and the *Heptateuchon* is the sole and unique instrument of philosophy: "Manifestum est Eptatheucon totius phylosophye unicum ac singulare esse instrumentum."[104] Thierry's *Heptateuchon* is the instrument of philosophy, because the seven liberal arts are the instrument of philosophy. In his *Commentum super Boethii librum De Trinitate,* Thierry states that "the ancients had the habit of learning mathematics before coming to an understanding of divinity."[105] In the commentary on the *De arithmetica,* it is said explicitly that no one can be a philosopher without the *quadrivium*[106] and that he who neglects it neglects wisdom.[107]

102 Jeauneau, "Note sur l'Ecole de Chartres," in *Lectio philosophorum,* p. 38. Cf. Vera Rodrigues, "La conception de la philosophie chez Thierry de Chartres," *Mediaevalia: Textos e Estudos* 11–12 (1997), 119–137.

103 Thierry of Chartres, *Commentum super Boethii librum De Trinitate* 2.2 (ed. Häring, p. 68).

104 Jeauneau, "Note sur l'Ecole de Chartres," in *Lectio philosophorum,* p. 38; the Latin is quoted above at p. 12.

105 Thierry of Chartres, *Commentum super Boethii librum De Trinitate* 2.15 (ed. Häring, p. 73, ll. 59–61): "... quia mathematicam solebant prius antiqui discere ut ad divinitatis intelligentiam possent pervenire." Cf. Agnieszka Kijewska, "Mathematics as a preparation for Theology: Boethius, Eriugena, Thierry of Chartres," in *Boèce,* ed. Galonnier, pp. 625–647.

106 Thierry of Chartres, *Commentum super Arithmeticam Boethii* I, 1, 6 (p. 110 below).

107 Thierry of Chartres, *Commentum super Arithmeticam Boethii* I, 1, 3 (p. 108 below). See also I, 1, 7 (p. 111 below), and I, 1, 8 (p. 113 below).

No clearer or more explicit articulation of Thierry's intellectual project can be found.

We can also adduce more factual data showing Thierry's adherence to the ideal of knowledge professed in Boethius's *De arithmetica*. Hubert L.L. Busard and Menso Folkerts demonstrated that in the *Heptateuchon* definitions VII.1–11 of Euclid's *Elements* have been replaced with definitions taken from Boethius's *De arithmetica*.[108] The Arab–Latin translation of the *Elements* preserved in the *Heptateuchon* is the "Adelard II" version that the editors attribute to the translator Robert of Chester. Definitions VII.1–11 were published, together with the variants rejected by the editors, in the critical apparatus:

1. <Unitas est> cuius multiplicatio numerum gignit.
2. Numerus vero est unitatum collectio cuius species sunt par atque impar.
3. Par est cuius est medietas.
4. Impar est cuius nulla est medietas propter abundantem unitatem.
5. Paris autem numeri tres sunt species: pariter par, pariter impar, impariter par.
6. Pariter par est numerus qui et ipse et quelibet sua pars in dimidia usque ad unitates dividitur.
7. Pariter impar est numerus qui et si in dimidia dividatur tamen nulla eius pars eandem sectionem recipit.
8. Impariter par[s] numerus est qui et ipse et alique sue partes sed non omnes dimidiantur.
9. Imparis vero numeri tres quoque species sunt: impar primus et incompositus, secundus et compositus, tercius eorum medius.
10. Primus et incompositus est quem solius unitatis multiplicatio gignit.
11. Secundus et compositus est quem eciam numeri multiplicatio procreat.
12. Medii vero cum per se sint secundi et compositi, tamen ad se invicem comparati primi et incompositi sunt quoniam nullam com munem mensuram habent preter unitatem.[109]

108 Busard and Folkerts, *Robert of Chester's (?) Redaction of Euclid's Elements*, 1: 24, 39. Euclid's *Elements*, integrated with the *De arithmetica*, are found on fols. 122rb-123va of MS 498.

109 Busard and Folkerts, *Robert of Chester's (?) Redaction of Euclid's Elements*, 2: 595.

According to Busard and Folkerts, the *De arithmetica* occupied a place of honor for Thierry of Chartres, and he transmitted his enthusiasm for this work to his disciple Hermann of Carinthia.[110] The latter dedicated his translation of Ptolemy's *Planisphaerium* to Thierry of Chartres (*diligentissime preceptor Theodorice*) and in order to do so, he used formulations taken from the Preface of Boethius's *De arithmetica*. Perhaps this is a mere coincidence, or perhaps it is a nod to his master Thierry. In any case, these correspondences deserve further attention:

> ... tibi, inquam, diligentissime preceptor Theodorice, quem haut equidem ambigam, Platonis animam celitus iterum mortalibus accommodatam. Quo factum est principaliter, ut non aliter quam **aureis culmis Cererem, maturo palmite Bacchum**, unum te Latini studii patrem astronomie **primitiis** donandum **iudicarim**, quippe cum nec ego quid offerrem melius, haberem, nec tibi sapientie dono quicquam acceptius cognoscerem; secundo vero, ut id quod solertiam tuam minime latere potest, aliis quoque per te innotescat interim: quanta presumptione astronomie nomen usurpant, qui necdum principium eius viderint, que sine tribus premissis ita recte possibilis est, ... qui novo freti ingenio conversis discipline gradibus a fine incipiunt ... .[111]

It is interesting to compare this passage to the Preface of the *De arithmetica*:

> Recte ergo, quasi **aureos Cereri culmos** et **maturos Baccho palmites**, sic ad te novi operis rudimenta transmisi. Tu tantum paterna gratia nostrum provehas munus. Ita et laboris mei **primitias** doctissimo **iudicio** consecrabis et non maiore censebitur auctor merito quam probator.[112]

Hermann completed the translation of the *Planisphaerium* in Toulouse on 1 June 1143. It is certainly tempting, if only as a hypothesis, to keep the year 1143 as the *terminus a quo* for the writing of Thierry's commentary on the *De arithmetica*. In his preface, Hermann also declares Thierry the father of astronomical studies among the Latins; however, no work by Thierry on astronomy has come down to us. Charles Burnett, who has examined

110 Hermann of Carinthia, *De essentiis* (ed. Burnett, p. 22); Busard and Folkerts, *Robert of Chester's (?) Redaction*, 1: 24–25.

111 Hermann of Carinthia, *De essentiis* (ed. Burnett, Appendix II, p. 349).

112 Boethius, *De arithmetica*, praefatio 4 (ed. Guillaumin, p. 3).

Hermann's preface, thinks that Thierry must have promoted the study of astronomy and perhaps encouraged the translation of Arabic texts, as is suggested by the presence of some of these texts in his *Heptateuchon* as well as in the library of the cathedral chapter at Chartres.[113] Indeed, the Stuttgart commentary on the *De arithmetica* contains a very interesting reference to the theory of astronomers (or, more accurately, astrologers) concerning the effects of the planets, the sun, and the moon on the natures and complexions of things, and to Firmicus Maternus's *Mathesis*, a text that was relatively rare before the twelfth century, and which was originally included, as we have seen, in the *Heptateuchon*.[114]

Other contemporary testimony also serves to highlight Thierry's propensity for the liberal arts. John of Salisbury's observation, "viri amatores litterarum utpote magister Theodoricus artium studiosissimus investigator," is one among several others collected by Alexandre Clerval.[115] Peter Abelard is also supposed to have secretly followed the *mathematicae lectiones* of master Thierry, little good though it did him. Although the authenticity of this anecdote is much debated, Constant Mews believes it is true, arguing that it was indeed with Thierry of Chartres that Abelard tried to initiate himself into the arts of mathematics.[116] Finally, the epitaph discovered by André Vernet, also praises Thierry's accomplishments in the arts of the *quadrivium* and the *trivium*:

> Quadruvium triviumque simul scrutando labore
> Pervigili cunctis fecit utrunque facet.[117]

In summary, all these attestations, while they certainly differ in origin and value, make it plausible that Thierry could have commented on the *De arithmetica* or read it in front of an audience of students.

113 Burnett, "La réception des mathématiques."

114 Thierry of Chartres, *Commentum super Arithmeticam Boethii* II, XLII, 1 (p. 198–199 below). This chapter is analyzed on pp. 67–70 below.

115 John of Salisbury, *Metalogicon* 1.5 (CCCM 98: 20). Cf. Alexandre Clerval, *L'enseignement des arts libéraux à Chartres et à Paris dans la première moitié du XIIe siècle d'après l'"Heptateuchon" de Thierry de Chartres* (Paris, 1889), p. 7.

116 On this anecdote, cf. David E. Luscombe, *The School of Peter Abelard* (Cambridge, 1970), p. 58, Maccagnolo, *Rerum universitas*, p. 2 (with the bibliography cited p. 16, nn 14–16), and more recently, Constant Mews, "In Search of a Name and Its Significance: A Twelfth-Century Anecdote about Thierry and Peter Abaelard," *Traditio* 44 (1988), 171–200.

117 Vernet, "Une épitaphe inédite," p. 670.

Thought, Language, Philosophy

There is little doubt that the philosophy of Thierry of Chartres evolved over time with regard to several doctrinal questions. An exhaustive essay, however, is beyond the scope of this introduction. I will limit myself to describing a few significant features that can serve as landmarks, with the goal of situating the commentary on the *De arithmetica* as precisely as possible. The discussion that follows relies in large measure on lexicographical analysis. It is, I believe, both an indispensable tool and consonant with its subject, since for Thierry philosophical writing entails the search for a language equal to the demands of thought, demands sometimes best fulfilled by the invention of neologisms.

It is generally accepted by scholars that the *Commentum super Boethii librum De Trinitate* and the *Tractatus de sex dierum operibus* share several common characteristics, notably a marked interest in natural philosophy.[118] When he composed these works, Thierry had not yet elaborated the theory that made him famous among both his contemporaries and historians of philosophy, namely the four modes of being of the *rerum universitas: necessitas absoluta,* that is God; *necessitas complexionis,* that is forms of every thing; *possibilitas absoluta,* that is prime matter; and *possibilitas determinata,* that is the sensible things in act. This theory would only be articulated explicitly in the *Lectiones in Boethii librum De Trinitate* and in the *Glosa super Boethii librum De Trinitate.* In the *Commentum,* Thierry explains that the *rerum universitas* can be considered in five (and not four) different modes: in God, in the created spirit, in number, in matter, and in act.[119] Only the first two modes are illustrated, for Thierry states that the *rerum universitas* needs more in-depth treatment, and that it would be necessary to dedicate an entire volume to it.[120] If we now want to seek, at any cost, a correspondence between the two versions of the doctrine of modes of being of the *rerum universitas,* we must note that when Thierry went from five modes to four, he deleted number, probably because it can be integrated into the second mode, the created spirit (in the terminology of the *Commentum*) or the *necessitas complexionis* (in the terminology of the *Lectiones* and of the *Glosa*). The strong Platonic and neo-Pythagorean character of the *Commentum* and the *Tractatus* means

118 See the section on Thierry's works, pp. 17–22 above.

119 Thierry of Chartres, *Commentum super Boethii librum De Trinitate* 4.7 (ed. Häring, p. 97): "Est enim rerum universitas in deo, est in spiritu creato, est in numeris, est in materia, est etiam rerum universitas in actu ...."

120 Ibid., 4.10 (p. 98): "Sed quoniam que de rerum universitate proposita sunt maioris indigent inquisitionis quam tractatus presens exigat dimittantur hec interim quorum explanationem voluminis integri series expectat."

that Thierry's philosophy is centered on the dialectic of *immutabilitas* and *mutabilitas, unitas* and *alteritas,* form and matter. The commentary on the *De arithmetica* shares these same interests and this same terminology, and in a manner analogous to the *Tractatus,* makes no mention of the four (or five) modes of the *rerum universitas.* It is also worth noting that in his *Cosmographia,* which we recall was dedicated to Thierry of Chartres, Bernard Silvestris does not mention the theory of the four modes of being, but does say he is basing his cosmology on the dialectic of two principles, *unitas* and *diversum,* in all likelihood in the wake of Thierry's early works.[121]

The *Commentum* represents Thierry's first reading of Boethius's *De Trinitate.* Its language is very close, in certain places, to the commented text; the *Lectiones* and the *Glosa,* in contrast, contain neologisms of Thierry's invention, and thus reflect an evolving, increasingly complex doctrine. Here are some key examples. On the subject of the much-studied division of philosophy, especially concerning the way in which one should proceed in the study of the three parts of the *speculativa,* Boethius had written: "In naturalibus rationabiliter, in mathematica disciplinaliter, in divinis intellectualiter."[122] The passage had sparked considerable debate from late antiquity onwards, and in the *Lectiones* Thierry unquestionably links the faculties of the human soul to the three parts of the *speculativa.* The *intellectibilitas* or *intelligentia* (which Thierry always pairs) is the faculty of the soul used in theology to understand the totality (*universitas*).[123] The intellect is used in mathematics; reason, the senses, and the imagination in physics.[124] *Intellectibilitas* allows us to apprehend undivided substance, that is to say divinity, which is the totality

121 Bernard Silvestris, *Cosmographia* 2.13 (ed. Dronke, p. 146): "Erant igitur duo rerum principia: unitas et diversum. Diversum, longe retro antiquissimum. Unitas non inceperat: simplex, intacta, solitaria, ex se in se permanens, infinibilis et eterna. Unitas deus, diversum non aliud quam yle, eaque indigens forma." On the two principles, see Tullio Gregory, *Anima mundi: La filosofia di Guglielmo di Conches e la Scuola di Chartres* (Florence, 1955), pp. 64–65, and Brian Stock, *Myth and Science in the Twelfth Century: A Study of Bernard Silvester* (Princeton, 1972), pp. 112–114.

122 Boethius, *De Trinitate* 2. The same Boethian terms are used by Thierry of Chartres, *Commentum super Boethii librum De Trinitate* 2.1 (ed. Häring, p. 68).

123 On the use of *intellectibilitas* and *intelligentia* in the work of Thierry of Chartres, cf. Dronke, "Thierry of Chartres," pp. 365–367.

124 Thierry of Chartres, *Lectiones in Boethii librum De Trinitate* 2.30 (ed. Häring, p. 164): "Et sciendum quod diversis anime viribus et conprehensionibus in phisica mathematica theologia utendum est ad conprehendendum universitatem ut subiecta est his tribus speculative partibus. Nam in theologia utendum est intellectibilitate sive intelligentia: in mathematica vero intellectu qui est disciplina: in phisica ratione sensu et imaginatione que circa materiam conprehendunt quicquid conprehendunt."

(*universitas*) in its simplicity.[125] *Intellectibilitas* is a faculty that is beyond reason; *intelligentia* (its synonym) is the highest faculty of the soul in Boethius's *Consolation of Philosophy*, which is explicitly cited by Thierry.[126] As for the term *intellectibile*, we encounter it in Boethius's first commentary on Porphyry's *Isagoge*, where it translates the Greek νοητόν and designates things that are separate from matter and which can only be grasped by the intellect.[127] The term *intellectibile* is Boethius's invention; it is also found once in his commentary on Aristotle's *Categories*,[128] but is used neither in the second commentary on the *Isagoge* nor in the *Opuscula sacra*. Thierry transforms the *intellectibile* found in the two Boethian logical commentaries (which he certainly knows) into a faculty of the soul pertaining to theology. The term

125 Ibid., 2.31 (p. 165): "Conprehendit enim anima intelligentia sive intellectibilitate individuam substantiam: scilicet divinitatem i.e. universitatem in simplicitate."

126 See Boethius, *Consolatio Philosophiae* 5.pr4.

127 Boethius, *In Isagogen Porphyrii Commenta, Editio prima* 1.3 (CSEL 48: 8–9): "Est enim una theoretices pars de intellectibilibus, alia de intellegibilibus, alia de naturalibus. ... Νοετά, inquam, quoniam Latino sermone numquam dictum repperi, intellectibilia egomet mea verbi compositione vocavi. Est enim intellectibile quod unum atque idem per se in propria semper divinitate consistens nullius umquam sensibus, sed sola tantum mente intellectuque capitur. Quae res ad speculationem dei atque ad animi incorporalitatem considerationemque verae philosophiae indagatione componitur: quam partem Graeci τεολογίαν nominant. Secunda vero est pars intellegibilis, quae primam intellectibilem cogitatione atque intellegentia comprehendit. Quae est omnium caelestium supernae divinitatis operum et quicquid sub lunari globo beatiore animo atque puriore substantia valet et postremo humanarum animarum. Quae omnia cum prioris illius intellectibilis substantiae fuissent, corporum tactu ab intellectibilibus ad intellegibilia degenerarunt, ut non magis ipsa intellegantur quam intellegant et intellegentiae puritate tunc beatiora sint, quotiens sese intellectibilibus applicarint. Tertia theoretices species est quae circa corpora atque eorum scientiam cognitionemque versatur: quae est physiologia, quae naturas corporum passionesque declarat. Secunda vero, intellegibilium substantia, merito medio collocata est, quod habeat et corporum animationem et quodammodo vivificationem et intellectibilium considerationem cognitionemque."

128 See Boethius, *In Categorias Aristotelis* 1 (PL 64: 183D–184A): "Sensibus vero obiiciuntur prima individua, merito igitur ea prima in divisione posuit. Eodem quoque modo illa quaestio solvitur quae dicit: Cum naturaliter primae intellectibiles sint substantiae, ut deus et animus, cur non has primas substantias nuncupaverit? Quoniam hic de nominibus tractatus habetur, nomina autem primo illis indita sunt quae principaliter sensibus fuere subiecta, posteriora vero in nominibus ponendis putantur quaecumque ad intelligibilem pertinent incorporalitatem; quare quoniam in hoc opera principaliter de nominibus tractatus est, de individuis vero substantiis quae primae sensibus subiacent prima sunt dicta vocabula in opere quo de vocabulis tractabatur, merito individuae sensibilesque substantiae primae substantiae sunt positae."

intellectibilitas enjoyed some success among Thierry's contemporaries; it is used again by his disciple Clarembald of Arras, by Arnold of Bonneval (who died in 1156),[129] as well as by Gundissalinus in his translation of Avicenna's *De anima* done between 1152 and 1166.[130] Moreover, in the *Lectiones* ontology and gnoseology meet and are mutually superimposed; the three parts of the *speculativa,* to which correspond the faculties of the soul, allow us to grasp the four modes of the *rerum universitas*:

> Sed SPECULATIVE TRES sunt PARTES: theologia que dat scientiam simplicitatis i.e. universitatis in simplicitate et mathematica que dat scientiam abstractorum scilicet necessitatis conplexionis et phisica que dat scientiam et cognitionem universitatis ut est tum in absoluta possibilitate tum in determinata.[131]

Neither these ideas nor the language used to explicate them appear in the *Commentum,* where the faculties of the soul, only briefly mentioned, are connected to the forms, according to whether they are abstract or in matter, immobile or in movement. The *Lectiones* show more advanced reflection on theology, which continues in the *Glosa,* where Thierry introduces *intelligibiliter* to replace *intellectualiter,* the word used by Boethius: "IGITUR

129 Arnold of Bonneval, *Commentarius in Psalmum* 132.5.4 (PL 189: 1595A): "... intellectibile est illud solum quod omni et soli bono intellectui praesto est ad fruendum. Huic soli intellectibilitas assignatur; nam caeterae incorporales substantiae intelligibiles nuncupantur." The use of the word *intellectibilitas* in the work of Thierry of Chartres, Arnold of Bonneval, and Clarembald of Arras was remarked on by Jacqueline Hamesse, "Un nouveau glossaire des néologismes du latin philosophique," in *Aux origines du lexique philosophique européen: L'influence de la "latinitas,"* ed. Jacqueline Hamesse (Louvain-la-Neuve, 1997), pp. 237–254. On the Thierry / Gundissalinus relationship, see Alexander Fidora, "La metodología de las ciencias según Boecio: Su recepción en las obras y traducciones de Domingo Gundisalvo," *Revista española de filosofía medieval* 7 (2000), 127–136. See also Tullio Gregory, "The Platonic Inheritance," in *Twelfth-Century Western Philosophy,* ed. Dronke, pp. 54–80, at pp. 71–76.

130 Avicenna, *Liber de anima seu Sextus de Naturalibus* 5.6, ed. Simone Van Riet (Louvain–Leiden, 1968–1972), p. 144: "Omnis autem apprehensio intelligibilis est similitudo aliqua ad formam separatam a materia et ab eius accidentibus materialibus, sicut praediximus. Sed anima habet hoc ex hoc quod est substantia recipiens impressa ab eo; intelligentia vero habet hoc ex hoc quod est substantia principium agens et creans. Quod autem proprium est eius ex principalitate eius ad animam, est eius intellectibilitas in effectu; quod vero proprium est animae ex hoc quod informatur ex ea et recipit ab ea, est eius intellectibilitas in effectu." For the date of the translation, see p. 95* of the introduction.

131 Thierry of Chartres, *Lectiones in Boethii librum De Trinitate* 2.18 (ed. Häring, p. 160).

IN NATURALIBUS RATIONABILITER, IN MATHEMATICIS DISCIPLINALITER, in theologicis VERSARI oportet intelligibiliter."[132] This lexical transformation continues and is refined from one work to another, and demonstrates conclusively that Thierry's interests changed between the writing of the *Tractatus* and that of the *Glosa*. Clarembald of Arras also uses Thierry's sophisticated vocabulary, adopting terms like *intellectibilitas* and even introducing the word *intellectibiliter* on his own.[133] Clarembald's commentaries on the *De Trinitate* and on the *De hebdomadibus* were written either during or after his stay in Laon, as is indicated in the dedicatory letter to Odo, that is to say at the end of the 1150s at the earliest. As for his *Tractatulus super librum Genesis*, if it is indeed addressed to the Empress Matilda, as scholars now seem to agree, it must have been written before 1164, the year of Matilda's death. The conclusion is inescapable: Clarembald's works are in the immediate aftermath of Thierry's teaching.

If we now approach the commentary on the *De arithmetica* preserved in Stuttgart, we might begin by noting that it clearly shows that his author, Thierry of Chartres, was familiar with Boethius's *De Trinitate*, from which he cites the division of philosophy and of the *speculativa*. During this phase of his career, however, he limits himself to linking the three parts of the *speculativa* to the three objects they treat, that is, theology, which deals with incorporeals that are beyond bodies, mathematics, with incorporeals that are around bodies, and physics, with bodies themselves. He does not speak of "forms," as he does in the commentaries on the *De Trinitate*, but of incorporeals and bodies, as he does elsewhere in his prologue to the *Lectiones*, which is very close to these passages in the commentary on the *De arithmetica*.[134] Here, Thierry insists on the median role played by mathematics

132 Thierry of Chartres, *Glosa super Boethii librum De Trinitate* 2.28 (ed. Häring, p. 274).

133 Clarembald of Arras, *Tractatus super librum Boetii De Trinitate* 2.19 (ed. Häring, p. 114); *Expositio super librum Boetii "De hebdomadibus"* 28 (ed. Häring, p. 206). See *The Boethian Commentaries of Clarembald of Arras*, ed. and trans. David B. George and John R. Fortin (Notre Dame, 2002), pp. xxii–xxiv.

134 Thierry of Chartres, *Commentum super Arithmeticam Boethii*, accessus (pp. 91–92 below, l. 18–32). Cf. Thierry of Chartres, *Lectiones in Boethii librum De Trinitate*, prol. 4 (ed. Häring, p. 126): "Sunt enim tres partes speculative: theologia cuius principium est de summo deo, de Trinitate, et inde descendit ad angelicos spiritus et animas et est de incorporeis que sunt extra corpora: et mathematica cuius est principium de numeris et inde descendit ad proporciones et ad magnitudines et est de incorporeis que sunt circa corpora sicut de linea superficie et de ceteris in hunc modum: et phisica que est de ipsis corporibus et habet principium a quatuor elementis."

which, through the study of the incorporeals around bodies, enables the knowledge both of bodies and of incorporeals that are outside them.[135] The faculties of the soul are not placed in relation with the parts of the *speculativa* or with the objects of understanding.

Let us return to the *Commentum super Boethii librum De Trinitate* which represents the result of Thierry's first readings of Boethius's *De Trinitate*. Like the *Tractatus de sex dierum operibus*, it also makes no mention of another famous theory that Thierry advances in the *Lectiones* and takes up again in the *Glosa*.[136] This theory, according to which names give their essence to things ("nomina res essentiant"), was admirably analyzed by Jean Jolivet, who coined the expression "grammatical Platonism" in order to refer to it.[137] Similarly, God is *yperusyon*, beyond substance (*supra substantiam*), in the *Lectiones* and the *Glosa*;[138] but this definition is not found in the *Commentum*. Several doctrinal developments that sealed the reputation of the Chartrian master, then, are lacking in the *Commentum*. Nevertheless, it is possible to point to the presence of the term *onitas* in the *Commentum*,[139] likely invented by Thierry himself, which he was to use again in the *Lectiones* and in the *Glosa*.[140] Unity is in fact *onitas* (formed from the Greek τὸ ὄν) that is, the entity of all things upon which it confers being. *Onitas* is not found in the *Tractatus* or in the commentary on the *De arithmetica*, though the latter speaks of unity as the entity of all things:

135 Thierry of Chartres, *Commentum super Arithmeticam Boethii*, accessus (p. 92 below).

136 Thierry of Chartres, *Glosa super Boethii librum De Trinitate* 4.14 (ed. Häring, p. 287).

137 Thierry of Chartres, *Lectiones in Boethii librum De Trinitate* 2.52 (ed. Häring, p. 172). See Jean Jolivet, "Quelques cas de platonisme grammatical du VIIe au XIIe siècle," in *Mélanges René Crozet*, ed. Pierre Gallais and Yves-Jean Riou, I (Poitiers, 1966), pp. 93–99; Luisa Valente, *Logique et Théologie: Les écoles parisiennes entre 1150–1220* (Paris, 2008), pp. 46–47, 64.

138 Thierry of Chartres, *Lectiones in Boethii librum De Trinitate*, 4.10 and *Glosa super Boethii librum De Trinitate* 4.16 (ed. Häring, pp. 189 and 288).

139 Thierry of Chartres, *Commentum super Boethii librum De Trinitate* 2.22 (ed. Häring, p. 75): "Actus igitur sine possibilitate necessitas est. Hec autem a Platone eternitas, ab aliis unitas, quasi *onitas* ab *on* Graeco i.e. entitas, ab omnibus autem usitato vocabulo appellatur deus."

140 Thierry of Chartres, *Lectiones in Boethii librum De Trinitate* 2.48 (ed. Häring, p. 170): "Unde et unitas dicitur i.e. *onitas* quasi entitas omnium rerum"; cf. ibid., 7.5 (p. 224); *Glosa super Boethii librum De Trinitate* 5.18 (ed. Häring, p. 297): "Unitas enim ipsa est *onitas* atque entitas que ante omnem alteritatem est."

> Dicitur autem unitas quasi entitas, quod nihil aliud est quam ipsam esse rerum intelligi. Deus autem rerum est entitas, quare et ipse vere est unitas.[141]

The term *onitas* thus seems to have been introduced later, and can be considered a key concept that captures the essence of Thierry's theory of being. It was to be used by Arnold of Bonneval who, however, we know did not like Thierry of Chartres's bold speculations,[142] as well as by the author of the commentary on the *De arithmetica* preserved in the Bern manuscript alluded to earlier.[143] Cardinal Nicholas of Cusa would also use it; I will return to this point below.

The rare terms invented by Thierry, such as *onitas* for example, allow us to delineate his philosophical journey diachronically. This approach is especially appropriate because Thierry himself accords much importance to the choice of words – which is not surprising in a philosopher for whom *nomina res essentiant*. The lexicographical analysis undertaken here also clearly shows that the commentary on the *De arithmetica* is prior to the grand speculations developed in the commentaries on the *Opuscula sacra*, and is probably contemporaneous with (or slightly later than) the *Tractatus*, with which it shares interest in natural philosophy and the liberal arts.

Accessus ad auctores

In the commentary on Cicero's *De inventione*, written in the 1130s, Thierry of Chartres develops a two-part schema to the *accessus ad auctores* that

141 Thierry of Chartres, *Commentum super Arithmeticam Boethii* I, XXXII, 1 (p. 146 below). See also n198 below.

142 Arnold of Bonneval, *Commentarius in Psalmum* 132.1.5 (PL 189: 1572A-B): "Et nota quod haec ipsa unitas, de qua hic agitur, longe ab illa unitate quae pars est numeri separatur; non enim haec unitas in numerum potest multiplicari; nec ex ea binarius, sive ternarius, vel quilibet alius numerus potest constitui, ut solet dici de numero, quod sit unitatum collectio. Trahit autem etymologiam de Graeco, et dicitur unitas quasi onitas, id est entitas sive essentialitas: unde et apud Graecos *on*, id est substantialis, videlicet in se et per se solum immutabiliter semper subsistens, deus vocatur. Est autem quaedam similitudo inter unitatem numeralem et unitatem substantialem. Nam quemadmodum dicitur unitas fons et origo omnium numerorum, ita quodammodo deus fons est et origo omnium rerum: fons, propter inexhaustae potentiae maiestatem; origo, propter creatricis substantiae sine principio dignitatem."

143 Commentary on the *Arithmetica*, Bern, Burgerbibliothek, Cod. 633, fol. 25rb, and fol. 25va: "*On*, quasi *onitas*, id est rerum entitas."

introduces the work: one *circa artem*, the other *circa librum*.[144] On the subject of rhetorical art, he asks ten questions: *genus, quid ipsa ars sit, materia, officium, finis, partes, species, instrumentum, artifex, quare rhetorica vocetur*. As for the book being commented, he observes that it is necessary to ask what its use is, as well as the intention of its author. Thierry adds that ancient authors (he is thinking of Marius Victorinus) divided the rhetorical art itself into two parts: the *ars extrinsecus*, which is propaedeutic and allows us to know all that is useful, via the ten questions on the subject of art, before moving to the *ars intrinsecus*, which is strictly speaking the *ars eloquendi*. For Thierry, these are not distinct or separate but two complementary methods of teaching the rhetorical art. Petrus Helias, often clearer than his master, will say that *extrinsecus* and *intrinsecus* are adverbs and not adjectives. As other historians before me have observed, in this type of *accessus* (D in Richard Hunt's classification), the part concerned with rhetorical art relies on several sources, such as the *De inventione* itself,[145] Boethius's *De differentiis topicis*,[146] Marius Victorinus's *Explanationes in Ciceronis Rhetoricam*[147] and,

144 Thierry of Chartres, *Commentarius super libros De inventione*, accessus (ed. Fredborg, pp. 49–55). On the origin of Thierry's *accessus*, cf. Fredborg, "The Commentary of Thierry of Chartres on Cicero's *De inventione*," 6–11 and *The Latin Rhetorical Commentaries*, pp. 9–12; Ward, "The Date of the Commentary on Cicero's *De inventione*," 247–263. Together with the now-classic study by Richard W. Hunt, "The Introductions to the *Artes* in the Twelfth Century," in *Studia mediaevalia in honorem Raymundi Josephi Martin*, ed. B.L. Van Helmond (Bruges, 1948), pp. 85–112, reprinted in Hunt, *The History of Grammar in the Middle Ages: Collected Papers*, ed. Geoffrey L. Bursill-Hall (Amsterdam, 1980), pp. 117–144, see also Edwin A. Quain, "The Mediaeval *accessus ad auctores*," *Traditio* 3 (1945), 215–264, and Hubert Silvestre, "Le schéma 'moderne' des *accessus*," *Latomus* 16 (1957), 684–689.

145 Cicero, *De inventione* 1.4: "Sed antequam de praeceptis oratoriis dicimus, videtur dicendum de genere ipsius artis, de officio, de fine, de materia, de partibus."

146 Boethius, *De differentiis topicis* 4 (PL 64: 1207A): "Dicemus itaque de generis artis, et speciebus, et materia, et partibus, et instrumento, instrumentique partibus, opere etiam officioque actoris et fine."

147 Marius Victorinus, *Explanationes in Ciceronis Rhetoricam* 1.4 (CCL 132: 32–33): "Necessario ergo priusquam eloquentiae praecepta traderet exhortatus est ab honesto, ab utili, a possibili, et debere ab hominibus et posse eloquentiam comprehendi. SED ANTE QUAM DE PRAECEPTIS ORATORIIS DICIMUS, VIDETUR DICENDUM DE GENERE IPSIUS ARTIS. Omnis ars duplex est, id est, duplicem faciem habet secundum praeceptum sententiam que Varronis, qui ait 'esse artem extrinsecus unam, aliam intrinsecus.' Ars extrinsecus talis est quae nobis scientiam solam tradit, intrinsecus quae ita dat scientiam, ut illud ipsum quod scientia dat quibus rationibus faciamus ostendat. Ita illa ad scientiam solam proficit, haec ad scientiam quae in actu sit. Ut puta si dicam 'grammatica ars est gnara partium orationis, gnara syllabarum, gnara litterarum; per hanc discimus omnia vitia

in all likelihood, the first redaction of William of Conches's *Glosae super Priscianum*, written around 1125, which begins with ten questions: "In principio huius artis haec sunt consideranda: quid sit ipsa ars, quod nomen artis, quae causa nominis, quod genus, quod officium, quis finis, quae materia, quae partes, quis artifex, quae intentio auctoris."[148]

For William grammar, unlike rhetoric and dialectic, has no instrument, or else its instrument is the *grammatica oratio*; grammar is also without types, contrary to those who wrongly claim that Latin grammar and Greek grammar are types of grammar.[149] These remarks seem to indicate that William of Conches was not the first to use such a schema for the *accessus*. However, the earlier commentaries on Priscian, that have come down to us, do not include a comparable *accessus*, and the interlocutors that William is addressing have not been identified.

The distinction *ars extrinsecus/ars intrinsecus* had already been introduced by Manegold of Lautenbach in his commentary on the *De inventione*, written in the last quarter of the eleventh century, and certainly before 1096.[150] Manegold stated that *genus*, *officium*, *materia*, and *partes* belong to the *ars extrinsecus* and that they allow the speaker to know (*ad sciendum*) the rhetorical art, while in the *ars intrinsecus* we find exactly the principles

devitare': haec cum dico cumque per hanc vitari vitia dico, non tamen quo modo vitentur ostendo. Artem illam extrinsecus doceo per quam sola scientia discitur; si autem dicam quae sint partes orationis, quae syllabae, quae litterae quibusque modis omnia illa constent, quo pacto vitia vitentur, tunc erit illa ars, cui est nomen intrinsecus, non quae ad scientiam nostram tantum proficit, sed quae in actu sit. Ergo et ars rhetorica duplex est; nam est extrinsecus et intrinsecus. Intrinsecus autem illa est quae nobis ad actus praecepta artis insinuat, extrinsecus porro quae nobis quid sit rhetorica demonstrat; quam cum demonstraverit, scimus tantum, sed exercere non possumus. Hanc itaque artem extrinsecus, quae solam scientiam parit, quinque rebus Tullius ostendit, genere artis, officio, fine, materia, partibus. Ipse etiam ostendit se de ea arte quae est extrinsecus esse dicturum."

148 Florence, Biblioteca Med. Laurenziana, MS San Marco 310, fol. 1ra; cited after Fredborg, *The Latin Rhetorical Commentaries*, p. 15 n66.

149 Ibid.: "Instrumentum vero non habet haec ars, velut dialectica argumentationem habet instrumentum, rhetorica orationem rhetoricam. Vel dicatur instrumentum eius esse grammatica oratio, quemadmodum <rhetoricae> rhetorica oratio, dialecticae dialectica oratio. Speciem iterum non potest habere. ... Et ideo non consentimus eis qui volunt quod grammatica in Latina lingua et grammatica in Graeca et ceterae sunt species huius artis ... ."

150 On *intrinsecus / extrinsecus*, see Ward, "The Date of the Commentary on Cicero's *De inventione*," 254–256. On the date of Manegold's commentary on the *De inventione*, see also Irene Caiazzo, "Manegold, *magister modernorum magistrorum*," in *Arts du langage et théologie aux confins des XIe–XIIe siècles*, ed. Rosier-Catach, pp. 317–349, at pp. 336–337.

that allow the speaker to practice (*ad agendum*).[151] Manegold of Lautenbach could well be the main source of inspiration for Thierry of Chartres. Whatever the case, the fact remains that Thierry conceived of and wrote down this *accessus* with the ten questions *circa artem* and the two questions *circa librum*; of his own invention is also the explicit association of the ten questions with the *ars extrinsecus* that, as I have said, is a prelude or complementary to the *ars intrinsecus*, the rhetorical art in the true sense, the *ars eloquendi*. Thierry proposes the same schema again in his commentary on the *Rhetorica ad Herennium*, where he states that he will not illustrate in detail the ten points on the subject of rhetoric, because he has already done so in the previous commentary (that is, the one on the *De inventione*), and that he will limit himself to two questions on the book, its utility and the intention of its author, since they are obviously specific to the *Rhetorica ad Herennium*.[152] There is no mention of the *ars extrinsecus* / *ars intrinsecus* distinction in this *accessus*. On the other hand, in commenting on the lemmata *neque ante rem neque praeter rem*, drawn from the preface of book 4 of the *Rhetorica ad Herennium*, Thierry explains that in the prologue *ante rem*, what is at issue is the *ars extrinsecus*, that is, the *materia* and the *officium* of the art, and so on, as is the case in the second prologue of Boethius's *De arithmetica*, while the prologue *praeter rem* is not about the *ars extrinsecus* but rather includes a *captatio benevolentiae* to the listener or a section in praise of the art.[153] A distant echo of these remarks is found in the commentary on the *De arithmetica* preserved in Stuttgart: "Est autem prologus iste 'ante rem,' cum praemissus fuerit 'praeter rem.' In isto enim praelibatur materia."[154] But it is especially the form of the *accessus circa artem* and *circa librum* from Thierry's

151 Manegold of Lautenbach's commentary on the *De inventione* is still unpublished; Filippo Bognini is preparing an edition (but I was not able to consult it). Cf. the passage in question cited by Ward, "The Date of the Commentary on Cicero's *De inventione*," 255: "Quid sit genus et officium et finis et materia et partes huius artis. Et sciendum quia hec pertinet ad extrinsecam artem. Ars alia intrinseca, alia extrinseca. Intrinseca ars dicuntur illa precepta que valent ad agendum; hoc est, que oratorem quomodo causa agere debeat instruunt. Extrinseca vero ars sunt ea que tantum valent ad sciendum non etiam ad agendum."

152 Thierry of Chartres, *Commentarius super Rhetoricam ad Herennium*, accessus (ed. Fredborg, p. 221): "Circa artem rhetoricam haec sunt inquirenda: quid ipsa rhetorica sit, quae eius materia, quod genus, quod officium, quis finis, quae partes, quae species, quod instrumentum, quis artifex, quare etiam rhetorica vocetur. Circa librum duo: quae auctoris intentio, quae libri utilitas."

153 Ibid. 4.1.1 (p. 313). The passage in question was already cited above among those that attest a familiarity with the *De arithmetica*.

154 Thierry of Chartres, *Commentum super Arithmeticam Boethii* I, 1, 1 (p. 102 below).

rhetorical commentaries, that is found in the commentary on the *De arithmetica*, as we can see in the following parallel passages:

Commentary on the *De inventione*

Circa artem rhetoricam decem consideranda sunt: quid sit genus ipsius artis, quid ipsa ars sit, quae eius materia, quod officium, quis finis, quae partes, quae species, quod instrumentum, quis artifex, quare rhetorica vocetur. Artem definiendi haec et dividendi et rationibus comprobandi antiqui rhetores artem extrinsecus vocant, eo quod extra et antequam ad doctrinam agendi perveniatur oportet ista praescire. Intrinsecus vero artem appellant ipsam artem eloquendi, quia ad eam prior scientia introductoria est. Non tamen idcirco haec distinguimus quod duae artes sint, sed quoniam his duobus modis una eademque ars docetur. Deinde circa librum Tullii, quem expositiuri sumus, consideranda duo sunt: quae sit in ipso auctoris intentio et quae libri utilitas. Horum unumquodque eo ordine quo proposuimus ostendendum est.[155]

Commentary on the *De arithmetica*

Quoniam ars ista, ars scilicet arithmetica, septem liberalium artium una et est et habetur – totius enim quadruvii fundamentum est –, ideo in eius principio decem inquirenda sunt, quippe cum in principio cuiuslibet artis decem inquirere iubeamur auctoritate Boetii in IIIIto *Topicorum*. Sunt autem haec decem: quae sit ars ista, quod genus artis, quae materia, quod officium, quis finis, quae partes, quae species, quod instrumentum, quis artifex, quae nominis interpretatio.[156]

… Ea quae prediximus circa artem versantur; circa librum vero, quae sit auctoris intentio, quae utilitas libri. Intentio auctoris est in hoc opere introducere lectorem in artem arithmeticam, utilitas est scientia artis.[157]

The precise textual correspondence between the two *accessus* cannot be a coincidence, but provides clear evidence that the same author wrote them. By contrast, when Thierry comments on *Genesis*, he limits himself to questions *circa librum*: *intentio auctoris, utilitas, titulus*,[158] and avoids the questions *circa artem*, no doubt because the text in question is biblical and thus does not belong among the liberal arts. In the same way, only the questions *circa librum* are treated in the commentaries on Boethius's *Opuscula*

155 Thierry of Chartres, *Commentarius super libros De inventione*, accessus (ed. Fredborg, pp. 49–50).
156 Thierry of Chartres, *Commentum super Arithmeticam Boethii*, accessus (p. 91 below).
157 Ibid. (p. 93 below).
158 Thierry of Chartres, *Tractatus de sex dierum operibus* 1 (ed. Häring, p. 555).

sacra, which are about theology; moreover, the questions are not the same in the *Commentum*, the *Lectiones*, and the *Glosa*.[159] These differences suggest that Thierry uses the *accessus circa artem* and *circa librum* only when he comments on works pertaining to the liberal arts. It is also important to mention here the *accessus* to the commentary on book 8 (*De astrologia*) of Martianus Capella's *De nuptiis Philologiae et Mercurii*, a commentary, as Edouard Jeauneau has shown, which bears the mark of Thierry's teaching, especially in the cosmological thesis it draws from the *Tractatus de sex dierum operibus*.[160] This commentary, which is preserved in two manuscripts, one of them from Chartres, but which still remains unpublished, begins with an *accessus* identical to those in Thierry's commentaries, on *De inventione*, on the *Rhetorica ad Herennium*, and on the *De arithmetica*:

> Quidam qui volunt hunc libellum esse artem inquirunt ea hic quae circa artes sunt inquirenda, scilicet quid ipsa ars sit, quod genus eius, quae materia, quod officium, quis finis, quae partes, quae species, quod instrumentum, quis artifex, unde sic dicatur. Haec circa artem inquiruntur. Circa hunc librum vero duo: intentio et utilitas.[161]

A table summarizing the schema Thierry used for the *accessus* in each of the four commentaries discussed here may prove useful:

159 Thierry of Chartres, *Commentum super Boethii librum De Trinitate*, Prologus 1 (ed. Häring, p. 57): "Inchoantibus librum hunc de Trinitate primo videndum est que sit auctoris intentio, que libri utilitas, ad quem scribat." Richard Hunt had already pointed out that Thierry had used two different *accessus* schemes, see "The Introductions to the *Artes*," p. 97. That the questions pertaining to the book are drawn from the type C *accessus*, the most common in the twelfth century, does not allow us to conclude anything.

160 Jeauneau, "Note sur l'Ecole de Chartres," pp. 17–23.

161 London, British Library, MS Harley 2510, fol. 132r. This manuscript, dated from the second half of the twelfth century, originates in Chartres, as the note on fol. 132r indicates: *Ista IIIIor folia sunt fratrum ordinis predicatorum Carnotensis*. According to Patricia Stirnemann at the IRHT (CNRS, Paris), to whom I am grateful for reviewing the microfilm with me and sharing her expertise, the folios in question were copied in Chartres. The contents of the manuscript are of considerable interest: Cicero's *De inventione*; the *Rhetorica ad Herennium*; Martianus Capella, *De nuptiis* 8: *De astrologia*, with a commentary on the text attributable to Thierry of Chartres; Boethius's *De arithmetica*. Chartres is also mentioned on fol. 123v: *Ista rethorica est fratrum predicatorum Carnotensis*; fol. 124r: *Ista astrologia et glose sunt domus fratrum predicatorum Carnotensis*; fol. 126r: *Ista astrologia est fratrum ordinis predicatorum Carnotensis*.

De inventione	***Rhetorica ad Herennium***	***De arithmetica***	***De nuptiis***
CIRCA ARTEM	CIRCA ARTEM	CIRCA ARTEM	CIRCA ARTEM
Genus	Quid	Quid	Quid
Quid	Materia	Genus	Genus
Materia	Genus	Materia	Materia
Officium	Officium	Officium	Officium
Finis	Finis	Finis	Finis
Partes	Partes	Partes	Partes
Species	Species	Species	Species
Instrumentum	Instrumentum	Instrumentum	Instrumentum
Artifex	Artifex	Artifex	Artifex
Quare vocetur	Quare vocetur	Nominis interpretatio	Unde sic dicatur
CIRCA LIBRUM	CIRCA LIBRUM	CIRCA LIBRUM	CIRCA LIBRUM
Intentio auctoris	Intentio auctoris	Intentio auctoris	Intentio
Utilitas libri	Utilitas libri	Utilitas libri	Utilitas

The schema for the *accessus* that Thierry applied to works concerning the seven liberal arts[162] was subsequently adopted and reworked by both Petrus Helias and Gundissalinus. In his *Summa* on Cicero's *De inventione*, Petrus Helias takes up the *circa artem / circa librum* and *ars extrinsecus / ars intrinsecus* distinctions, the ten questions about rhetoric, to which he adds an order to follow in teaching and in learning (*ordo docendi et discendi*), and explains the four questions (not two, like his master) pertaining to the *De inventione*: intention, utility, whether it is part of philosophy, and the title.[163] In the *Summa super Priscianum,* he applies to grammar the same

162 The fragmentary commentary on the *De arithmetica* preserved in Bern also contains this schema (lacking *quare vocetur*): *quid sit ars ista, genus, materia, officium, finis, partes, species, <instrumentum>, artifex,* and *circa librum*: *auctoris intentio* and *utilitas libri*; the distinction *ars extrinsecus / ars intrinsecus* is also found.

163 See Fredborg, *The Latin Rhetorical Commentaries,* pp. 13–20; Fredborg, "Petrus Helias's *Summa,*" 164: "Sicut ordo nostre doctrine exigit, ita quoque circa artem rethoricam consideranda sunt hec: primo quid sit rethorica, deinde quod genus rethorice, que materia, quod officium, quis finis, que partes, que species, quod instrumentum, quis artifex et quare rethorica dicatur. Ad ultimum quo ordine docenda sit et discenda. Dicit autem Victorinus quod docere hec hoc est artem extrinsecus docere, ideo scilicet quia extra et antequam ad litteram artis perveniatur oportet hec omnia prescire. Quis enim audeat se profiteri rethoricem nisi sciat quid sit rethorica, de quo scientiarum sit genere et cetera que enumerata sunt? Vel ideo extrinsecus quod totus artis vigor et virtus in officio exercendo consistit. Illa autem que enumerata sunt extra officium artis sunt. Non enim informant oratorem ut per ea melius suum officium complere sciat; ideoque docere illa est artem extrinsecus docere. Intrinsecus vero docere artem est tradere precepta que

eleven questions, but omits both the *ars extrinsecus / ars intrinsecus* and the *circa artem / circa librum* distinctions, and will say nothing further about the book he is paraphrasing, Priscian's *Institutiones grammaticae*.[164] Gundissalinus also uses Thierry's schema in the *De divisione philosophiae*, which he most likely knew through Petrus Helias's *Summa* on the *De inventione*, as we can surmise from the eleven questions he includes *circa artem*. Gundissalinus does make some changes inspired by works that were not available to his predecessors, such as al-Fārābī's *Catalogue of Sciences* and Avicenna's *Logic*. The question of Gundissalinus's sources has provoked considerable scholarly debate.[165] To appreciate Thierry's distinctiveness it may be relevant to observe that Gundissalinus applies the *ars extrinsecus / ars intrinsecus* distinction to grammar, and not to rhetoric, as Thierry and Petrus Helias do.[166] Second, Gundissalinus associates this distinction with the fact that each art is divided into theory and practice, a division that he finds in al-Fārābī's *Catalogue of Sciences*, a work unknown in the Latin West before the translations

oratorem ad persuadendum informant. Nec est dicendum ars extrinseca vel intrinseca, ne due artes esse putentur, sed ars extrinsecus et intrinsecus ut duo diversi modi docendi unam et eandem artem per adverbia illa designentur. Omnis enim ars duobus modis docetur, extrinsecus scilicet et intrinsecus, sicut iam supra dictum est." Ibid., p. 175: "Hoc ordine docenda et discenda est. Primo enim agendum est de inventione, secundo de dispositione, tertio de pronunciatione, quarto de memoria, ad ultimum de elocutione. Ratio autem huius ordinis in *Rhetorica ad Herennium*, quam *Secundam* uocant *Rethoricam* assignabitur. Et hec quidem circa artem extrinsecus inquirenda sunt. Circa librum vero Tullii consideranda sunt ista. Que sit eius intentio in hoc opere, que libri utilitas, an opus hoc ad philosophiam pertineat, quis sit operis titulus. Intentio ergo Tullii est in hoc opere tractare primam partem rethorice artis, id est inventionem. Non enim in hoc opere totam intendit rethoricam tradere, sed de parte eius prima solummodo agere. Utilitas libri est scientia facile inveniendi et copiose que ad persuadendum sunt utilia."

164 Petrus Helias, *Summa super Priscianum* (ed. Reilly, pp. 61–65).

165 Nikolaus M. Häring argued, wrongly, that Thierry of Chartres relied on Gundissalinus: see "Thierry of Chartres and Gundissalinus," *Mediaeval Studies* 26 (1964), 271–286. On the other hand, Richard Hunt correctly situated Thierry before Petrus Helias and Gundissalinus; see "The Introduction to the *Artes*," pp. 86–93, as well as Richard McKeon, "Rhetoric in the Middle Ages," *Speculum* 17 (1942), 17. Cf. on this subject John Ward's article as well as Karin M. Fredborg's introduction to the rhetorical commentaries of Thierry of Chartres, cited above. See also Charles Burnett, "A New Source for Dominicus Gundissalinus's Account of the Science of the Stars?" *Annals of Science* 47 (1990), 361–374; Alexander Fidora, *Die Wissenschaftstheorie des Dominicus Gundissalinus: Voraussetzungen und Konsequenzen des zweiten Anfangs der aristotelischen Philosophie im 12. Jahrhundert* (Berlin, 2003), pp. 67–76.

166 Dominicus Gundissalinus, *De divisione philosophiae: Über die Einteilung der Philosophie*, ed. and trans. Alexander Fidora and Dorothée Werner (Freiburg, 2007), p. 110.

from the Arabic.[167] At the very end of the chapter on rhetoric, Gundissalinus states that the (eleven) points that he just illustrated go back to the *ars extrinsecus*.[168] But Gundissalinus does not forget the *accessus circa librum*, and at the end of his *De divisione philosophiae*, he explains that it is necessary, when discussing books concerning the arts, to ask *seven* questions[169] (instead of Petrus Helias's four and Thierry's two).

The Trinity and the Four Causes

In the *Tractatus de sex dierum operibus*, Thierry shows that in the account of the creation of the world set forth in *Genesis* it is possible to find a correspondence between the Trinity and the four (Aristotelian) causes. The Father is the efficient cause that creates matter *ex nihilo*; the Son, the co-eternal wisdom of God, is the formal cause that informs and arranges matter; the Holy Ghost, the goodness of God, is the final cause that governs and loves the matter informed by Wisdom; the four elements (the "sky" and the "earth" of *Genesis*), that is, the matter created by the efficient cause, are themselves the material cause.[170] In the words of Enzo Maccagnolo, we can state that for Thierry of Chartres "the creative act is a simultaneous act of the Trinity."[171]

167 Ibid., p. 110. However, the division of medicine into *theorica* and *practica* was known in the Latin West by way of translations by Constantine the African; see Danielle Jacquart, "*Theorica* et *practica* dans l'enseignement de la médecine à Salerne au XIIe siècle," in *Vocabulaire des écoles et des méthodes d'enseignement au Moyen Age, Actes du colloque, Rome 21–22 octobre 1989*, ed. Olga Weijers (Turnhout, 1992), pp. 102–110. Moreover, Hugh of Saint Victor divides geometry into *theorica* and *practica* in his *Practica geometriae*. The issue merits extended, independent discussion.

168 Dominicus Gundissalinus, *De divisione philosophiae* (ed. Fidora and Werner, p. 148).

169 Ibid., pp. 262–266. This was also remarked upon by Hunt, "The Introductions to the *Artes*," p. 129, who explains that in the case of both Petrus Helias and Gundissalinus, it is a combination of the C and D types of *accessus*.

170 Thierry of Chartres, *Tractatus de sex dierum operibus* 2–3 (ed. Häring, pp. 556–557): "Si quis igitur subtiliter consideret mundi fabricam, efficientem ipsius causam deum esse cognoscet formalem vero dei sapientiam finalem eiusdem benignitatem materialem vero quatuor elementa que et ipse creator IN PRINCIPIO de nichilo CREAVIT. ... In materia igitur que est quatuor elementa operatur summa Trinitas ipsam materiam creando in hoc quod est efficiens causa: creatam informando et disponendo in eo quod est formalis causa: informatam et dispositam diligendo et gubernando in eo quod est finalis causa. Nam pater est efficiens causa filius vero formalis spiritus sanctus finalis quatuor vero elementa materialis. Ex quibus quatuor causis universa corporea substantia habet subsistere."

171 Maccagnolo, *Rerum universitas*, pp. 227–229.

In the last lines of the *Tractatus*, which is unfinished, Thierry of Chartres explains that the *unitatis aequalitas*, that is, the Son, is the *forma essendi* or the formal eternal cause of all existing things.[172] This sentence summarizes the philosophy explained in the *Tractatus* by evoking at the same time the four causes, the generation of the Trinity *ad intra*, God as *forma essendi*, as well as a beginning of the theory that historians of philosophy have called the *Trinitas perpetuorum* – the presence and perpetual operation of the three persons of the Trinity in the created world – that Thierry expounds in the *Commentum super Boethii librum De Trinitate*.[173]

The commentary on the *De arithmetica* develops a version of the "Trinity of perpetuals" (although the term is not used) that is more complex than that of the *Tractatus* and less accomplished than the one in the *Commentum*, as well as the theory of four causes. It also includes important remarks on physics and mathematics: mathematics, conceived as knowledge of equality and inequality, is useful in physics.[174] This, in my view, sheds light on the philosophical project of the *Tractatus*, making it more comprehensible. In the commentary on the *De arithmetica*, we learn that physics studies "the cause of the sensible world" and the formation of sensible bodies; the reading of *Genesis* "according to physics" that Thierry proposes to carry out in the *Tractatus* means in fact that he will insist on the objects studied by physics, in particular the formation of the sensible world and of bodies. But we also see more clearly the role of mathematics here: the knowledge of equality and inequality, studied by mathematics, is useful for physics, because it allows us to understand the different weights of the elements and, consequently, to explain their place in the cosmos, and to recognize that the equilibrium in sensible bodies can be broken by the inequality caused by an excess of warmth or of one of the four humors.[175]

172 Thierry of Chartres, *Tractatus de sex dierum operibus* 45 (ed. Häring, p. 574): "… non est dubium quin ipsa unitatis equalitas sit rebus omnibus forma essendi eterna ac formalis causa secundum quam artifex eternus modum existendi omnibus rebus constituit."

173 Thierry of Chartres, *Commentum super Boethii librum De Trinitate* 2.39 (ed. Häring, pp. 80–81): "Ab hac igitur sancta et summa Trinitate descendit quedam perpetuorum Trinitas. Unitas enim secundum hoc quod unitas est materiam creat. Formas vero rerum secundum hoc quod est unitatis equalitas. Ex eo quod amor est et conexio creat spiritum."

174 Thierry of Chartres, *Commentum super Arithmeticam Boethii* I, XXXII, 1 (pp. 152–153 below).

175 Thierry of Chartres, *Commentum super Arithmeticam Boethii* I, XXXII, 1 (p. 153 below): "Quatuor enim humoribus animantium vegetatur natura: sanguine, flegmate, colera nigra, quae melancolia dicitur, et rubea. Qui quidem humores quamdiu adaequantur in corporibus, viget complexio. Cum vero subrepit humoris

Unitas *and* aequalitas

These theories of the Trinity and the four causes are linked to the doctrine of unity and equality, a doctrine that met with much success in the twelfth century and beyond to become perhaps one of the best known and most studied themes in Thierry's writings.[176] In the *Lectiones,* Thierry explains that one can treat the Trinity *theologice, mathematice,* and *ethice,*[177] and also that it is precisely *mathematice* that Augustine spoke of the Trinity when he stated that the Father is unity, the Son equality, and the Holy Ghost the conjunction of unity and equality. In the *Glosa,* Thierry explains that it is possible to discover the Trinity in all sensible things as Augustine did in the *De Trinitate* because, according to physical doctrine, we find "weight, measure, and number" in sensible things; he repeats that we can discuss the Trinity theologically, mathematically, and ethically.[178] In the *Tractatus*

et caloris inaequalitas, infirmatur animal et plerumque moritur. Ecce quid scientia aequalitatis et inaequalitatis valet ad hanc partem phylosophiae." On medicine in twelfth century, see Richard McKeon, "Medicine and Philosophy in the Eleventh and Twelfth Centuries: The Problem of Elements," *The Thomist* 24 (1961), 211–256, and Danielle Jacquart, *La science médicale occidentale entre deux renaissances (XIIe s.–XVe s.)* (Aldershot, 1997).

176 In addition to the studies cited in n89, see Tullio Gregory, *Anima mundi*, pp. 59–63, 73–74; Stephen Gersh, *Concord in Discourse: Harmonics and Semiotics in Late Classical and Early Medieval Platonism* (Berlin and New York, 1996), pp. 115–118; Fernand Brunner, "*Creatio numerorum, rerum est creatio,*" in *Mélanges offerts à René Crozet à l'occasion de son soixante-dixième anniversaire,* ed. Pierre Gallais and Yves-Jean Riou, 2 vols. (Poitiers, 1966), 2: 719–725.; Efrem Betton, "Avicebron è l'unica fonte dell'ilemorfismo universale?" in *Actas del V Congreso internacional de filosofia medieval,* 2 vols. (Madrid, 1979), 1: 619–629; Carlo Riccati, *"Processio" et "Explicatio": La doctrine de la création chez Jean Scot et Nicolas de Cues* (Naples, 1983), pp. 69–72; Jean-Michel Counet, "Les mathématiques au service d'une théologie de la matière chez Thierry de Chartres et Clarembaud d'Arras," in *Vie spéculative, vie méditative et travail manuel à Chartres au XIIe siècle,* pp. 103–114; Christian Trottmann, "*Unitas, aequalitas, conexio*: Alain de Lille dans la tradition des analogies trinitaires arithmétiques," in *Alain de Lille, le docteur universel. Philosophie, théologie et littérature au XIIe siècle,* ed. Jean-Luc Solère, Anca Vasiliu and Alain Galonnier (Turnhout, 2005), pp. 401–427.

177 Thierry of Chartres, *Lectiones in Boethii librum De Trinitate* 7.5 (ed. Häring, p. 224): "Tribus enim modis de Trinitate loquimur: theologice scilicet mathematice et ethice. Et Augustinus quidem mathematice dicit quod in patre est unitas in filio equalitas in spiritu sancto unitatis equalitatisque conexio."

178 Thierry of Chartres, *Glosa super Boethii librum De Trinitate* 5.17 (ed. Häring, pp. 296–297): "Hanc Trinitatem in rebus singulis investigat Augustinus in libro *De Trinitate* et secundum phisicam doctrinam invenit pondus mensuram et numerum: trinitatem in rebus singulis. Quod evidenter docet Claudianus Mamertus in eo

de sex dierum operibus, he proposes to demonstrate briefly, "in this theology," in what manner the *probationes arithmeticae, musicae, geometriae,* and *astronomicae* lead man to knowledge of the creator. We may wonder what is meant here by *hac theologia.*[179] Based on what he states a few lines earlier, it is clear that Thierry does not want to linger on mathematical considerations and the Trinity in the framework of a literal commentary on *Genesis,* unless it is to show the creator's action on created things. He details his arguments in his commentaries on Boethius's *De Trinitate,* as well as in his commentary on the *De arithmetica.* In fact, in the *Tractatus,* only the arithmetical proofs (*probationes arithmeticae*) are explained, and these only partially. Thierry begins his demonstration by postulating the unity / alterity distinction: unity, he states, precedes all alterity.[180] Unity, he adds, is divinity itself and divinity the *forma essendi* of all things, a concept and a formulation borrowed from Boethius's *De hebdomadibus,* the interpretation of which differs among medieval as well as modern commentators.[181] However, it seems

capitulo in quo de numero et pondere et mensura anime et dei agit. Unde huc quia transferre doctrinam potes quomodo hoc sit pretermitto. Tractatur autem tripliciter de Trinitate preter hunc modum: theologice mathematice ethice. Ethica autem consideratio fidem spem caritatem pro trinitate attendit cuius et theologice studiosi sunt investigatores theologi. Mathematicam super hanc rem doctrinam non addit nisi Augustinus dicens unitatem esse in patre et eiusdem unitatis equalitatem in filio atque unitatis equalitatisque conexionem et amorem in spiritu sancto."

179 Thierry of Chartres, *Tractatus de sex dierum operibus* 30 (ed. Häring, p. 568): "Adsint igitur quatuor genera rationum que ducunt hominem ad cognitionem creatoris: scilicet arimethice probationes et musice et geometrice et astronomice. Quibus instrumentis in hac theologia breviter utendum est ut et artificium creatoris in rebus appareat et quod proposuimus rationabiliter ostendatur." *Hac theologia,* in Maccagnolo's rational theology (see *Rerum universitas,* pp. 202–203) is not very persuasive.

180 Thierry of Chartres, *Tractatus de sex dierum operibus* 30 (ed. Häring, p. 568): "Omnem alteritatem unitas praecedit quoniam unitas pracedit binarium qui est principium omnis alteritatis."

181 Boethius, *De hebdomadibus* 2: "Diversum est esse et id quod est; ipsum enim esse nondum est, at vero quod est accepta forma essendi est atque consistit." See Pierre Hadot, "*Forma essendi:* Interprétation philologique et interprétation philosophique d'une formule de Boèce," *Les études classiques* 38 (1970), 143–156; Fernand Brunner, "*Deus forma essendi,*" in *Entretiens sur la renaissance du 12e siècle,* ed. Maurice de Gandillac and Edouard Jeauneau (Paris, 1968), pp. 85–102; Bruno Maioli, *Gilberto Porretano: Dalla grammatica speculativa alla metafisica del concreto* (Rome, 1979). For an account of the different interpretations, medieval and modern, of the *forma essendi,* see Alain Galonnier's Introduction to the *De hebdomadibus* in Boethius, *Opuscula sacra,* ed. Claudio Moreschini, trans. and comm. Alain Galonnier, 2 vols. (Louvain and Paris, 2007–2013), 1: 337–348.

certain that Thierry of Chartres was the first to identify the *forma essendi* with God, and becomes something of a "trademark" of the Chartrian philosopher. It follows that *unitas* is the *forma essendi* of all singular things since: "Omne quod est, ideo est, quia unum est." Thierry relies on Boethius's second commentary on Porphyry's *Isagoge*: "... omne enim quod est, idcirco est, quia unum est."[182] In the *Contra Eutychen et Nestorium*, Boethius states that: "Quod enim non est unum, nec esse omnino potest. Esse enim atque unum convertitur, et quodcumque unum est, est."[183] Boethius, we should note, is explicitly cited in the commentary on the *De arithmetica* preserved in Bern, and contains theories that unquestionably derive from Thierry of Chartres's teaching:

> Omne enim quod est participatione unitatis habet esse. Unde Boethius dicit: *Omne quod est, ideo est, quia unum numero est.* Hac igitur ratione quidam numerum non esse volebant, quod quia a philosophia remotum est numerum non esse, ostendendum est. Et prius quid sit unitas, et quomodo ex ea numerus descendat videamus. Unitas est ipsa rerum entitas, id est existentia, scilicet id quod ex se existit, et per quod omnia existere habent. Unde eam essendi formam nominat, quae forma ex se existere habet, et est actu ab aeterno. Haec quoque sola forma non indiget materia, immo actu est extra omnem materiam, et ut aperte dicatur: haec forma nichil aliud est quam ipsa sapientia artificis, per quam omnia existere habent. Sed vero appareat quomodo per ipsam omnia existant, altius incipiendum est. Volunt philosophi duo prima rerum esse principia, actum scilicet divinum et potentiam, id est unitatem quae est forma essendi ex se existens et quae ex sui natura una est, et materiam primordialem quae nec ex se est nec ex se non est, nec ex se iterum una

182 Thierry of Chartres, *Tractatus de sex dierum operibus* 31 (ed. Häring, p. 569); cf. Boethius, *In Isagogen Porphyrii commenta editio prima* 1.10 (CSEL 48: 162).

183 Boethius, *Contra Eutychen et Nestorium* 4, ed. Claudio Moreschini in *Opuscula theologica* (Munich, 2000), p. 220. This is not the place to launch into an extended discussion of the transcendentals and their convertibility, a much-discussed subject over the past few years. For the twelfth century, see Luisa Valente, "*Illa quae transcendunt generalissima*: Elementi per una storia latina dei termini trascendentali (XII secolo)," *Quaestio: Annuario di storia della metafisica* 5 (2005), 217–239; Luisa Valente, "*Ens, Unum, Bonum*: Elementi per una storia dei trascendentali in Boezio e nella tradizione boeziana del XII secolo," in *Ad Ingenii Acuitionem: Studies in Honour of Alfonso Maierù*, ed. Stefano Caroti, Ruedi Imbach, Zénon Kaluza, Giorgio Stabile, and Loris Sturlese (Louvain-la-Neuve, 2006), pp. 483–545; Luisa Valente, "Names That Can Be Said of Everything: Porphyrian Tradition and 'Transcendental' Terms in Twelfth-Century Logic," *Vivarium* 45 (2007), 298–310.

est nec ex se plura est, sed potens esse una, potens esse plura. Potest enim esse terra, aqua, aer, ignis et quidlibet aliud secundum susceptionem diversarum formarum. Quod ex se non sit una materia apparet, quia tunc ex se unitate quae forma est essendi participaret. Iterum si ex se esset eadem forma participaret ex se, quod ab omni philosophia remotum est. Nec iterum ex se plura est, quia tunc ex se pluribus formis participaret. Non tamen ideo concedemus non esse ex se. Haec igitur talis potentia principium numquam habuit. Non tamen potest dici eam fuisse, quia haec verba esse, fuisse, fore idem significant quod unitas. Omnia enim ista vocabula unum, ens et unitas idem significant, et dicitur unitas a greco quod est *on* quasi *onitas,* id est rerum entitas.[184]

However, the Bern commentary seems to have been composed in the wake of Thierry's commentaries on the *Opuscula sacra.* The fragmentary form in which we now have it does not give any indication of the author's name.

In the *Tractatus de sex dierum operibus,* after having postulated the identity between unity, divinity, and the *forma essendi,* Thierry explains how unity engenders equality. He shifts the discourse from the *ad extra* activity of the divine Trinity, where the question at issue was the relation between unity and created things, to the *ad intra* activity, since he is now speaking of the generation of the persons of the Trinity. In the *De doctrina christiana,* Augustine stated that in the Father there is unity, in the Son equality, and in the Holy Ghost concord, and that the three are one thanks to the Father, equal thanks to the Son, and united thanks to the Holy Ghost.[185] Augustine does explore the underlying mathematical properties of the triune person, but in relying on this text, Thierry notes that, according to the rules of arithmetic, numbers are generated either through the multiplication of one number by another number, or through powers, that is, the repeated multiplication of one number by itself. Unity multiplied by itself engenders

184 Bern, Burgerbibliothek, Cod. 633, fol. 25va.

185 Augustine, *De doctrina christiana* 1.5.12 (CCL 32: 9): "In patre unitas, in filio aequalitas, in spiritu sancto unitatis aequalitatisque concordia, et tria haec unum omnia propter patrem, aequalia omnia propter filium, conexa omnia propter spiritum sanctum." This passage from the *De doctrina christiana* is also evoked by Otloh of Saint Emmeram in the *Dialogus de tribus quaestionibus,* as Stephen Gersh has remarked in *Concord in Discourse,* pp. 115–120. However, Thierry's 'mathematical' approach is very different from Otloh's. On Otloh, see also Thierry Lesieur, *Sciences des nombres et spiritualité entre Danube et Meuse (IXe–XIIe siècles)* (Paris, 2009), pp. 24, 25.

equality, while unity multiplied by other numbers engenders all the numbers.[186] But unity is eternal and precedes all numbers. Thus the equality of unity, engendered by unity, also precedes all numbers itself. It is an eternal generation (*generatio aeterna*), but since two or more eternal things cannot exist, the result is that "unitas igitur et aequalitas unitatis unum sunt."[187] Unity remains identical to itself and, through this engendering, it also becomes equal to itself. The equality of unity is the equality of the existence of things, that is, the definition or the eternal determination of all things beyond and beneath which existence (*esse*) is not possible. It is this frontier of being, as it were, that Thierry says the ancient philosophers called equality of unity or *mens divina*, providence or the wisdom of the creator.[188] If the equality of unity is the equality of the existence of things, by determining it almost like a law and an eternal rule of its existence, then the equality of unity is the eternal *forma essendi* and the formal cause according to which the creator constituted the mode of existence of all things.[189] The *Tractatus* is unfinished and stops abruptly just after having announced the explanation of the way in which the "conexio aequalitatis et unitatis ab utraque earum procedat."[190] Fortunately, Thierry continues the demonstration in the three commentaries on Boethius's *De Trinitate*, explaining that the *conexio*, which he plainly identifies with the Holy Ghost, is the link and the love that unites equality and unity and that it proceeds from both.[191] However, the *Commentum* as well as the commentary on the *De arithmetica* devote more space to the generation of equality from unity, using very similar reasoning and

186 Thierry of Chartres, *Tractatus de sex dierum operibus* 38–39 (ed. Häring, p. 571): "Hec igitur duplex generatio circa unitatem potest reperiri. Et quidem per alios numeros multiplicata omnes numeros generat. Ex se autem et ex sua substantia nichil aliud generare potest nisi equalitatem cum alii numeri ex se multiplicati inequalitates producant. Unitas enim semel nichil aliud est quam unitas. Est igitur gignentis et geniti una et eadem substantia quoniam utraque vera unitas. Unitas enim per se nichil aliud gignere potest nisi eiusdem unitatis equalitatem. Nam cum equalitas inequalitatem precedat necesse est generationem equalitatis precedere. Cum igitur unitas gignat utrumque et per quemlibet numerum multiplicata non possit gignere nisi inequalitatem necesse est ut gignat equalitatem per illud quod naturaliter omnes numeros precedit. At illud est unitas. Unitas igitur ex se et ex sua substantia nichil aliud gignere potest nisi equalitatem."

187 Ibid., 40 (p. 572).

188 Ibid., 42 (pp. 572–573).

189 Ibid., 45 (p. 574).

190 Ibid., 47 (p. 575).

191 Thierry of Chartres, *Commentum super Boethii librum De Trinitate* 2.31–38 (ed. Häring, pp. 78–80); *Lectiones in Boethii librum De Trinitate* 5.16 (ed. Häring, p. 218); *Glosa super Boethii librum De Trinitate* 16–19 (ed. Häring, p. 307).

formulations, in particular the eternal tetragon, the reference to Paul and the prophecy of the Spanish Sibyl, and the idea that equality is the integrity of being (*essendi videlicet integritas*).[192] It is the engendering of the Son that especially interests Thierry of Chartres in this phase. The similarities between the *Commentum* and the commentary on the *De arithmetica* do not stop here: both works share the idea that the *conexio* proceeds from both unity and equality, in very close terms that are worth highlighting, despite the fact that in the *Commentum* the developments on the Trinity are, as is normal, more important:

> Amor autem hic et conexio nec gignitur nec gignit sed ab unitate et ab unitatis equalitate procedit: non ab uno scilicet illorum sed ab utroque.[193]

> Unitas autem, ut saepe diximus, una est. Est enim unitas, est unitatis aequalitas, est et unitatis conexio et aequalitatis, quae ab unitate et unitatis aequalitate procedit.[194]

In the commentary on the *De arithmetica*, we find the generation of equality and *conexio* from unity again, inserted to illustrate chapter XXXII of the first book of the *De arithmetica*, in which Boethius demonstrates how inequality proceeds from equality. Compared to the *Tractatus*, we find a supplementary identification: unity is now called the *entitas* of things, since it confers being upon them:

> Dicitur autem unitas quasi entitas, quod nihil aliud est quam ipsam esse rerum intelligi. Deus autem rerum est entitas, quare et ipse vere est unitas. Eadem vero unitas entitatis est aequalitas. Haec enim est essendi forma, perfectio et integritas … .[195]

This idea is expressed in the three commentaries on the *De Trinitate*, where another explanation is advanced: *entitas* is in fact *onitas*, a word

192 Thierry of Chartres, *Commentum super Boethii librum De Trinitate* 2.31 (ed. Häring, p. 78).
193 Thierry of Chartres, *Commentum super Boethii librum De Trinitate* 2.38 (ed. Häring, p. 80).
194 Thierry of Chartres, *Commentum super Arithmeticam Boethii* II, XLII,1 (p. 199 below).
195 Thierry of Chartres, *Commentum super Arithmeticam Boethii* I, XXXII, 1 (p. 146 below).

invented by Thierry discussed above.[196] In the commentary on the *De arithmetica*, we read also that the equality of the entity is the *forma essendi* from which are derived the forms of all things, such as humanity and "lapidity," which are the equality of being of man and of stone respectively.[197] God is the entity of all things; humanity is the entity of man and divinity is the entity of humanity.[198] These same formulations and concepts are found in the commentaries on the *De Trinitate*.[199]

As already suggested, the philosophy of Thierry of Chartres shows a striking resemblance with the thought of Nicholas of Cusa. Although extensive comparison of the two figures is outside the limits of this introduction, and the "historiographical legitimacy" of comparing two thinkers separated by three centuries in any case must seem doubtful, a number of studies, some very recent, nevertheless, have tried to clarify the cardinal's debt to Thierry.[200] The discovery of the commentary on the *De arithmetica* should serve to anchor further analysis in specific philosophical and textual sources. For example, a passage from the *De docta ignorantia* shows remarkable similiarity with the excerpt from Thierry's commentary on the *De arith-*

196 See above pp. 39–40, 53.

197 Thierry of Chartres, *Commentum super Arithmeticam Boethii* I, xxxii, 1 (pp. 146–147 below): "Haec est ergo illa essendi forma et entitatis aequalitas ... ."

198 Ibid., I, xxxii, 1 (p. 148 below): "Deus enim omnium rerum est entitas ... ."

199 The parallels are all catalogued in the notes to the edition of the *Commentum* that follows.

200 Pierre Duhem, "Thierry de Chartres et Nicolas de Cuse," *Revue des sciences philosophiques et théologiques* 2 (1909), 525–531; Riccati, *"Processio" et "Explicatio"*; Werner Beierwaltes, "Unità e Uguaglianza: Una formulazione del problema nel platonismo di Chartres e la sua recezione attraverso Nicola Cusano," in *Pensare l'Uno: Studi sulla filosofia neoplatonica e sulla storia dei suoi influssi* (Milan, 1991, Italian trans. of *Denken des Einen: Studien zum Neuplatonismus und dessen Wirkungsgeschichte* [Frankfurt, 1985]), pp. 315–328; Jean-Michel Counet, *Mathématiques et dialectique chez Nicolas de Cuse* (Paris, 2000); Marteen J. Hoenen, "'Ista prius inaudita': Eine neuentdeckte Vorlage der *Docta ignorantia* und ihre Bedeutung für die frühe Philosophie des Nikolaus von Kues," *Medioevo: Rivista di storia della filosofia* 21 (1995), 375–476. See also the contributions of David C. Albertson, especially "'Our Boethius': Traditions of Thierry of Chartres and the Christology of Nicolaus Cusanus in *De docta ignorantia* (PhD diss., University of Chicago, 2008); "A Learned Thief? Nicholas of Cusa and the Anonymous *Fundamentum Naturae*: Reassessing the *Vorlage* Theory," *Recherches de théologie et philosophie médiévales* 77 (2010), 351–390; "A Late Medieval Reaction to Thierry of Chartres's (d. 1157) Philosophy: The Anti-Platonist Argument of the Anonymous *Fundamentum Naturae*," *Vivarium* 50 (2012), 53–84; "Achard of St. Victor (d. 1171) and the Eclipse of the Arithmetic Model of the Trinity," *Traditio* 67 (2012), 101–144. Albertson's *Mathematical Theologies: Nicholas of Cusa and the Legacy of Thierry of Chartres* (Oxford, 2014) appeared after my own study had been completed.

metica just cited, and the question of whether Nicholas of Cusa consulted it directly becomes more legitimate:

> De generatione aeterna. Ostendamus nunc brevissime ab unitate gigni unitatis aequalitatem, conexionem vero ab unitate procedere et ab uni tatis aequalitate. Unitas dicitur *quasi ontas* ab *on* Graeco, quod Latine ens dicitur; et est unitas quasi entitas. Deus namque ipsa est rerum entitas; forma enim essendi est, quare et entitas. Aequalitas vero unitatis quasi aequalitas entitatis, id est aequalitas essendi sive existendi. Aequalitas vero essendi est, quod in re neque plus neque minus est; nihil ultra, nihil infra. Si enim in re magis est, monstruosum est; si minus est, nec est. Generatio aequalitatis ab unitate clare conspicitur, quando quid sit generatio attenditur. Generatio est enim unitatis repetitio vel eiusdem naturae multiplicatio a patre procedens in filium. Et haec quidem generatio in solis rebus caducis invenitur. Generatio autem unitatis ab unitate est una unitatis repetitio, id est unitas semel; quod, si bis vel ter vel deinceps unitatem multiplicavero, iam unitas ex se aliud procreabit, ut binarium vel ternarium vel alium numerum. Unitas vero semel repetita solum gignit unitatis aequalitatem; quod nihil aliud intelligi potest quam quod unitas gignat unitatem. Et haec quidem generatio aeterna est.[201]

We will return to these possible affinities between the writings of Nicholas of Cusa and the commentary on the *De arithmetica* when we explore the theory of *complicatio* and *explicatio*.

The Prophecy of the sybilla hyspana

The generation of equality from unity, which we have just analyzed in detail, was also announced by a prophecy. While commenting on *De arithmetica* I, XXXII, which deals with equality and inequalities (*aequalitas / omnes species inaequalitatis*), Thierry compares the generation of the Son to that of the first tetragon. We find the same reasoning that he had developed in the *Tractatus*, by which one multiplied by itself (1×1) produces equality;[202] however, now he calls the Son "the first tetragon" (that is, the first square), and adds that this eternal generation of the tetragon was announced by the prophecy of the Spanish Sibyl:

201 Nicholas of Cusa, *De docta ignorantia* 1.8.
202 Thierry of Chartres, *Tractatus de sex dierum operibus* 40 (ed. Häring, pp. 571–572); cf. above pp. 50–54.

> ... hoc est aequalitas illa quae ad concordiam redigit omnem naturaliter discordiam. Haec est tetragona<tura> prima, quam ex se gignit unitas, quae multiplicata per se primum efficit tetragonum. Semel enim unum, primus tetragonus est. Haec est aeterna illa tetragoni generatio, de qua hyspana sybilla in suo dicit vaticinio: "Cum pervenitur ad costam tetragoni sedentis aeterni, et ad costas tetragonorum stantium aeternorum, et caetera." Tetragonum quidem sedentem et aeternum appellat dei filium, quem unitatis natura ex sua genuit substantia.[203]

Thierry evokes this prophecy a second time when explaining that it is the force of the square that makes equality the *forma essendi* of all things, and that, further, divinity is the cause or the form of all things. It is precisely for these reasons, Thierry concludes, the Spanish Sibyl spoke of the Son of God as the "eternal tetragon."[204] The prophecy also appears in the *Commentum super Boethii librum De Trinitate,* where he explains that the Son is the first tetragon, playing on the double meaning of this term, that is to say the first number squared and the figure of the square, the most perfect figure, since it has all equal sides; this is why, Thierry adds, we say that the Son is the "figure" of the substance of the Father.[205] The generation of the Son is eternal and the Spanish Sibyl herself calls the Son "eternal tetragon":

> Vel filium ideo figuram substantie patris appellat quia substantia patris est unitas. Sed unitas semel tetragonatura prima est. Sicut enim binarius bis vel quaternarius quater arismethica ratione tetragonum constituit ita quoque unitas semel tetragonum primum efficit. Sed hec tetragonatura est generatio filii et, ut verum fatear, tetragonatura hec prima est generatio. Hec autem ab eterno est. Generatio autem filii eterna est. Et quoniam tetragonatura prima generatio filii est, et filius tetragonus primus est. Tetragonatio vero figura est. Merito ergo filium figuram substantie patris appellat. Quem tetragonum eternum sibilla hispana

203 Thierry of Chartres, *Commentum super Arithmeticam Boethii* I, XXXII, 3 (p. 152 below).

204 Thierry of Chartres, *Commentum super Arithmeticam Boethii* II, XXXI, 1 (p. 186 below): "Ex vi namque quadrati aequalitas et forma est existendi aequalitas, ut iam saepe dictum est. Dico itaque quod ex vi quadraturae et unitatis forma habet materiam determinare. Divinitas ergo ex vi quadrati forma vel causa rerum est. Et hoc est quare sybilla hyspana in vaticinio suo filium dei tetragonum aeternum nominat."

205 On the *tetragonatura* and the double meaning of the first tetragon in the *Commentum super Boethii librum De Trinitate,* see the illuminating contribution by Vera Rodrigues, "Thierry de Chartres," pp. 656–660.

> nominat dicens: "Cum perveneris ad costam tetragoni sedentis eterni, et ad costam tetragonorum stantium eternorum." Bene autem tetragonus filio attribuitur quoniam figura hec perfectior ceteris propter laterum equalitatem iudicatur.[206]

In the commentary on the *De arithmetica*, Thierry explains that quadrilaterals (*tetragoni*) are of the substance of unity because they are engendered by odd numbers that participate in unity.[207] Elsewhere in the same work, he affirms that the equality of the entity is precisely the figure of unity, and the Apostle (that is, Saint Paul) called the Son of God "the figure and the substance of the Father."[208] This constellation of ideas is linked to the prophecy in the *Commentum*, Thierry's first commentary on Boethius's *De Trinitate*. This not only confirms that Thierry is indeed the author of the commentary on the *De arithmetica* but also demonstrates it was written before the commentaries on the *Opuscula sacra*. In the *Tractatus de sex dierum operibus*, the tetragon is merely mentioned, following Boethius's *De arithmetica*, because it has equal sides,[209] and the equality of unity itself is described as almost the figure and the splendor of unity.[210] In the *Lectiones* and the *Glosa*, there are no references to the eternal tetragon. However, Thierry speaks explicitly of

206 Thierry of Chartres, *Commentum super Boethii librum De Trinitate* 2.34 (ed. Häring, pp. 78–79). The Apostle Paul is mentioned in *Commentum super Boethii librum De Trinitate* 2.32 (ed. Häring, p. 78).

207 Thierry of Chartres, *Commentum super Arithmeticam Boethii* II, XXXVI, 1 (p. 193 below): "CONSTAT IGITUR. Quia tetragonus eiusdem essentiae est per medios impare, impares vero eiusdem propter unitatis participationem, ergo unitas in primo loco est eiusdem substantiae"; ibid., II, XXXI, 3 (p. 188): "VEL AB AEQUALITATE FOR<MANTUR> TETRAGONI: quia sicut tetragonus per aequales dimensiones crescit, vel quia ex imparibus nascitur, ut supra monstratum est. ET QUONIAM. Vult ostendere non solam unitatem esse eiusdem substantiae, sed et omnes impares quorum ipsa effectrix est, et omnes tetragonos qui ab imparibus procedunt."

208 See Heb. 1:3: "Qui cum sit splendor gloriae et figura substantiae eius ... ." See Thierry of Chartres, *Commentum super Arithmeticam Boethii* I, XXXII, 1 (p. 147 below): "Est ergo haec illa entitatis aequalitas quae vere et proprie dicitur unitatis ipsius figura. Quod Apostolus, subtilissime inspiciens, filium dei figuram substantiae patris appellat."

209 Thierry of Chartres, *Tractatus de sex dierum operibus* 37 (ed. Häring, p. 571); passage cited p. 27 above.

210 Thierry of Chartres, *Tractatus de sex dierum operibus* 41 (ed. Häring, p. 572): "Est igitur ipsa unitatis equalitas eiusdem unitatis quasi quedam figura et splendor. Figura quidem quia est modus secundum quem ipsa unitas operatur in rebus. Splendor vero quia est id per quod omnia discernuntur a se invicem. Fine enim modoque proprio cuncta invicem a se discreta sunt."

the "eternal generation" of unity in the *Tractatus*,[211] as well as in the *Glosa*: "Apparet quod unitas de unitate generatur. Sed quia hoc a principio fuit – ex quo enim si dici potest fuit unitas mox fuit similiter unum quod est unitas – ideo hanc generationem eternam appello."[212] This doctrine appears to be unique to Thierry and does not seem to be found elsewhere in the twelfth century.[213] Note that the *generatio aeterna* is common to the commentary on the *De arithmetica*, to the *Tractatus de sex dierum operibus*, to the *Commentum*, and to the *Glosa* on Boethius's *De Trinitate*.

The prophecy of the Spanish Sibyl is also mentioned in the *Tractatus super librum Boetii De Trinitate* by Clarembald of Arras who, by all accounts, recopied and rearranged pieces of Thierry's *Commentum super Boethii librum De Trinitate*.[214] Finally, the prophecy is evoked in an anonymous commentary on the *De Trinitate*, known by its incipit "In titulo," attributed to Bede by its first publisher, Johannes Heerwagen, reprinted in the Patrologia Latina, and finally republished by Nikolaus Häring under the title *Commentarius Victorinus*. There is a single known manuscript witness for this text, Paris, BnF, MS lat. 14489, which contains, among other works, Thierry's *Lectiones* on the *De Trinitate*.[215] The author is very much influenced by Thierry of Chartres, even if he has more frequent recourse to patristic authors and seems to have consulted the *Summa sententiarum*, composed in 1139. He writes after the council of Reims in 1148, since he openly criticizes some of the theses

211 Thierry of Chartres, *Tractatus de sex dierum operibus* 40 (ed. Häring, p. 572): "Equalitas igitur unitatis et eius ab unitate generatio eterna est." Cf. above p. 54.

212 Thierry of Chartres, *Glosa super Boethii librum De Trinitate* 5.18 (ed. Häring, p. 297).

213 On *generatio aeterna* in the twelfth century, see Albertson, "Achard of St. Victor."

214 Clarembald of Arras, *Tractatus super librum Boetii De Trinitate* 37–38 (ed. Häring, pp. 121–122): "Vel filium iccirco figuram substantiae patris appellat quia patris substantia unitas est. Sed unitas semel, licet non opere et actu, tamen potestate et natura est primus tetragonus. Sicut enim binarius bis positus vel quaternarius quater, arithmethica ratione tetragonum constituit ita quoque unitas semel primum natura tetragonum efficit. ... Quia ergo unitatis per suam denominationem, hoc est per semel, multiplicatae figura ad divinam generationem aliquam similitudinis umbram quodammodo retinet, Apostolus filium figuram patris fortasse appellat. ... Sic etiam hyspana sybilla 'tetragoni' nomine filium patris designans: 'Cum perveneris, inquit, ad costam tetragoni sedentis aeterni et ad costam tetragonorum stantium aeternorum.' Bene autem tetragonum primum natura filio attribuit quoniam eius tetragoni figura ceteris perfectior per laterum aequalitatem indicatur. Ea enim aequalitas identitas est."

215 See Bede, *Opera omnia*, 8 vols. in 4 (Basel, 1563), 8:1088–1120; partially reprinted in PL 95: 391–411, with Nikolaus M. Häring, "A Short Treatise on the Trinity from the School of Thierry of Chartres," *Mediaeval Studies* 18 (1956), 125–134. On this treatise, see also Ermenegildo Bertola, "Il *De Trinitate* dello pseudo Beda," *Rivista di filosofia neo-scolastica* 48 (1956), 316–333.

supported by Gilbert of Poitiers and subsequently condemned in Reims. The author depends on Thierry's *Commentum* for his citation of the Spanish Sibyl, which he curtails; in identifying the eternal tetragons of the prophecy with angels, however, he differentiates himself both from Thierry and from Clarembald of Arras. The mathematical interpretation of the prophecy loses ground to an interpretation that is more allegorical and mystical.[216] Note that neither Clarembald of Arras nor the author of the *Commentarius Victorinus* speak of the *generatio aeterna* pertaining to generation from unity.

The prophecy of the Spanish Sibyl is thus cited in a small number of texts from the twelfth century, but its exact origin remains unknown and has, surprisingly, eluded even specialists.[217] Raymond Klibansky, the only scholar to examine the matter in any detail, established a link between the prophecy and the second crusade in an article that appeared in 1961, promising a study that was never published.[218] Little of direct relevance has appeared since.

Let us focus for a moment on the political context in France in the 1140s. Bernard of Clairvaux and his chancellery produced propaganda in favor of the second crusade. Bernard himself is known to be the author of at least two letters written in 1146 and 1147, the first of which is known to historians as the "encyclical letter" on the crusade.[219] The crusade was then proclaimed

216 *Commentarius Victorinus De Trinitate* 95–96 (*Commentaries on Boethius*, ed. Häring, p. 501): "Et sciendum est has similitudines valere ad exponendum illud Apostoli *figura* et *splendor*. Quod enim dixit *figura* retulit ad hoc quod filius dicitur unitatis equalitas i.e. primus tetragonus. Est namque semel unum primus tetragonus qui rationibus predictis filio dei attribuitur. Unde sibilla hispana: 'Cum perveneris, inquit, ad costas primi tetragoni sedentis ad costas tetragonorum stantium eternorum et cetera.' Sic per primum tetragonum filium dei designans: per tetragonos angelos. Ideoque hunc sedentem illos vero stantes asseverans quasi huius regnare illorum vero utilium famulantium astare est proprium. Quod autem dixit splendorem tulit ad hoc quod filius dicitur lumen de lumine."

217 The Spanish Sibyl is not cited in Peter Dronke, "*Hermes* and the *Sibyls*: Continuations and Creations," in *Intellectuals and Poets in Medieval Europe* (Rome, 1992), pp. 219–244. I have consulted several works and also enquired with various experts on the religions of antiquity.

218 Raymond Klibansky, "Peter Abailard and Bernard of Clairvaux: A Letter by Abailard," *Mediaeval and Renaissance Studies* 5 (1961), 1–27, at 3–5, followed by Marie-Thérèse d'Alverny, *Alain de Lille: Textes inédits, avec une introduction sur sa vie et ses oeuvres* (Paris, 1965), p. 176 n62; Häring, *Commentaries on Boethius*, p. 24. Cf. also Albertson, "Achard of St. Victor," 114–116.

219 Bernard's letters 363 and 457 have a very complex textual tradition. For the details, I refer to Jean Leclercq, "L'encyclique de saint Bernard en faveur de la croisade," and "Pour l'histoire de l'encyclique de saint Bernard sur la croisade," both in *Recueil d'études sur saint Bernard et ses écrits*, 5 vols. (Rome, 1962–1992), 4: 227–246, and 4: 247–263.

by Conrad III of Hohenstaufen and King Louis VII of France on 20 June 1147. The connection between the crusade and the prophecy of a Sibyl is established in the annals of the church of Saint Jacques of Liège: in 1146, it is indicated that the *libri sibillini* had been used to justify the reconquest of Jerusalem by the king of France.[220] The annals are undoubtedly using the version of events related by Otto of Freising, who wrote the *Gesta Friderici I Imperatoris* in the years 1148–1149, precisely on his return from the second crusade that ended in a terrible military defeat for the crusaders. Otto associates the prophecy with the crusade: he is very harsh and critical toward the French, who embarked on a catastrophic crusade because of a naive belief in the prophecy of the *libri sibillini*.[221] In another chapter of the *Gesta*, Otto reproduces a large portion of Saint Bernard's "encyclical letter" on the crusade.[222]

220 *Annales Sancti Iacobi Leodiensis: Chronicon breve Leodiense, ex codice Aureaevallis, cum proemio et notis L.C. Bethmann, recogn. J. Alexandre* (Liège, 1874), p. 17: "AD 1146 ... ex libris Sibillinis ad votum interpretatis regi Francie ituro Ierosolimam magnifice falsa promittuntur."

221 Otto of Freising, *Gesta Friderici I Imperatoris*, 1.proem (MGH SS rer. Germ. 46: 10–11): "Quod autem dixi illo in tempore spiritu peregrini dei occidentales populos afflatos, nemo sic intelligat, acsi quisquam a nobis peregrinus deus putetur, sed ab illo scripto, quod illis in diebus in multis Galliae locis lectitabatur, nos hoc dictum mutuasse sciat, quod tale fuit: 'Tibi dico L pastor corporum primo elemento materiae silvae tuae, quem inspiravit spiritus diei peregrini dei.' In cuius scripturae tenore sub quodam verborum involucro de expugnatione regiae urbis necnon et antiquae Babylonis et ad instar Cyri regis Persarum vel Herculis totius orientis triumphus prefato Ludewico Francorum regi promittebatur. Unde talia ibidem dicta reperta sunt: 'Cum perveneris ad costam tetragoni sedentis aeterni et ad costam tetragonorum stantium aeternorum et ad multiplicationem beati numeri per actualem primum cubum, surge ad eam, quam promisit angelus matris tuae visitare et non visitavit, et pertinges de ea usque ad penultimum, primum cuius cum ascendit promissor, deficit promissio propter optimam mercem, et figantur vexilla tua rosea usque ad extremos labores Herculis, et aperietur tibi porta civitatis B. Nam erexit te sponsus arthemonem, barcha cuius pene cecidit, in capite cuius triangulare velum, ut sequatur te qui precessit te. Tuum ergo L vertetur in C, qui dispersit aquas fluminis, donec pertransirent illud, qui student in procuratione filiorum.' Quod scriptum tantae auctoritatis a probatissimis et religiosissimis Galliarum personis tunc putabatur, ut a quibusdam in Sibillinis libris repertum, ab aliis cuidam Armenio divinitus revelatum affirmaretur. Sed quisquis fuit ille propheta seu trotannus, qui hoc promulgavit, videat, si in futura adhuc aliqua expeditione implendum expectetur, aut tamquam iam non impletum conculcandum Gallicanae levitati, quod fidem aliquam habere potuit, imputetur; hoc tantum sciens, quod non sine rationis proportione spiritus ille omnes pene occidentales in peregrinationem mittens spiritus peregrini dei tam a nobis quam ab illo vocatus est."

222 Ibid., 43 (MGH SS rer. Germ. 46: 61–63).

Klibansky noted that the prophecy existed in two previously published versions, a long one and a short one, and stated that it had been transmitted in four manuscripts: Munich, Bayerische Staatsbibliothek, MSS lat. 5254 and lat. 9516, Cambridge, Mass., Harvard University, Houghton Library, MS Riant 36, and Heidelberg, Universitätsbibliothek, Hs. 71.[223] Ultimately, Klibansky only examined the last manuscript in detail, written by a German scribe around 1150, and containing, among other things, Saint Bernard's letters on Peter Abelard, the principal object of his article. Indeed, in the Heidelberg manuscript, the prophecy follows Saint Bernard's letter 363 without interruption on fols. 9r–v, but it has been erased and is now almost illegible. I was not able to consult the Munich manuscripts, but the Houghton's Riant 36 manuscript is fortunately online. Dating from the third quarter of the twelfth century and of French origin, it contains, on fol. 84r, the short version of the prophecy, also erased, but which I partly succeeded in transcribing:

> Ex prophetia sybille. Tibi dico L. pastor corporum, quem inspiravit spiritus peregrini Dei: Cum ascenderis ad costam tetragoni sedentis aeterni et costam tetragonorum stantium aeternorum, et descendes ad eam, quam promisit filius matris tuae visitare et non visitavit, et aperientur tibi portae civitatis et figes vexilla tua rosea usque ad extremos labores Herculis, et erexit te Christus artemonem in capite cuius est triangulare velum, ut serviant ubi dom …†. Tuum ergo L vertetur in C, …† fluminis, ut transirent …† in procuratione filiorum.[224]

This is not the version cited by Thierry, Clarembald of Arras, and by the anonymous author of the *Commentarius Victorinus*. Thierry seems to have had access to the long version of the prophecy used by Otto of Freising, since it uses the verb *perveneris*.[225] But, as we have seen, Thierry and the other

223 Both the long and short versions of the prophecy were published by Wilhelm von Giesebrecht, *Geschichte der deutschen Kaiserzeit*, 6 vols. in 8. (Leipzig, 1873–1895), 4: 504.

224 Cambridge, Mass., Harvard University, Houghton Library, MS Riant 36, fol. 84r. Cf. the short version (in Giesebrecht, *Geschichte der deutschen Kaiserzeit* 4: 504): "Tibi dico L pastor corporum, quem inspiravit spiritus peregrini dei. Cum ascenderis ad costam tetragoni sedentis aeterni et ad costam tetragonorum stantium aeternorum, tunc aperientur tibi portae civitatis, quam promisit filius matris tuae visitare et non visitativit, et pones vexilla tua rosea usque ad extremos labores Herculis, quia erexit te Christus in artemonem navis, in capite cuius est velum triangulatum. Tuum ergo L vertatur in C, qui divisit aquas fluminis, ut transirent per eas, qui student in procreatione filiorum." No manuscript source is provided.

225 Aside from a few close variants, it corresponds to the long version (a) in Giesebrecht, *Geschichte der deutschen Kaiserzeit* 4: 504.

authors who follow him do not link the prophecy to the reconquest of Jerusalem and, above all, they attribute it to the Spanish Sibyl, a specific detail that is not provided in the prophecy (again, I should note I have not seen the Munich manuscripts). It seems reasonable to suppose that the prophecy was already circulating before it was used for political propaganda. Based on our current knowledge, then, we can state that Thierry was the first to cite the prophecy, free of any political implication: it is the generation of the first tetragon that interests him and nothing else. This lends even greater support to the view that Thierry of Chartres wrote his commentary on the *De arithmetica* as well as his *Commentum super Boethii librum De Trinitate* before 1147.

Complicatio *and* explicatio

The discovery of the commentary on the *De arithmetica* also helps clarify the origins and development of Thierry's doctrine of *complicatio* and *explicatio* (later adopted in the fifteenth century by Nicholas of Cusa). Two chapters of Boethius's *De arithmetica* can be said now to be its foundation.[226] Thierry's *Lectiones in Boethii librum De Trinitate* represent the most mature version of his theory. To begin, unity is the "complication" of plurality, and plurality is the "explication" of unity. Since unity is God, He "complicates" within himself, in His simplicity, the totality of things (*rerum universitas*).[227] The "explication" of this "complication" is none other than the totality of all that has been, all that is, and all that will be. Just as unity precedes plurality, God precedes the diversity and plurality of all things. Moreover, the

226 Boethius, *De arithmetica* II, IV, 4 (ed. Guillaumin, p. 89): "Est igitur unitas vicem obtinens puncti, intervalli longitudinisque principium; ipsa vero nec intervalli nec longitudinis capax, quemadmodum punctum principium quidem lineae est atque intervalli, ipsum vero nec intervallum nec linea"; ibid., II, IV, 6 (p. 90): "Ex hoc igitur principio, id est ex unitate, prima omnium longitudo succrescit quae a binarii numeri principio in cunctos sese numeros explicat, quoniam primum intervallum linea est." See ibid., II, V, 1 (p. 92).

227 Thierry of Chartres, *Lectiones in Boethii librum De Trinitate* 2.4–5 (ed. Häring, p. 155): "Sicut enim unitas conplicatio est omnis pluralitatis et non est tamen pluralitas sed unitas nisi vi – unitas enim vi et potestate pluralitas. Et pluralitas vero explicatio est unitatis et unitas est principium et origo pluralitatis. Ab unitate enim pluralitas. Ab unitate descendit omnis alteritas – sic deus est unitas conplicans in se rerum uniuersitatem in simplicitate quadam. Cuius conplicationis explicatio est omnia que fuerunt que erunt et que sunt. Nam sicut unitas precedit pluralitatem ita simplicitas que deus est in qua conplicata est universitas precedit rerum diversitatem et pluralitatem."

four modes of being of the *rerum universitas* (*necessitas absoluta, necessitas complexionis, possibilitas absoluta, possibilitas determinata*) are related to the theory of "complication" and "explication."[228] Absolute necessity is the "complication" of the totality of things, while the necessity of their complexion constitutes the "explication" (also called *fatum* by natural philosophers – Thierry says). As for absolute possibility, it is the "complication" of all things, as a possibility, while determinate necessity is the "explication" of these same things in act. The *rerum universitas,* in its four modes of being, is subject (*subjecta*) to the three parts of the *speculativa*: theology, mathematics, and physics. Theology deals with absolute necessity, mathematics with the necessity of complexion, and physics with both absolute possibility and determinate possibility.[229] It is perhaps worth repeating that the *Lectiones* contain the mature version of Thierry's philosophy and is not found either in the *Tractatus* or the *Commentum,* where the theory of the four modes of being is not explained. However, in the *Commentum,* we have an interesting reference concerning the divine form that "complicates" (*conplicat*) in itself the forms that the mutability of matter "explicates" (*explicat*).[230]

228 Ibid., 2.10 (pp. 157–158): "Absoluta enim necessitas rerum omnium conplicatio est in simplicitate. Necessitas conplexionis earum rerum explicatio in quodam ordine. Qui ordo a phisicis fatum dicitur. Absoluta autem possibilitas est eiusdem universitatis rerum conplicatio in possibilitate tantum de qua veniunt ad actum. Et vocatur a phisicis primordialis materia sive caos. Determinata vero possibilitas est explicatio possibilitatis absolute in actu cum possibilitate. Sic eadem rerum universitas quatuor modis est." In the *Glosa,* the *necessitas complexionis* is also called *necessitas determinata*; see *Glosa super Boethii librum De Trinitate* 2.21 (ed. Häring, p. 273): "Hec vero determinata dicitur necessitas vel necessitas complexionis eo quod cum aliquam eius materiam incurrimus causarum reliquarum seriatam conexionem vitare non possumus." For *fatum* in Thierry of Chartres, cf. Paolo Lucentini, "L'*Asclepius* ermetico nel secolo XII," in *From Athens to Chartres: Neoplatonism and Medieval Thought,* ed. Haijo J. Westra (Leiden, 1992), pp. 397–420. There is a great deal of scholarship on *complicatio / explicatio* and the four modes of being. See, among others, Maccagnolo, *Rerum universitas,* pp. 40–42, 59–60; Kijewska, "Mathematics as a preparation for Theology," pp. 644–646.

229 Thierry of Chartres, *Lectiones in Boethii librum De Trinitate* 2.11 (ed. Häring, p. 158): "Considerat enim theologia necessitatem que unitas est et simplicitas. Mathematica considerat necessitatem conplexionis que est explicatio simplicitatis. Mathematica enim formas rerum in veritate sua considerat. Phisica vero considerat determinatam possibilitatem et absolutam. Absoluta autem necessitas et absoluta possibilitas in his extrema sunt."

230 Thierry of Chartres, *Commentum super Boethii librum De Trinitate* 2.49 (ed. Häring, p. 84): "Quas igitur formas per pluralitatis diversitatem possibilitatis i.e. materie mutabilitas explicat, easdem quodam modo in unum forma divina conplicat et ad unius forme simplicitatem inexplicabili modo revocat."

The commentary on the *De arithmetica* itself offers an early reflection on the doctrine of *complicatio* and *explicatio* that is closer to the concerns of the *Commentum* than of the *Lectiones*. Here, Thierry states that unity is the complication of number and number is the explication of unity, in the same way that the point is the complication of each magnitude and magnitude is the explication of the point. The point is simultaneously the limit and the perfection of the thing and is the totality itself. The thing in its perfection is thus contained in the point. The point, unity and totality, and plurality coincide.[231] Totality is the complication of magnitude; and totality explicates itself: that is, it is divided, into parts and through the parts that are the totality itself.[232] Pressing Thierry's reasoning even further we could argue that unity, or the point, as the perfection of each thing and the limit of its being, is its *forma essendi,* thus returning to the theory by which all that is, is precisely because it is one. In the commentary on the *De arithmetica,* unity is identified with God, while in the *Lectiones,* it is clearly stated that "Deus est enim unitas in se conplicans universitatem rerum in simplicitate."[233] Accordingly, we can only surmise an evolution in a theological direction in Thierry's thought and that his interests changed in the course of the commentaries on the *Opuscula sacra.*

231 Thierry of Chartres, *Commentum super Arithmeticam Boethii* II, IIII, 4 (pp. 163–164 below): "Unitas vero locum obtinet puncti, et merito, quia sicut unitas est complicatio numeri, et numerus explicatio unitatis, ita quoque punctum est complicatio omnis magnitudinis, et magnitudo est explicatio puncti. ... Haec est ipsa rei totalitas, quae rem totam integre terminat et perfecte, nec est nisi unius rei proprie terminus. Cum enim terminos dicimus, abutimur, quemadmodum cum unitates dicimus. Hanc ergo totalitatem, quam terminum nominamus, explicat ipsius lineae extensio; linea namque est puncti evolutio. Et quoniam magnitudo a puncto incipit, ideo ab ipsa totalitate merito decrescere in infinitum dicitur, sicut pluralitas in infinitum crescere, eo quod ab unitate incipiat. Quoniam enim magnitudo a totalitate inchoat, id est ab ipsa rei perfectione, ultra totalitatem vero rei eiusque perfectionem nihil de ipsa est, eo scilicet quod totalitati et perfectioni rei ipsius nihil addi possit ut ipsa perfecta sit, ideo non crescere sed decrescere in infinitum potest."

232 Thierry of Chartres, *Commentum super Arithmeticam Boethii* II, IIII, 4 (pp. 164–165 below): "Ultra perfectionem namque rei nihil est, sed infra. Unde merito dictum est quod magnitudo decrescit in infinitum, eo quod a perfectione et ipsa rei totalitate incipit, ut praemonstratum est, ultra quam eius nihil est, sed infra. ... Nec tamen est ubi una rei perfectio, una totalitas quae ipsam magnitudinem includit; ipsa enim est magnitudinis complicatio. Explicare vero totalitatem hoc nihil aliud est quam totum per partes dividere. In unaquaque vero partium est eadem totalitas, quae et in ipso toto, quamvis totum non remaneat idem."

233 Thierry of Chartres, *Lectiones in Boethii librum De Trinitate* 2.4 (ed. Häring, p. 155).

With the exception of the commentary on the *De arithmetica,* Thierry's other works adduce only the mathematical example of unity that complicates within itself all numbers. Nicholas of Cusa, for his part, speaks of the point as a complication of magnitude, and of magnitude as the explication of the point; he also has other examples of complication / explication, such as God / beings, rest / movement, equality / inequality. This might reasonably lead to the conclusion that "the concept of complication has become more extensive, compared to what it was in Thierry of Chartres."[234] Now, however, and without diminishing the genius of Nicholas of Cusa, it is possible to suppose that he made use of Thierry's commentary on the *De arithmetica,* which develops the theory of the point and magnitude, of *complicatio* and *explicatio* in precisely the ways that previously seemed unprecedented.

Aequalitas *and* conexio

The concluding chapters of book II of Boethius's *De arithmetica,* devoted to the theory of proportions, were the starting point for bold speculations by several medieval mathematicians, especially in the fourteenth century. Thierry's exegesis of these chapters in his commentary on the *De arithmetica* is shaped by his own reflections on the fundamental place of mathematics in his philosophy. Patiently decoded, these densly written reflections about the Trinity and astrology allow us better to understand his conception of the *necessitas complexionis,* one of the four modes of being of the *rerum universitas,* which are developed further in the *Lectiones* and in the *Glosa.* Theology, physics, and mathematics encompass the four modes of being of the *rerum universitas,* and it is mathematics that Thierry links to the necessity of complexion.[235]

According to Boethius, a proportion (*proportio,* a translation of the Greek *logos*) is a reciprocal relationship between two quantities.[236] This relationship is presented in five types: multiple, superparticular, superpartient, multiple-superparticular, and multiple-superpartient. Proportion-

234 Counet, *Mathématiques et dialectique,* p. 82. See also the discussion in Markus L. Fuehrer, "*Cusanus Platonicus:* References to the Term 'Platonici' in Nicholas of Cues," in *The Platonic Tradition in the Middle Ages: A Doxographic Approach,* ed. Stephen Gersh and Maarten J.F.M. Hoenen (Berlin, 2002), pp. 345–371.

235 Thierry of Chartres, *Lectiones in Boethii librum De Trinitate* 2.11 (ed. Häring, p. 158); *Glosa super Boethii librum De Trinitate* 2.26 (ed. Häring, p. 274).

236 Boethius, *De arithmetica* II, XL, 3 (ed. Guillaumin, p. 140): "Proportio est duorum terminorum ad se invicem quaedam habitudo et quasi quodammodo continentia … ."

ality (*proportionalitas*, a translation of the Greek *analogia*) is the reunion into a single relationship of relationships that exist between at least three quantities[237] – exactly what we call "proportion" today – of which there can be three types, that is, arithmetical, geometric, and harmonic. A proportionality with three terms is called continuous (*continua*), while one with four or more terms is discontinuous (*disiuncta*).[238] We should note, nevertheless, that the distinction is not always maintained, and in some cases Boethius uses *proportio* where we would expect *proportionalitas*. At the beginning of book II, chapter XL, Boethius states that the proportions that follow will be useful for understanding the subtleties of astronomy, among other things.[239] Thierry, for his part, explains that the proportionalities have a maximal number in the nature and composition of things, which are studied, above all, by astronomy (by which he means astrology). The signs are positioned among themselves in opposition or according to figures (square, triangular, and many others); this produces a proportionality that has a maximal effect on things. According to the astronomers, among them Firmicus Maternus, Thierry continues, planets have affinities with the sun and the moon that in turn influence our natures, which are thus also influenced by the planets. As previously noted, the text of Firmicus Maternus's *Mathesis* enjoyed only limited circulation in the Middle Ages; it nevertheless left its mark on two texts from the twelfth century linked to Chartres, the *Cosmographia* and *Mathematicus* of Bernard Silvestris, as well as the *Liber de sex rerum principiis* and was also included in Thierry's

237 Boethius, *De arithmetica* II, XL, 2 (ed. Guillaumin, p. 140): "Est igitur proportionalitas duarum vel trium vel quotlibet proportionum adsumptio ad unum atque collectio. Ut etiam communiter definiamus, proportionalitas est duarum vel plurium proportionum similis habitudo, etiamsi non eisdem quantitatibus et differentiis constitutae sint"; ibid., p. 140: "Ex iunctis enim proportionibus proportionalitas fit. In tribus autem terminis minima proportionalitas invenitur. Fit etiam in pluribus, sed longior; ut binarius ad unum, quoniam duo sunt termini, duplam obtinet proportionem. Sin vero quattuor contra duo compares, hic quoque dupla proportio est. Quos tres terminos si continue consideres, ex duabus proportionibus fit proportionalitas et est proportionalitas unum ad duo et duo ad quattuor. Est enim proportionalitas, ut dictum est, collectio proportionum in unumque redactio."

238 Ibid., II, XLIII, 2 (p. 144).

239 Ibid., II, XL, 1 (p. 140): "... nunc res admonet quaedam de proportionibus disputantes, quae nobis vel ad musicas speculationes vel ad astronomicas subtilitates vel ad geometricae considerationis vim vel etiam ad veterum lectionum intellegentiam prodesse possint, arithmeticam introductionem commodissime terminare."

Heptateuchon.[240] The *Mathesis* does not contain the theory of astral influence to which Thierry refers. Still, it was undoubtedly the source of inspiration for his system: since we have at least three terms, that is, a planet, a luminary, and a nature, a proportionality proceeds (*procedit*) from these three terms – a proportionality that does not engender and is not engendered, because it does not entail multiplication. Thierry then returns to the *unitas*, the *aequalitas*, and the *conexio* that simultaneously proceed from unity and the equality of unity.[241] This echoes what Thierry writes in the theological commentaries, for example the *Commentum*, where unity engenders, the equality of unity is engendered, and the *conexio* proceeds from both;[242] or again in the *Glosa*, where Thierry clearly states the Father engenders, the Son is engendered, and the Holy Ghost proceeds from both.[243] Clarembald of Arras, we might note, puts it differently: God the Son proceeds from God the Father, and God the Holy Ghost proceeds from God the Father and God the Son.[244]

By placing several distinct passages of the commentary on the *De arithmetica* side-by-side, we learn that the *conexio*, the Holy Ghost, is the proportionality that descends from equality, that it is the link between proportion

240 On Bernard Silvestris see Stock, *Myth and Science in Twelfth Century*, pp. 79–86, 206–207. See also the recent edition of Hermes Trismegistus, *Liber de sex rerum principiis*, ed. Mark Delp and Paolo Lucentini (CCCM 142).

241 Thierry of Chartres, *Commentum super Arithmeticam Boethii* II, XLII, 1 (pp. 198–199 below): "NUNC VERO DE PROPORTIONALITATIBUS. Nota quod proportionalitates maximum numerum habent in rerum naturis et complexionibus, ut maxime consideratur in astronomia. Signa namque respiciunt se vel trigona ratione, vel tetragona, vel diametra, vel aliqua alia; et est secundum hoc inter ea aliqua proportionalitas, scilicet quod maximum in rebus habent effectum. Dicunt praeterea astronomici, et inter caeteros Iulius Firmicus, quod planetae quidam habeant respectum et affinitatem ad solem et lunam, sol autem et luna ad nostras naturas, et sic caeteri planetae sole et luna mediantibus. Ecce igitur quod secundum proportionalitatem planetae in nostris habent vim maximam naturis. Illud quoque notandum quod proportionalitas non gignit nec gignitur; per nullam enim fit multiplicationem, sed tantum procedit. Sicut enim aequalitas est complexio inaequalitatis, et ab ea inaequalitas omnis descendit, ita quoque conexio complexio est proportionalitatis et proportionis, et a conexione haec et illa descendit. Unitas autem, ut saepe diximus, una est. Est enim unitas, est unitatis aequalitas, est et unitatis conexio et aequalitatis, quae ab unitate et unitatis aequalitate procedit."

242 Thierry of Chartres, *Commentum super Boethii librum De Trinitate* 2.38 (ed. Häring, p. 80).

243 Thierry of Chartres, *Glosa super Boethii librum De Trinitate* 5.27 (ed. Häring, pp. 298–299).

244 Clarembald of Arras, *Tractatus super librum Boetii De Trinitate* 17 (ed. Häring, p. 181); ibid., 19 (p. 182).

and proportionality, or, again, that it has the force of proportionality.[245] In fact, since three terms are necessary for a proportionality, we can see how the *conexio* may be useful in explaining the connection between the three persons of the Trinity. This may be an application of the *probationes arithmeticae* and *astronomicae* of which Thierry speaks briefly in the *Tractatus de sex dierum operibus,* intended to lead man to knowledge of the creator.[246] In the same work, the equality of unity, the second person of the Trinity, produces from itself all the proportions and inequalities.[247] In the *Lectiones,* Thierry asserts, clearly and concisely, that unity engenders number, equality proportion, and connection proportionality: "Unitas autem numerum facit: equalitas vero proportionem: conexio proportionalitatem."[248] Although scholars had noticed these ideas in the *Lectiones,* the discovery of the commentary on the *De arithmetica* sheds light on their origin and makes them more intelligible. To put the theory in its most succinct form: unity, alone, is a number; with a second term we have a proportion (a relation), and with a third term we have a proportionality (a proportion).

Traditions of Commentary on the De arithmetica

All the texts from Antiquity and Late Antiquity traditionally associated with the school of Chartres and with Chartrian authors are cited in Thierry of Chartres's commentary on the *De arithmetica,* among them: Boethius's *Consolation of Philosophy, Opuscula sacra,* and *De differentiis topicis,* Plato's *Timaeus* and Calcidius's *Commentary on the Timaeus,* Macrobius's *Commentary on Scipio's Dream,* Martianus Capella's *De nuptiis Philologiae et Mercurii,* Augustine's *Retractationes,* Symmachus's letters to Ausonius, and, especially, Firmicus Maternus's *Mathesis.* Medieval authors are also cited,

245 Thierry of Chartres, *Commentum super Arithmeticam Boethii* II, XXXI, 1 (p. 186 below): "Finalis enim causa est utilitas ipsa; hoc est conexio, quae alio nomine bonitas, alio spiritus sanctus dicitur, cui inserta est vis proportionalitatis, id est numerorum proportionalium, quibus ista vis est et figurarum." Ibid., I, XXXII, 3 (p. 152 below): "SINT ENIM: in quo diligenter intuendum est quod, cum debuisset ostendere quomodo inaequalitas descendat ab aequalitate, ostendit quomodo ab aequalitate descendat proportionalitas … ."

246 Thierry of Chartres, *Tractatus de sex dierum operibus* 30 (ed. Häring, p. 568); the passage is cited in n179 above.

247 Thierry of Chartres, *Tractatus de sex dierum operibus* 43 (ed. Häring, p. 573): "Et sicut ipsa unitas omnes numeros ex se procreat ita ipsa unitatis equalitas omnes proportiones et inequalitates omnium rerum ex se producit." On *aequalitas* and *conexio,* see Maccagnolo, *Rerum universitas,* pp. 31–34.

248 Thierry of Chartres, *Lectiones in Boethii librum De Trinitate* 7.7 (ed. Häring, p. 225).

but not always explicitly: Isidore of Seville's *Etymologies*; Gerbert of Aurillac's scholia on *De arithmetica* II, 1, known as the *Saltus Gerberti* from the eleventh century onward; the *De utilitatibus astrolabii* (J), a work associated with Gerbert of Aurillac's scientific writings in some manuscripts and sometimes attributed to him.[249] Thierry also very likely consulted Remi of Auxerre's *Commentary on Martianus Capella*. Special mention should be made of Hugh of Saint Victor's *Didascalicon*, particularly because of the etymology of the word *mechanicus*, which Thierry cites in the commentary on the *De arithmetica*.[250] Hugh of Saint Victor himself used the false etymology of *mechanicus* (from *moechus*, adulterer), found already in the *Scholica graecarum glossarum*, whose attribution to Martin of Laon, is now contested.[251] However, it seems likely that Thierry borrowed this etymology from Hugh, whose *Didascalicon* was circulating before the summer of 1121.[252] It is

249 The text, found in Chartres, Bibliothèque municipale, MS 214, was destroyed in the bombardment of 1944; no fragments have been found to date for the moment. Hermann of Reichenau may have rewritten the *De utilitatibus astrolabii*, but there is no in-depth study of the entire manuscript tradition that would allow us to distinguish the two texts. Nikolaj Bubnov published the *De utilitatibus astrolabii* (J), a version in 21 chapters, among the *opera dubia* of Gerbert of Aurillac, without concern for the quality or correctness of the manuscript witnesses he used. I thank David Juste for enlightening me on this point; for the astrolabe literature before the twelfth century, see David Juste, *Les Alchandreana primitifs: Etude sur les plus anciens traités astrologiques latins d'origine arabe (Xe siècle)* (Leiden, 2007), pp. 1–8. See the description of the Stuttgart manuscript below (pp. 81–85) for full bibliographic references.

250 Thierry of Chartres, *Commentum super Arithmeticam Boethii*, praefatio (p. 98 below): De mechanicis, id est adulterinis artibus..." Cf. Hugh of Saint Victor, *Didascalicon de studio legendi* 2.21, ed. Charles H. Buttimer (Washington, 1939), p. 39: "Hae mechanicae appellantur, id est adulterinae, quia de opere artificis agunt, quod a natura formam mutuatur."

251 See *Scholica graecarum glossarum* as cited in M.L.W. Laistner, "Notes on Greek from the Lectures of a Ninth-Century Monastery Teacher," *Bulletin of the John Rylands Library* 7 (1923), 421–456, at 439: "Moechus est adulter alterius thorum furtim polluens. Inde a moecho dicitur mechanica ars, ingeniosa atque subtilissima et paene quomodo facta vel administrata sit invisibilis in tantum, ut etiam visum conspicientium quodam modo furetur, dum non facile penetratur eius ingeniositas." For a recent development, see Patrizia Lendinara, "The *Scholica graecarum glossarum* and Martianus Capella," in *Carolingian Scholarship and Martianus Capella: Ninth-Century Commentary Traditions on 'De nuptiis' in Context*, ed. Mariken Teeuwen and Sinéad O'Sullivan (Turnhout, 2011), pp. 301–361.

252 Dominique Poirel, "*Tene fontem et totum habes*: L'unité du *Didascalicon* d'Hugues de Saint-Victor," in *Universitas scolarium: Mélanges offerts à Jacques Verger par ses anciens étudiants*, ed. Cédric Giraud and Martin Morard (Geneva, 2011), pp. 293–328.

interesting to observe that a very rare definition of the *officium* found in Thierry's commentary on the *De inventione* ("... officium, quod est congruus actus uniuscuiusque secundum morem patriae"[253]) is repeated in the commentary on the *De arithmetica*: ("Vel officium est congruus actus uniuscuiusque personae secundum morem patriae"[254]).

Numerous additional sources and parallel passages with Thierry's other works can be found in the notes to the edition below. Nevertheless, the main medieval source used in Thierry of Chartres's commentary on the *De arithmetica* is another commentary on the *De arithmetica*, in all likelihood written in the tenth century, on which I will now focus.

Two principles guide this undertaking. First, philology in its most inclusive sense is critical to understanding medieval philosophy. It is no longer necessary to insist on this point: every historian of medieval philosophy is convinced that the history of texts, codicology, and paleography can help grasp doctrines, their origin, diffusion, and evolution. Second, appreciating the literary genre of commentaries, studying the entire tradition in order to situate any given commentary as precisely as possible in a chain of commentaries, is fundamental. This principle is difficult to follow in practice. We typically risk attributing to an author, known or anonymous, ideas professed by others before him, since as we know, commentators repeat one another. They do not conceal this fact; William of Conches sketches in detail the portrait of the good commentator who cuts, adds, shifts, corrects, and imitates: "... aliorum superflua recidentes, praetermissa addentes, obscura elucidantes, male dicta removentes, bene dicta imitantes."[255] Such a confession might alarm any historian of philosophy with a professional conscience!

The commentaries on the *De arithmetica* as a whole, however, have never yet been subjected to comprehensive and systematic enquiry; Boethius's work inhabits the shifting boundary between philosophy and history of science and has been wrongly neglected, as a consequence, by most medievalists (Gillian Evans and Alison Peden are notable exceptions). Convinced of their importance, I have undertaken an inventory of these commentaries myself[256] and published the first complete lemmatic commentary preserved in three manuscripts, with the incipit "Ne subesse possit similitudo

253 Thierry of Chartres, *Commentarius super libros De inventione* 1.1.1 (ed. Fredborg, p. 59).

254 Thierry of Chartres, *Commentum super Arithmeticam Boethii*, praefatio (p. 95 below).

255 William of Conches, *Glosae super Platonem*, Prologus 1 (ed. Jeauneau, p. 5).

256 See Appendix, pp. 209–210 below.

falsitatis."[257] If I emphasize this point it is because Thierry's commentary on the *De arithmetica,* as it has come down to us in the Stuttgart manuscript, includes sizable excerpts of this commentary dating from the end of the tenth century. The fragments from this "old" commentary are easily recognizable, and I have indicated them in the apparatus to the edition, since they are often separate and stand alongside Thierry's explanations without influencing them. Now, our current state of knowledge does not allow us to establish whether Thierry himself integrated these passages or whether, as sometimes happens, a scribe took the initiative of combining the "old" commentary with the new. Fortunately, since the publication of the edition of the "old" commentary, new discoveries have been made that afford us greater insight into Thierry's own commentary on the *De arithmetica.*

To begin, the use of the word "filiolitas" in the "old" commentary unquestionably links its anonymous author to the circle of Johannes Scotus Eriugena, who is the inventor of this neologism that did not circulate widely (even Remi of Auxerre, who disseminated Eriugena's doctrines, uses *filietas*).[258] The word *filiolitas* is also found in Thierry's commentary.[259] More important, however, is the fact that by a fortunate coincidence, I have been able to reconstitute a manuscript that was considered acephalous at the time I published my edition: Leiden, Bibliotheek der Rijksuniversiteit, MS Voss. lat. Fol. 70 A^II^, fols. 74r-78v, from the end of the tenth century.[260] Section II of codex 70 A, written by several French scribes from the tenth and eleventh

257 Irene Caiazzo, "Un commento altomedievale al *De arithmetica* di Boezio," *Archivum Latinitatis Medii Aevi* 58 (2000), 113–150. A partial edition based on the acephalous Leiden manuscript was previously published by Michael Bernhard, "Glossen zur Arithmetik des Boethius," in *Scire litteras: Forschungen zum mittelalterlichen Geistesleben,* ed. Sigrid Krämer and Michael Bernhard (Munich, 1988), pp. 23–34.

258 Concerning *filiolitas,* see Ernesto S. Mainoldi, "Le fonti del *De praedestinatione liber* di Giovanni Scoto Eriugena," *Studi Medievali* 45 (2004), 651–697, at 694; Paul E. Dutton, "*Filiolitas*: The Short History of One of Eriugena's Inventions," *American Catholic Philosophical Quarterly* 79 (2005), 549–566.

259 See Thierry of Chartres, *Commentum super Arithmeticam Boethii* I, 1, 2 (p. 106, l. 134 below).

260 The information found in Karel A. de Meyier, *Codices Vossiani Latini,* Pars I: *Codices in folio* (Leiden, 1973) pp. 144–145 needs to be thoroughly revised. Here is a description based on my own examination of the manuscript. Section A^II^: (membr.; mm. 290 × 225); fols. 74r–78v [saec. X/XI]: *Commentarius in Boethii De arithmetica;* fol. 78v: *Notae de tono;* fols. 79r–80v [saec. X/XI]: *Adnotationes grammaticales;* fol. 81r: blank; fol. 81v [saec. X/XI]: *Formulae theologicae;* fol. 82r [saec. XI]: *Tractatus theologicus;* fol. 82v [saec. X/XI]: *Catalogus codicum monasterii S. Petri Resbacensis;* fol. 83r–v [saec. XI]: *Alter catalogus codicum monasterii S. Petri Resbacensis.*

centuries, contains a quire of eight folios (fols. 74–81) that is of interest to us here:

> Commentary on the *De arithmetica,* acephalous, inc. "asserunt. Continua: unita pacta," uninterrupted; notes on the tone: inc. "Tonus est quod vocula voculam [*corr. ex* vocabula vocabulam MS] … pone arculum tange"; grammatical notes: *Incipiunt quedam excerpta utilium verborum,* inc. "Aurio diversos sensus habet … vilis et pro nichilo ductus"; prenuptial contract between one Gauscelinus and one Ascelina, inc. "Summus et ineffabilis pater."[261]

This last text was written by another scribe, perhaps in the twelfth century. The last two folios originally came from Saint-Pierre de Rebais,[262] but ended up in Saint-Maur-des-Fossés.[263] The manuscript came into the hands of the bibliophile Alexander Petau, then passed to Queen Christine of Sweden, and then to Isaac Vossius; hence its entry into the Leiden library. The eight folios that interest us are in fact the continuation of the now-destroyed manuscript Chartres, Bibliothèque municipale, MS 46, originating in the cathedral chapter, of which no fragments remain. Thanks to the detailed description provided in the nineteenth-century catalogue, we learn that MS 46 was damaged, and that fol. 28v contained a commentary on Boethius's *De arithmetica,* with the incipit "Ne subesse possit similitudo falsitatis in auctore," and the explicit "alii non circa corpus, sed intus ea."[264] The continuation of the sentence is in fact in the Leiden manuscript at fol. 74r: "asserunt. CONTINUA: unita pacta".[265] This Chartres/Leiden manuscript is the antigraph of another manuscript witness, Reims, Bibliothèque municipale,

261 fol. 81v: "… ego miles Gauscelinus te o amantissima sponsa mea nomine Ascelina in matrimonium volo procreandorum amore filiorum ducere … ."

262 Bernhard Bischoff, *Katalog der festländischen Handschriften des neunten Jahrhunderts,* Vol. 2: Laon–Paderborn (Wiesbaden, 2004), p. 53 no. 2197: "Frankreich (Rebais?), IX. Jh., 3. Viertel."

263 Ex-libris fol. 83v: "Iste liber est monasterii fossatensis [s. XII?]."

264 *Catalogue général des manuscrits des bibliothèques publiques de France: Départements,* Vol. 11: *Chartres,* p. 23: "46 (38). Boetii arithmetica. Même texte que dans le manuscrit 45. — Mêmes tableaux à la fin, fol. 28, avec le même fragment à la suite. Fol. 28v. 'Glose super arithmeticam Boetii. Ne subesse possit similitudo falsitatis in auctore … alii non circa corpus, sed intus ea …' XIe siècle. Parchemin. 29 feuillets. 292 sur 230 millim. Rel. peau. (Chapitre)." The difference in dimensions of the Chartres and Leiden manuscripts (29 mm. versus 292 and 23 versus 225) is of no importance.

265 See the full text in Caiazzo, "Un commento altomedievale," 132: "… alii non circa corpus, sed intus ea asserunt. CONTINUA: unita pacta."

MS 975, dating from the twelfth century, which I had previously used for my edition.[266] This latter manuscript gives us access to that portion of text which is missing from the manuscript of Chartres today. It would have been marvelous to put the manuscript of Chartres and the manuscript of Leiden side by side and to prove irrefutably that the quire preserved in Leiden today was taken from the Chartrian codex 46. Unfortunately, no reproduction of MS 46 is available, and the reconstruction of *membra disiecta* proposed here will remain forever hypothetical.[267]

Let me return to the main point. The commentary inserted into the commentary by Thierry of Chartres was thus in the library of the Chapter of Chartres. It is not possible to say much more, even if Thierry did indeed bequeath his volumes to the Chapter before his departure.[268] The inventory of the library drawn up at the end of the sixteenth century further supports my argument. Two copies of the *De arithmetica* are mentioned: "68. Boetii arithmetica; fol., optimae notae" and "69. Boetii arithmetica; fol., optimae notae," very likely MSS 45 and 46 according to the *Catalogue général* of 1890.[269]

Detailed examination of excerpts from the "old" commentary does not allow us to situate with any certainty the copy used by Thierry (or by the scribe of the Stuttgart manuscript) within its manuscript tradition, and to state definitively that the Chartres/Leiden manuscript was in front of Thierry. However, the study of a few key passages suggests there was a slight rewriting of the text, which is difficult to attribute to a simple scribe, and leads us to conclude that Thierry himself inserted the passages of the "old" commentary. Here is an example of such rewriting where the word *filiolitas* is precisely at issue as it appears in all known witnesses of the "old" commentary. I have compared Thierry's commentary with all the known manuscript witnesses of the "old" commentary, viz. Reims, Bibliothèque municipale, MS 975 (twelfth-century copy of Chartres/Leiden manuscript, that is lacking at this point), Melk, Stiftsbibliothek, Cod. Mell. 740, and

266 *Catalogue général des manuscrits: Départements*, Vol. 39: *Reims* (Paris, 1904–1906), 2.1: 232–234. Summary description: (membr.; fols. 1–33: mm. 308 × 202, fols. 34–75: mm. 207 × 195); fols. 1r–28r [s. XI]: Boethius, *De arithmetica*; fol. 28v [s. XIIin]: Prenuptial contract; fols. 29r–32v [s. XIIin]: *Commentarius in Boethii De arithmetica*; fols. 32v–33r [s. XIIin]: *Notae de tono*; fol. 33r–v: *Adnotationes grammaticales;* fols. 34r–75v [s. XIIin]: Cassiodorus, *Institutiones*.

267 Some fragments from the now-destroyed Chartres manuscripts have yet to be catalogued.

268 See p. 8 above.

269 See the introduction to *Catalogue général des manuscrits des bibliothèques publiques de France: Départements*, Vol. 11: *Chartres*, p. xiv.

Munich, Bayerische Staatsbibliothek, MS lat. 18764 (an indirect witness, because it preserves only few extracts of the "old" commentary).

MS Stuttgart, fol. 5va	**MS Reims, fol. 30r (copy of Chartres/Leiden)**	**MS Melk, fol. 40r**	**MS Munich, fol. 5va**
Qualitas vero in suo genere consonanter et perpetuo manet, sic et caetera. Et quicquid adunatum corporibus invenitur: scilicet participatione non generalitate, ut longitudo, paternitas, filiolitas. Ideo dicit: "corporibus adunatum quodammodo," quia alii dicunt quod circa corpus sint, alii non circa corpus sed intus **inesse** asserunt, **sicut habere et pati, quae sunt duo praedicamenta**.[270]	Qualitas vero in suo genere qualitatis constanter et perpetuo manet, sic et cetera: et quicquid adunatum quodammodo corporibus invenitur, scilicet participacione non generalitate, ut longitudo, paternitas, filiolitas; ideo dicit "corporibus adunatum quodammodo," quia alii dicunt quod circa corpus sunt, alii non circa corpus sed intus ea asserunt.[271]	Qualitas vero in suo genere qualitas constanter et perpetuo manet, sic contra: quicquid adunatum quodammodo corporibus invenitur, scilicet participatione non generalitate, ut longitudo, paternitas, filiolitas. Ideo dicit "corporibus adunatum quodammodo," quia alii dicunt quod circa corpus sint.	Qualitas vero in suo genere constanter et perpetuo manet, sicut cetera: quicquid adunatum quodammodo corporibus invenitur, scilicet participatione non generalitate, ut longitudo, paternitas, filiolitas.[272]

The textual lacuna in the manuscripts of Melk and of Munich appears to show on the one hand that they could not have been used for Thierry's commentary and, on the other, that the rewriting of the "old" commentary on the *De arithmetica* in the Stuttgart manuscript cannot be the work of a simple scribe. Two solutions are possible: either Thierry himself produced the additions, or another writer, drawing from both the "old" commentary and Thierry's commentary, conflated them. In light of the date of the Stuttgart manuscript (toward the middle or second half of the twelfth century) and the dates of the texts that it contains (the most recent of which date to the 1140s), it seems reasonable to me to suppose that Thierry of Chartres used the "old" commentary on the *De arithmetica* to write his own commentary.

Another passage seems to indicate that Thierry did not use precisely the Chartres/Leiden manuscript. It clearly shows Thierry's interest in the

270 See Thierry of Chartres, *Commentum super Arithmeticam Boethii* I, 1, 2 (p. 106, ll. 132–136 below).

271 The Leiden manuscript, which in all likelihood came from Chartres, begins with *asserunt*.

272 The Munich manuscript (an indirect witness) adds a gloss: "(fol. 5va) Nam IIIIor matheseos disciplinae nec intentione crescunt, nec diminutione minuuntur nec aliis varietatibus permutantur, si in [in] propria permanentes regulas suas inconvertibili firmitate custodiunt."

division of the dialectic into *diaretike* and *analitike*, an idea introduced by Eriugena.[273] However, he does not copy the entire passage from the "old" commentary and he adds new remarks, for example introducing the concept of "inferentia."[274]

MS Stuttgart, fol. 8ra	**MS Chartres/ Leiden, fol. 74r**	**MS Melk, fol. 40v**	**MS Munich, fol. 5vb**
... ubi tanguntur duae species dialecticae, diareticae et analiticae. Diaretice est quae a generalissimo in specialissimum dividitur, ut ab animali ad unum solum. Analitice est quae a specialissimo in generalissimum resolvitur, ut: quid est Cicero? Animal.[275]	... ibi tanguntur duae speties dialecticae: diaretike et analitike, id est resolutoria. Diaretike, id est qua a generalissimum dividitur, ut ab animali ad unum solum hominem. Analitike, quae a spetialissimo in generalissimum resolvitur: quid est Cicero? Animal.	... ubi tanguntur duae species dialecticae: diarethike et analetice. Diaretice, quae a generalissimo in specialissimum dividitur, ut ab animali ad unum sonum hominem. Analetice, quae a specialissimo in generalissimum resolventur. Quid est Cicero? Animal.	... ubi tanguntur duae species dialecticae: diaretike et analetike. Diaretike, quae a generalissimo in specialissimum dividitur, ut ab animali ad unum hominem solum. Analitike, quae a specialissimo in generalissimum solvitur, ut: quid est Cicero? Animal.

The Chartres/Leiden manuscript contains a textual variant, *Diaretike, id est qua a generalissimum dividitur* not found in the Stuttgart manuscript: *Diaretice est quae a generalissimo in specialissimum dividitur*. Should we presuppose the existence of another now lost manuscript, that Thierry of Chartres might also have consulted? Whatever the case, the text clearly proved easy enough to correct, despite a defective antigraph. It is for this reason that I refrain from making a definitive pronouncement on the question.

Since only two commentaries on the *De arithmetica* have been published at present, and also since the history of the medieval reception of Boethius's work has yet to be written, it is premature to attempt a comparative study with the aim of establishing a typology of hermeneutic trends.[276] It is nevertheless possible to sketch the principal characteristics of the two commentaries.

273 Johannes Scotus Eriugena, *Expositiones in Hierarchiam coelestem* VII, 578–603 (CCCM 31: 106–107).

274 Thierry of Chartres, *Commentum super Arithmeticam Boethii* I, 1, 8 (p. 114, l. 358 below).

275 Thierry of Chartres, *Commentum super Arithmeticam Boethii* I, 1, 8 (p. 114, ll. 354–357 below).

276 See Alison M. White, "Boethius in the Medieval *Quadrivium*," in *Boethius: His Life, Thought and Influence*, ed. Margaret Gibson (Oxford, 1981), pp. 162–205; Jean-Yves Guillaumin, "Boethius's *De institutione arithmetica* and its Influence on Posterity," in *A Companion to Boethius in the Middle Ages*, ed. Noel H. Kaylor, Jr. and Philip E. Phillips (Leiden, 2012), pp. 135–161.

The author of the "old" commentary dwells on the *essentiae* enumerated in the prologue of the *De arithmetica*, which he identifies with Aristotle's categories. He carefully glosses the paragraphs on the four disciplines of the *quadrivium* and their respective objects. Like commentaries on the *Consolation of Philosophy*, which are often preceded by the *Vitae Boetii*, his commentary also evinces an interest in Boethius's biography.[277] He neglects the definition of number and unity. The chapters on equality and inequality receive comment, but no mathematical rule is stated or applied to explain the return of inequalities into equality. We know that at least three rules were invented to reduce inequalities (that is, plurality) to equality (unity): Gerbert of Aurillac's rule, the most often used, also known as *saltus Gerberti*, Abbo of Fleury's rule, the rule of Notker of Liège, as well as other anonymous rules found in various manuscripts.[278] The return of inequalities into equality has a moral meaning, suggested by Boethius himself, in that the vices are brought back into virtue, as well as a metaphysical meaning, in that evil is brought back to good. In keeping with the view propounded in the *Consolatio*, evil does not exist in itself and is only a failure of the good.

If we now turn to Thierry's commentary, it is clear that all the crucial points of the *De arithmetica* glossed in the "old" commentary receive a great deal of attention. Of course, as we have seen, Thierry introduces a very complex *accessus*, of which he is most likely the inventor. Above all, he insists on the fundamental role of mathematics as a preparation for the study of philosophy and, subsequently, of theology. He gives much more space, accordingly, to the strictly mathematical chapters and explains the *saltus Gerberti* rule in considerable detail. The chapters on unity, on proportions and geometric figures like the tetragon, which is simultaneously a number and a figure, give rise to a long excursus on the nature of God and of the Trinity. The same holds true for the chapters on equality and inequality, that is, good and evil, the latter being nothing other than distance from the good, through excess or lack. Against the dualism of the Manicheans, Thierry argues that mathematics (the four disciplines of the *quadrivium*) allow us to prove that evil does not exist as a principle opposed to good, and that it is possible to reduce evil to good. Thierry also refers to rithmomachy, a game that resembles chess, invented at the beginning of the eleventh century and popular from the twelfth century onward, the rules of which were inspired by the *De*

277 The different *Vitae* were published in Boethius, *Philosophiae Consolationis Libri Quinque*, ed. Rudolf Peiper (Leipzig, 1871), pp. xxx–xxxv.

278 Cf. White, "Boethius in the Medieval *Quadrivium*," pp. 169–171; Gillian R. Evans, "The *Saltus Gerberti*: The Problem of the 'Leap,'" *Janus* 67 (1980), 261–268.

arithmetica.[279] Thierry ponders the discontinuous and the continuous, the point and the line (is the point merely the limit of the line, or is it part of the line?). The subject is much debated from the twelfth century onward, in commentaries on the exposition of quantity in Aristotle's *Categories* and on Macrobius's *Commentary on Scipio's Dream*, as well as in the *Dialectica* of both Peter Abelard and Garlandus Compotista.[280] Thierry's answer is exact: the point is the limit of the line; but the point is identified with unity, the *aequalitas essendi*, the *forma essendi*. In sum, Thierry's answer is more that of a metaphysician than that of a logician or a mathematician. Finally, Thierry shows that he knows and has understood the theory of the elements in Plato's *Timaeus*, which he cites from Calcidius's commentary on the *Timaeus*.[281]

Although he takes the "old" commentary as a starting point, Thierry of Chartres distances himself from it considerably, partly because of his "metaphysical" approach to the *De arithmetica*, and partly because of the themes and subjects he dealt with, which were in the air at the time. Thierry of Chartres reads the *De arithmetica* through a Platonic, neo-Pythagorean lens, considering mathematics and the disciplines of the *quadrivium* as a whole as fundamental to the study of philosophy and theology.

Furthermore, even a cursory comparison of the commentary on the *De arithmetica* with the treatises on "the symbolism of numbers" written in the twelfth century reveals the wide gap that separates Thierry from the Victorines and from authors such as Odo of Morimond, Thibaut of Langres, and Geoffrey of Auxerre.[282] These authors consider all numbers to be symbols that need to be decoded. For example, for Geoffrey of Auxerre, the number five designates the five books of Law and the five senses, while seven represents the seven days of the week, the creation of the world, and the gifts of the Holy Spirit.[283] Thibaut of Langres associates three with the theological

279 In addition to the works of Arno Borst and Menso Folkerts on rithmomachy, cited above in n79, see Menso Folkerts, "La rithmomachie et le manuscrit Avranches 235," in *Science antique, science médiévale* ed. Louis Callebat and Olivier Desbordes (Hildesheim and New York, 2000), pp. 347–357. Specialists agree that manuscript 235 of the Bibliothèque municipale d'Avranches is a copy of a manuscript from Chartres that has not been preserved.

280 These texts are quoted in the apparatus fontium of the edition.

281 Irene Caiazzo, "La forme et les qualités des éléments: Lectures médiévales du *Timée*," in *Il Timeo: Esegesi greche, arabe, latine*, ed. Francesco Celia and Angela Ulacco (Pisa, 2012), pp. 307–345.

282 These treatises were published by Hanne Lange in several issues of the *Cahiers de l'Institut du Moyen Age grec et latin de Copenhague*.

283 Hanne Lange, "Les données mathématiques des traités du XIIe siècle sur la symbolique des nombres," *Cahiers de l'Institut du Moyen Age grec et latin de Copenhague* 32 (1979), 1–128, at 43.

virtues and the powers of the soul; as for thirteen, it is a theophanic number, since the Epiphany happened thirteen days after Christmas.[284] We could multiply the examples, but it should be clear the intellectual approaches and goals of these authors are very different from those of the Chartrian master.[285] For completeness, I should mention also William of Lucca's *De arithmetica compendiose tractata*,[286] that is a mere summary of Boethius's *De arithmetica*, and doesn't show original doctrinal development nor echos of Thierry's teaching. Given the current state of scholarship, it is still too early to say definitively if Thierry's commentary on the *De arithmetica* had an influence on later thought, beyond the very particular case of Nicholas of Cusa explored above. The publication and the study of other unpublished commentaries on the *De arithmetica* will no doubt provide answers to these questions.

Some Concluding Reflections

The commentary on the *De arithmetica* preserved in Stuttgart, Württembergische Landesbibliothek, Cod. math. 4° 33 can be attributed with certainty to Thierry of Chartres. It belongs to a phase when the Chartrian master's interests were mainly directed toward the liberal arts; however, we can also distinguish in it themes and problematics that were abundantly developed in his commentaries on Boethius's *Opuscula sacra*. Furthermore, thanks to the discovery of the commentary on the *De arithmetica*, some of the more difficult or enigmatic pages in the theological commentaries as well as from the *Tractatus de sex dierum operibus* have become more intelligible, and Thierry's whole philosophical project has gained greater clarity. The decision to publish an integral edition of the commentary on the *De arithmetica* seemed both obvious and necessary. This is an important work that will permit us, especially through comparison with other medieval commentaries on Boethius's *De arithmetica*, to refine our knowledge of philosophy in the twelfth century and its intersection with the arts of the *quadrivium*. An initial inventory of the medieval commentaries on the *De arithmetica* seems to indicate that a strong interest in this work developed precisely in the twelfth century. Indeed, from the Carolingian renaissance onward, the scholarly *curriculum* was based on the seven liberal arts; Boethius's *De arithmetica* soon became a canonical text for the study of arithmetic. At the beginning of the

284 Ibid., 53.
285 Cf. Marie-Dominique Chenu, "Platon à Citeaux," *AHDLMA* 29 (1954), 99–106.
286 William of Lucca, *De arithmetica compendiose tractata: Dal codice 614 (sec. XII) della Biblioteca Capitolare Feliniana di Lucca*, ed. Gino Arrighi (Pisa, 1964).

thirteenth century, the birth of universities was to revolutionize the educational structures and the *curricula* in the Latin West; however, Boethius's *De arithmetica* never diminished in importance, and remained a foundational text, at least for the study of speculative arithmetic, until the modern era.[287] The history of the influence of Boethius's *De arithmetica*, and consequently of neo-Pythagoreanism, in the Latin West has yet to be written. But it is my hope that this edition of Thierry's commentary, taken together with my earlier edition of the "old" commentary on the *De arithmetica*, can serve those willing to take up the challenge.

The Present Work: From Manuscript to Edition

Stuttgart, Württembergische Landesbibliothek, Cod. math. 4° 33 is a miscellany of 86 folios containing texts on the *quadrivium* and computistical material, for some of which it is the sole, known manuscript witness. Given that the collection of mathematical manuscripts at the Württembergische Landesbibliothek has never been the subject of a printed catalog, a comprehensive and precise description of this manuscript should prove useful.[288] The manuscript was written in southern Germany, probably in the middle or in the second half of the twelfth century;[289] its binding also dates from the

287 See Guy Beaujouan's pioneering essays collected in *Par raison de nombres: L'art du calcul et les savoirs médiévaux* (Aldershot, 1991). On the use and influence of Boethius's text for the study of arithmetic in the middle ages, see Michael Masi, "Boethius' *De institutione arithmetica* in the Context of Medieval Mathematics," in *Atti del congresso internazionale di studi boeziani*, ed. Luca Obertello (Rome, 1981), pp. 263–272, and "The Influence of Boethius' *De Arithmetica* on Late Medieval Mathematics," in *Boethius and the Liberal Arts: A Collection of Essays*, ed. Michael Masi (Bern, 1981), pp. 81–95; Pearl Kibre, "The Boethian *De institutione arithmetica* and the *Quadrivium* in the Thirteenth Century University Milieu at Paris," in *Boethius and the Liberal Arts*, pp. 67–80. See also Guy Beaujouan, "Le *quadrivium* et la Faculté des arts" and Max Lejbowicz, "Les disciplines du *quadrivium*," in *L'enseignement des disciplines à la Faculté des arts (Paris et Oxford XIIIe–XVe siècles)*, ed. Olga Weijers and Louis Holtz (Turnhout, 1997), pp. 185–194, and pp. 195–216, respectively.

288 A handwritten catalogue from the nineteenth century has been preserved and can only be consulted in situ at the library. I thank Carmen Kämmerer, director of the department of manuscripts, for sending me bibliographical information on this manuscript.

289 Bernhard Bischoff, "Ostertagtexte und Intervalltafeln," in *Mittelalterliche Studien: Ausgewählte Aufsätze zur Schriftkunde und Literaturgeschichte*, 3 vols. (Stuttgart, 1966–1981), 2: 192–226 at 195 n20: "Süddeutschland, XII. Jh." Some information on the content and material characteristics of Cod. math. 4° 33 can also be found in the works of Arno Borst cited in n299 below.

twelfth century, something of a rarity.[290] Before being acquired by the state library in Stuttgart, it was at the Benedictine abbey of Komburg, founded in 1078 and belonging to the diocese of Wurtzburg.[291] The manuscript was most likely written in the scriptorium of the abbey itself. The miscellany is very homogenous, from both a paleographic and a codicological point of view; it was written by several scribes whose handwriting is so similar, it is difficult to tell them apart with any certainty. Three codicological elements, however, can be distinguished. The first (fols. 1r–40v), written in two columns, contains lemmatic commentaries on mathematical texts; the second (fols. 41r–54v) contains calendars and tables of computus and astronomy; the third (fols. 55r–86v), written in a single column, comprises three of the four texts commented on in the first codicological element.

Parchment, 86 numbered folios, 205 × 140 mm
Origin: Southern Germany; monastery of Komburg?
Owner: Komburg

I: folios 1–40

1. fols. 1ra–34ra: Thierry of Chartres, *Commentary on the "De arithmetica" of Boethius*, inc. "In dandis accipiendisque muneribus." Title written by the scribe: "Glossae super arithmeticam"; inc: "Quoniam ars ista, ars scilicet arithmetica." No other known manuscript witnesses.
2. fols. 34ra–37va: anonymous (1143 or 1150 AD), *Commentary on the "Computus" of Gerlandus*, inc. "Sepe volumina domini Bedae." Title written by the scribe: "De compoto"; inc. "Ista ars vocatur compotus, id est computatio vel numerus." The *accessus*, on its own, is also found divided across Troyes, Bibliothèque municipale, MS 1558, fol. 2r and fol. 28r–v, from the twelfth century.[292]
3. fols. 37va–38ra: anonymous, *On the astrolabe*. Title written by the scribe: "De astrolabio"; inc. "Videndum est in primis de nomine

290 Andreas Fingernagel, "Der Romanische Bucheinband," in *Romanik*, ed. Andreas Fingernagel, 2 vols. (Graz, 2007), 1: 407 n116.

291 G. Michels, "Komburg (S.-Nicholas)," in *Dictionnaire d'histoire et de géographie ecclésiastiques*, ed. André Baudrillart et al. (Paris, 1912–), 29: 533–534.

292 The incipit is listed in Lynn Thorndike and Pearl Kibre, *A Catalogue of Incipits of Mediaeval Scientific Writings in Latin*, rev. and augm. ed. (Cambridge, MA, 1963), col. 778, following the Troyes manuscript. However, there is no indication that the *accessus* continues on fol. 28r. Furthermore, the identification of the text is wrong: "Gerlandus, Prol. Computus." On texts 2 through 5, see Caiazzo, "Sur l'astrolabe et le comput au XIIe siècle."

huius instrumenti."[293] This is a proper *accessus* to the two lemmatic commentaries that follow. No other known manuscript witnesses.[294]

4. fols. 38rb–39rb: anonymous, *Commentary on the "De mensura astrolabii" of Hermann of Reichenau* (**h**), inc. "Herimannus Christi pauperum." Title written by the scribe: "Glossae de mensura astrolabii"; inc. "In instrumento precisa est spera." No other known manuscript witnesses.[295]
5. fols. 39va–40vb: anonymous, *Commentary on the "De utilitatibus astrolabii"* (**J**), inc. "Quicumque astronomicae peritiam." Title written by the scribe on the last line of fol. 38rb: "De libro astrol<apsus>"; inc. "Incipiunt quasi hic initium capiunt." No other known manuscript witnesses.[296]

II: folios 41–54

6. fol. 41r: Verses. No title; inc. "Quisquis erit amans."[297]
7. fols. 41v–42r: Table of movable feasts and of Easter. No title.[298]
8. fols. 42v–43r: Computus tables. No title. In fol. 42v, the table illustrating the solar cycle of 28 years begins with the year 1064 and ends with the year 1568.
9. fols. 43v–46v: Calendar-martyrology for the Benedictine monastery of Komburg. No title; inc. "Principium Iani sancit tropicus Capricornus."[299]
10. fol. 47r: blank

293 Ernst Zinner, *Verzeichnis der astronomischen Handschriften des deutschen Kulturgebietes* (Munich, 1925), p. 44, no. 933 (treatise on the astrolabe, fol. 37v; inc. "Videndum est in primis").

294 The "De astrolabio" is published in Caiazzo, "Sur l'astrolabe et le comput," 36–37.

295 See Caiazzo, "Sur l'astrolabe et le comput," 38, for the transcription of the beginning of *Commentary on the "De mensura astrolabii."*

296 The opening of *Commentary on the "De utilitatibus astrolabii"* (**J**) is published in Caiazzo, "Sur l'astrolabe et le comput," 38.

297 Wolfgang Irtenkauf, "Der *computus ecclesiasticus* in der Einstimmigkeit des Mittelalters," *Archiv für Musikwissenschaft* 14 (1957), 1–15, at 6–7. The verses are transcribed on page 6; illustration 5 is a reproduction of fol. 41r.

298 The same table is contained in Oxford, St. John's College, MS 17, fol. 31v. See Faith Wallis's remarks on the site http://digital.library.mcgill.ca/ms-17: "Key letter method of Paschal reckoning: Three tables, fols. 30v–31v."

299 Arno Borst, *Die karolingische Kalenderreform* (Hannover, 1998), p. xxvi: "Komburg nach 1150, fols. 43v–46v und fols. 51r–53r"; cf. pp. 516, and 521. See also *Der karolingische Reichskalender und seine Überlieferung bis ins 12. Jahrhundert*, ed. Arno Borst, 3 vols. (Hannover, 2001), 1: 332–333.

11. fols. 47v–48r: *Tabula Gerlandi*. No title.[300]
12. fol. 48v: Raymond of Marseilles, *Liber de astrolabio* (excerpt from chapter 7). No title; inc. "Quod si in qua parte."[301]
13. fols. 48v–49r: Raymond of Marseilles, *Liber de astrolabio* (table of fixed stars according to the moderns). Title written by the scribe: "Tabula fixarum stellarum secundum sententiam modernorum."
14. fol. 49r: Ibn al-Saffār, *Liber de astrolabio,* in the translation of John of Seville (excerpt). No title; inc. "Cognitio stellarum quas ignoraveris."[302]
15. fol. 49r: Anonymous, short glosses on Martianus Capella, *De nuptiis Philologiae et Mercurii* 8.860–861. No title; inc. "Lunaris circulus centies."[303]
16. fol. 49r: Anonymous, notes on the months, the sun and the Zodiac. Title written by the scribe: "Initia XII mensium secundum *librum cursuum* super Massiliam factum"; inc. "Martius super VIIII gradum."[304]
17. fol. 49v: blank
18. fols. 50r–53r: Calendar-martyrology for the Benedictine monastery of Komburg (continuation of fols. 43v–46v).[305]
19. fols. 53v–54r: Computus table illustrating the lunar cycle of 19 years. No title. The table begins with the year 1140 and ends with the year 1653.[306]
20. fol. 54v: blank

300 On this table, see Pieter Verbist, *Duelling with the Past: Medieval Authors and the Problem of the Christian Era (c.990–1135)* (Turnhout, 2010), pp. 150–161.

301 This manuscript was not used by the editors of the text. See Raymond of Marseilles, *Opera omnia*, Vol. 1: *Traité de l'astrolabe – Liber cursuum planetarum*, ed. and trans. Marie-Thérèse d'Alverny, Charles Burnett and Emmanuel Poulle (Paris, 2009), p. 72. On items 12–13, see Caiazzo and Burnett, "New Manuscripts of *On the astrolabe* by Raymond of Marseilles," *Scriptorium* 65 (2011), 338–349.

302 See "Tratado sobre el uso del astrolabio planisférico, de Maslama al-Maŷrītī, de Córdoba, en la traducción latina de Johannes Hispanus (perdido el original árabe)," in *Las traducciones orientales en los manuscritos de la Biblioteca Catedral de Toledo*, ed. José M. Millás Vallicrosa (Madrid, 1942), pp. 261–284, at 271–272. On the subject of this excerpt, see Caiazzo and Burnett, "New Manuscripts of *On the astrolabe*," 342–343.

303 Ibid., 343 (transcription of the glosses).

304 This short text should be compared with the *Liber cursuum planetarum* of Raymond of Marseilles. See Caiazzo and Burnett, "New Manuscripts of *On the astrolabe*," 343.

305 Cf. the works of Arno Borst cited above, n299.

306 Irtenkauf, "Der *computus ecclesiasticus*," 6: "Ob damit eine aktuelle Jahresangabe – entstand die Hs. um oder kurz vor 1140? – beabsichtigt war, kann nicht entschieden werden, paläographisch wäre eine derartige Festlegung ohne weiteres möglich."

III: folios 55–86

21. fols. 55r–73r: Gerlandus, *Computus*. Title written by the scribe: "Incipit prologus Gerlandi in Compotum suum"; inc. "Sepe volumina domini Bedae."[307] Some marginal and interlinear glosses.
22. fols. 73v–78r: Hermann of Reichenau, *De mensura astrolabii* (**h**) [308] Title written by the scribe on the last line of fol. 73r: "Prologus Hermanni De mensura astrolabii"; inc. (on fol. 73v) "Herimannus Christi pauperum." Some marginal and interlinear glosses.
23. fols. 78v–86r: *De utilitatibus astrolabii* (**J**).[309] Title written by the scribe on the last line of fol. 78r: "Incipiunt regulae astrolabii"; inc. (on fol. 78v) "Quicumque astronomicae peritiam." The titles of the 21 chapters of the treatise are transcribed on folio 78r, without interruption, after the last words of *De mensura astrolabii*. The scribe skipped the title of chapter XVI ("In quo signo sint stellae horarum"). The titles are displaced in relation to the content of the chapters. The title of chapter XXI ("In quo signo et gradu signi invenienda sit luna") is in fact the title of chapter XXII. Bubnov had published the version of the *De utilitatibus astrolabii* in 21 chapters, but he did not use this manuscript, which he did not know. Our manuscript's version of *De utilitatibus astrolabii* is composed of the 21 chapters of the Bubnov edition, in the same order, with a twenty-second chapter added.[310] Some marginal and interlinear glosses.
24. fol. 86v: blank

307 On the work of Gerlandus, see Verbist, *Duelling with the Past*, pp. 147–171. Cf. now *Der Computus Gerlandi: Edition, Übersetzung und Erläuterungen*, ed. Alfred Lohr (Stuttgart, 2013).

308 The text has been edited by Julius Drecker, "Hermannus Contractus Über das Astrolab," *Isis* 16 (1931), 200–219, at 203–212.

309 The *De utilitatibus astrolabii*, in the version with 21 chapters, usually indicated by the initial **J** by specialists of astrolabe literature, has been published among the *opera incerta* of Gerbert d'Aurillac by Nikolaj Bubnov, *Opera mathematica* (Berlin, 1899), pp. 114–147. For a historical development of the *De utilitatibus astrolabii*, see Arianna Borrelli, *Aspects of the Astrolabe: Architectonica Ratio in Tenth- and Eleventh-Century Europe* (Stuttgart, 2008), pp. 83–87.

310 fol. 86r: "XXII. Si vis invenire in quo signo et gradu signi sit luna, sole in die lucente et luna posita super hemisperium, cape solem et pone eum in reperta altitudine, sive ex parte almarech, sive ex parte almagrib, et deinde luna prende in sua altitudine, et nota ubi eam ceperis. Posito igitur sole in sua altitudine, vide qui gradus cuius signi iaceat in altitudine pro luna reperta. In eo nimirum gradu signi est luna."

The text of the commentary on the *De arithmetica* edited here is based on Stuttgart, Württembergische Landesbibliothek, Cod. math. 4° 33, fols. 1ra–34ra. Although this manuscript is the sole surviving witness to the work, what follows is not, strictly speaking, a diplomatic edition, since it both corrects obvious errors in the original and also introduces small emendations deemed essential to the understanding of the text. Anyway, the quality of the text is excellent.

The edition adopts the division into chapters and paragraphs instituted by Jean-Yves Guillaumin in his Budé edition of Boethius's *De arithmetica*.[311] This practice aims to facilitate convenient identification of those parts of the *De arithmetica* on which Thierry has commented, since it is practically impossible to follow the sequence of the text without also consulting Boethius's work alongside. However, in certain cases, the lemmata in the commentary do not correspond to the text as edited by Guillaumin or to the variants provided in the critical apparatus to his edition. I have consulted both the edition of the *De arithmetica* by Henri Oosthout and Johannes Schilling in the Corpus Christianorum[312] and the text of Boethius's *De arithmetica* found in Thierry's *Heptateuchon*, preserved in manuscript 498 of the library of Chartres; but this systematic study of variants has proved inconclusive: it is impossible to demonstrate definitively which manuscript of the *De arithmetica* Thierry used to write his commentary.

The apparatus fontium, which collects all parallel passages from the "old" commentary and the many other sources Thierry consulted and used, should allow readers to appreciate the scope of his borrowings and to situate the commentary more precisely in the exegetical tradition of Boethius's *De arithmetica*.

The edition preserves the orthography of the manuscript, but in the interest of greater textual uniformity some aberrant forms have been corrected, either where there is an obvious mistake or where it risks possible confusion. We know that from the twelfth century onward, and following contemporary pronunciation, the written form *e* begins to appear everywhere for the diphthongs *ae* and *oe*.[313] The manuscript contains the *e cau-*

311 See Boethius, *Institution arithmétique*, ed. Jean-Yves Guillaumin (Paris, 1995).

312 See Boethius, *De institutione arithmetica*, ed. Henri Oosthout and Johannes Schilling, CCSL 94A (Turnhout, 1999).

313 On the subject of dipththongs, it is interesting to mention the opinion of Guillaume of Conches, cited by Edouard Jeauneau in his "Deux rédactions des gloses de Guillaume de Conches sur Priscien," *Recherches de théologie ancienne et médiévale* 27 (1960), 212–247, at 242: "A titre documentaire, nous transcrivons maintenant quelques lignes où il est question de diphtongues. Nous les livrons à la réflexion de ceux qui s'intéressent à l'édition des textes médiévaux. *Glose sur Priscien* XVII,7:

data, simple *e* being very rare. Following classical practice, the diphthong *ae* has thus been systemically restored in the edition. Certain uses of *e* are clearly the result of scribal errors: accordingly, *ipse* has been corrected to *ipsae*, *certe* to *certae*, *ille* to *illae*. In a few cases, such as *caera* for *cera*, the diphthong *ae* was mistakenly used. In order to facilitate reading, the letters *v* and *u* have been distinguished, even though *v* remains an exception in the manuscript: *veracissime*, *quamvis*. Words of Greek origin show alternations between *y* and *i*: for instance, in most cases we find *phylosophia* and *physica*, but the alternate forms *philosophia* and *phisica* are also used. *H*, a letter that in the twelfth century had long since lost its phonetic value, shows considerable variability, especially in words derived from Greek, such as *rethorica* for *rhetorica*, *dyathesseron* (used once) for *dyatesseron*, or even *Nichomachus*, *Thimaeus*. On the other hand, we find *h* within *ch* in medial position, as in *nichil*, which follows common medieval usage, although the classical form *nihil* is generally used in the manuscript. *C* and *g* are sometimes confused, as in the word *flegma* which is found in the form *flecma*. The forms *ci*/*ti* are generally well differentiated in the manuscript; the form *ratiocinacio* has thus been corrected to *ratiocinatio*. The manuscript also reflects the widespread practice of not distinguishing the pronunciation of double consonants in an intervocalic position: hence *conexio* and *connexio* (normalized here as *conexio*), *paralellus* for *parallelus*, *paralellepipedus* for *parallelepipedus*. The adverbial particles *tamen* and *tantum* have been replaced according to the meaning of the sentence, since the scribe is not always consistent in the abbreviations used for them.

It should be noted that the scribe sometimes identifies the case of a word (nominative, ablative) above the line; these minor interventions are all noted in the apparatus criticus. In one case, the scribe writes above the word *indago* its vernacular equivalent *hage*; but the text also includes

'Et notandum quod, quamvis omnis ditongus duas obtineat vocales, non tamen in omni utraque sonat, sed in quibusdam ultima tantum ut ae, oe; in quibusdam utraque sed altera magis ut au, eu. Si querat aliquis quare in istis utraque sonet, in illis altera tantum, dicimus quod 'u', que est ultima in istis, multos habet sonos: aliquando enim per se potest pronunciari, aliquando cum alio. Sed 'a' et 'e' unum solum habent: unde in eadem sillaba non possunt simul pronunciari. Sed, quamvis una pronunciatur sola, tamen debet utraque scribi. Propter imperitos qui volunt quicquid vident scriptum pronunciare, instituerunt moderni ut illa sola que ibi habet sonum scribatur, sed tamen ei virgula que sit ditongi nota subscribatur (*P*, fol. 88ra).' *Glose sur Priscien* 1.30: 'Celare, in prolatione, æquivocum est ad duo, et actum sculpandi et ad actum abscondendi; et in scriptura non, quia habet ditongum cum significat actum sculpandi (*P*, fol. 15rb).' Des règles ainsi formulées, les copistes ne semblent avoir tenu aucun compte. Et lorsqu'à notre tour nous éditons des textes médiévaux, nous faisons comme eux."

the phrase “quod teutonice dicitur *smalzet*,” which suggests it was likely an addition made by the scribe (the same is perhaps also the case in the gloss “hage” for “indago”). Capitalization and punctuation have been normalized following modern usage but always taking the peculiarities of the manuscript into account.

Finally, in quotations from published editions, the letters *u* and *v* have been distinguished, in line with orthographical conventions underlying the edition of the commentary.

Teodoricus Carnotensis

COMMENTUM SUPER ARITHMETICAM BOETHII

Sigla

MS	Stuttgart, Württembergische Landesbibliothek, Cod. math. 4° 33
Chartres 498	Chartres, Bibliothèque municipale, MS 498
ed. Guillaumin	Boethius, *Institution arithmétique,* ed. Jean-Yves Guillaumin (Paris, 1995)
ed. Oosthout and Schilling	Boethius, *De institutione arithmetica,* ed. Henri Oosthout and Johannes Schilling, Corpus Christianorum: Series Latina 94A (Turnhout, 1999)
Caiazzo, "Un commento altomedievale"	Irene Caiazzo, "Un commento altomedievale al *De arithmetica di Boezio,*" *Archivum Latinitatis Medii Aevi* 58 (2000), 113–150.

<ACCESSUS AD ARITHMETICAM BOETHII>

Quoniam ars ista, ars scilicet arithmetica, septem liberalium artium una et est et habetur – totius enim quadruvii fundamentum est –, ideo in eius principio decem inquirenda sunt, quippe cum in principio cuiuslibet artis decem inquirere iubeamur auctoritate Boetii in IIIIto *Topicorum*. Sunt autem haec decem: quae sit ars ista, quod genus artis, quae materia, quod officium, quis finis, quae partes, quae species, quod instrumentum, quis artifex, quae nominis interpretatio.

Horum singula, ut proposuimus, exequamur modo. Quid ipsa sit primitus. Arithmetica est scientia virtutis numerorum contemplativa ad cognitionem naturae. Cum enim, ut in huius libri principio legitur, omnium rerum exemplar sit numerus, et omnia secundum numerum constructa sint, merito de ipsius natura, vi et virtute in hac arte docemur, ut qualiter secundum numerum res factae sint contemplemur.

Genus huius artis est mathematica ars, quae sub speculativa continetur. Phylosophiae enim sunt tres partes, quarum una dicitur speculativa; alia vero activa, quae docet quid sit agendum et quid non; tercia logica, quae est de rationibus. Speculativa iterum tres habet partes: theologiam quae agit de incorporeis, mathematicam quae de incorporeis circa corpora, phisicam

2 Glossae super arithmeticam *add. supra lineam* MS

3–5 Boethius, *De differentiis topicis* 4 (PL 64: 1207A, 1211C).

5–8 See the accessus to Thierry of Chartres's *Commentarius super libros De inventione* (ed. Fredborg, pp. 49–50) and to his *Commentarius super Rhetoricam ad Herennium* (ed. Fredborg, p. 221). On the accessus ad auctores more generally, see above, pp. 40–48. See also the section "De arithmetica" in Dominicus Gundissalinus, *De divisione philosophiae* (ed. Fidora and Werner, p. 180): "Circa arithmeticam quoque haec eadem inquirenda sunt: quid sit ipsa, quod genus eius, quae materia, quae partes, quae species, quod instrumentum, quis artifex, quod officium, quis finis, quare sic dicatur, quem ordinem habeat inter scientias."

10–14 See the overview of Thierry of Chartres as commentator of *De arithmetica*, above pp. 22–33. Cf. Dominicus Gundissalinus, *De divisione philosophiae* (ed. Fidora and Werner, p. 188): "Arithmetica autem dicitur scientia de virtute numeri. 'Ares' enim 'virtus' interpretatur. Inde 'arithmetica' de virtute numeri scientia vocatur. Ipsius est enim omnem virtutem numeri considerare et quid in creaturis valeat demonstrare ... quod mundi conditor deus numerum habuit ratiocinationis exemplar. Quaecumque enim constituit fabricante ratione, secundum numeros colligavit."

15–16 Boethius, *De Trinitate* 2. See the remarks on the development of Thierry of Chartres's thought and language above, pp. 34–39.

18–20 Cf. Thierry of Chartres, *Lectiones in Boethii librum De Trinitate*, prol.4 (ed. Häring, p. 126); the text is cited above, p. 38, n134.

quae de ipsis corporibus. De theologia et physica nihil ad praesens opus. Mathematica vero quatuor habet species: arithmeticam, musicam, geometriam, astronomiam. Et dicitur mathematica, id est doctrinalis, hac de causa, quia tria sunt genera mathesis nec plura. Unum genus est incorporeorum extra corpora, ad quorum doctrinam theologia inventa est. Aliud genus est incorporeorum circa corpora, et de his agit mathematica. Tercium genus est ipsa corpora, causa quorum reperta est phisica. Sed quamvis ad uniuscuiusque generis doctrinam disciplina propria inventa sit, per incorporea tamen quae circa corpora sunt, de quibus mathematica tractat, habemus corporeorum cognitionem et incorporeorum quae sunt extra corpora, id est divinorum. Ex figuris corporeorum, ex ordine, qualitatibus et quantitatibus ipsorum, quae omnia corporibus adiacent, perpenditur res omnis creata esse ab auctore sapientissimo, id est deo. Et quia ex his, de quibus mathematica tractat, perpenduntur divina, merito mathematica doctrinalis appellatur. Mathesis autem sine aspiratione significat divinationem, aspiratum significat doctrinam. Dicuntur autem artes istae quadruvii mathematicae, id est doctrinales, quia maior in eis docendis doctrina adhibetur quam in trivio. Datis enim praeceptis, subponuntur figurae probantes praemissa.

Materia huius artis est numerus, quia secundum hanc artem tractat numerum arithmeticus. Quod autem et de relato numero hic tractat, incidenter hoc facit. Numerus per se consideratus est unus, duo, tres et reliqui. Numerus relatus sive comparatus ad alium est duplus, sesqualter, superpartiens, de quo specialiter agit<ur> in musica, quia tales faciunt simphonias.

Officium huius artis est dividere, multiplicare, coacervare virtutesque numerorum perpendere ad speculationem naturae numerorum.

Finis eius est divisum esse, multiplicatum esse, coacervatum esse, et hominem fieri divisorem et scire multiplicare.

Partes eius sunt par et impar numerus. Vel hae IIII: prima de numero per se, secunda de numero relato, tercia de numero geometrico, quarta de proportionalitate.

33–35 Cf. Hugh of Saint Victor, "De mathematica," *Didascalicon* 2.3 (ed. Buttimer, p. 25): "Mathematica autem doctrinalis scientia dicitur. Mathesis enim quando 't' habet sine aspiratione, interpretatur vanitas, et significat superstitionem illorum, qui fata hominum in constellationibus ponunt. Unde et huiusmodi mathematici appellati sunt. Quando autem 't' habet aspiratum, doctrinam sonat." Cf. William of Conches, *Glosae super Platonem,* accessus.5 (ed. Jeauneau, p. 10): "Mathematica vero quadruvium continet, dicta mathematica, id est doctrinalis. Mathesis enim cum aspiratione est doctrina, sine ea est vanitas. Et dicitur doctrinalis antonomasice, scilicet quia perfectior fit doctrina in quadruvio quam in aliis artibus." Cf. Thierry of Chartres, *Commentum super Boethii librum De Trinitate* 2.15 (ed. Häring, p. 73): "... quia mathematicam solebant prius antiqui discere ut ad divinitatis intelligentiam possent pervenire. Unde etiam mathematica i.e. doctrinalis dicitur. Mathesis namque doctrina interpretatur."

Species, ut quibusdam placet, nullas habet sed, secundum alios, species eius sunt scientia de dispositione figurarum et scientia de inventione medietatum.

Instrumentum eius est, ut quidam volunt, abacus vel, ut aliis placet, mensa Pythagorea, quae dicitur rithmachia, quia in istis maximum huius artis habetur exercicium. Vel ipsae figurae, quae subponuntur, sunt eius instrumentum.

Artifex artis dupliciter dicitur. Vocatur enim arithmeticus tam ille qui dat praecepta quam ille qui secundum praecepta operatur dividendo, multiplicando.

Nominis interpretatio haec est: "ares" graece, "virtus" latine, "rithmus" "numerus," "ika" "scientia." Inde arithmetica: scientia virtutis numerorum.

Ea quae praediximus circa artem versantur; circa librum vero, quae sit auctoris intentio, quae utilitas libri. Intentio auctoris est in hoc opere introducere lectorem in artem arithmeticam, utilitas est scientia artis.

Ne subesse possit similitudo falsitatis in auctore, tria requiruntur in initio libri, quae nobis certa sunt: persona, tempus et locus. Persona fuit Boetius, romanus et consul tempore Iustini imperatoris. Cuius tempore Theodericus, rex Gothorum, Italiam contra Romanos invasit, et per XXX annos et eo amplius obtinuit potestatem. Locus autem Romae aut certe in Italia fuit. Locus ideo forsitan quaeritur, ut si mendosi habeantur libri, illuc revertatur, ubi certiores esse arbitrantur. Causa arithmeticae disciplinae

53–55 On rithmomachy, see the commentary below *ad Arithmeticam* II, XXIIII, 5. See also Dominicus Gundissalinus, "De arithmetica," *De divisione philosophiae* (ed. Fidora and Werner, p. 186): "Instrumentum vero theoricae est demonstratio. Quae quid sit et ex quibus fiat, superius dictum est. Instrumentum vero practicae est tabula abaci seu rhythmomachia, quae est ludus de pugna numerorum secundum proportiones eorum."

62–64 Cf. Thierry of Chartres, *Commentarius super Rhetoricam ad Herennium*, accessus (ed. Fredborg, p. 221): "Est igitur in hoc libro auctoris intentio plenarie de tota rhetorica disputare, id est de omnibus partibus eius ... . Utilitas vero libri est partium rhetoricae artis omnium cognitio." See also above, pp. 40–48.

65–76 Cf. Caiazzo, "Un commento altomedievale," 126, ll. 2–12: "Ne subesse possit similitudo falsitatis in auctore, tria requiruntur in initio libri, quae nobis certa sunt: persona, tempus et locus. Persona fuit Boetius, romanus et consul tempore Iustini imperatoris. Cuius tempore Theodoricus, rex Gothorum, Italiam et Romanos invasit, et per triginta annos et eo amplius potestatem exercuit Romae; et certe in Italia fuit. Ideo forsitan quaeritur, ut si mendosi habentur libri, illuc revertatur, quo certiores esse arbitrantur. Causa arithmeticae disciplinae scribendae haec fuit, ut numeri et proportiones numerorum scirentur a sedulis lectoribus, qui ipsam sollerti indagine perlecturi erant. Nam in numeris maxima versatur dubitatio, quam iste Boetius contendit auferre, dum obscure a Nicomacho et breviter dicta, plana producto sermone reddit, et diffuse dicta, moderata brevitate collegit."

scribendae haec fuit, ut numeri, proportiones numerorum scirentur a sedulis lectoribus, qui ipsam sollerti indagine perlecturi erant. Nam in numeris maxima versabatur dubitatio, quam iste Boetius contendit auferre, dum obscure a Nichomacho et breviter dicta, plana producto sermone reddit, et diffuse dicta, moderata brevitate colligit. Arithmeticae disciplinae apud Graecos Pythagoras primus fuit auctor, post Nichomachus; apud Latinos vero Apuleius, deinde Boetius. Nam Latini imitati sunt Graecos, fontem scientiae ab ipsis haurientes, et in proprium sermonem transferentes. Et licet quaedam disciplinae ab Aegyptiis originem videantur habere, ipsas tamen viciniores Graeci ab eis accipientes post tradiderunt Romanis.

76–78 Isidore of Seville, *Etymologiae* 3.2.1.

76–81 Cf. Caiazzo, "Un commento altomedievale," 126–127, ll. 12–20: "Causa insuper requiritur, quae ea est, ut sciat videlicet naturam numerorum, quemadmodum ex unitatibus ipse numerus crescat ac multiplicetur, et ut quaelibet summa eius facile valeat deprehendi. Ut verbi causa: si proponatur octies octo quot sunt, et si qua similia. Rithmon dicunt Graeci 'numerum,' inde arithmetica 'disciplina numerorum'; cuius Pytagoras apud Graecos primus fuit auctor, postea Nicomachus; apud Latinos vero Apuleius, deinde Boetius. Nam Latini imitati sunt Graecos, fontem scientiae ab ipsis haurientes, et in proprium sermonem transferentes. Et licet quaedam disciplinae ab Aegyptiis originem videantur habere, ipsas tamen viciniores Graeci ab eis accipientes, postea tradiderunt Romanis." Thierry of Chartres leaves out three lines of the old commentary.

79–81 For the birth of the "disciplinae" in Egypt, see Thierry of Chartres, *Prologus in Heptateuchon*, in Jeauneau, "Note sur l'Ecole de Chartres," *Studi Medievali*, 3rd ser., 5 (1964), 821–865, at 855.

<PRAEFATIO>

1. IN DANDIS. More recte scribentium duplicem praemittit prologum: primum praeter rem, in quo nulla subsequentis materiae mentio fit, secundum vero illum qui est ante rem, in quo praelibatur materia cum trium captatione.

Scribit autem hoc prohemium ad Symmachum patricium, in quo captat eius benivolentiam, tam a persona ipsius Symmachi quam a re ipsa, et a persona sua propria. Symmachus enim interpretatur compugnans: "sym" "com," "machia" "pugna" dicitur. Boetius quoque interpretatur adiutor, quod nomen ei non proprium fuit, sed Severinus. Nam ex accidenti Boetius dictus est. Plurimis enim prodesse satagebat, maxime pupillos et inopes suis fovens stipendiis. Consul enim fuit, cuius dignitatem sui liberi post eum sunt adepti. Symmachus quoque patricius fuit et suo tempore clarus, et fuit socer Boetii. Cui hanc epistulam loco prohemii et praefationis illi ponere placuit. Daturus autem hanc artem Symmacho quasi pro munere, ostendit quomodo ab amicis dari debeant et accipi munera. Et est sententia litterae talis: quod si aliquis det munus amico, ille officium suum exequitur si ei munus offerat quo nihil inveniat honestius, accipiens iterum suum conservat officium si munus oblatum iocunde suscipiat. OFFICIA dicuntur ab efficiendo, quae pro amicicia ponuntur nonnumquam. Vel officium est congruus actus uniuscuiusque personae secundum morem patriae. Magni facit

2–5 Cf. Thierry of Chartres, *Commentarius super Rhetoricam ad Herennium* 4.1.1 (ed. Fredborg, p. 313). See above, p. 43.

6–8 See the following *loca parallela* from Thierry of Chartres, *Commentarius super Rhetoricam ad Herennium* 1.1.1: "Hic captat benivolentiam a persona sua" (ed. Fredborg, p. 222, l. 10); "Hic a persona auditoris captat benivolentiam" (p. 222, ll. 19–20); and "Hic captat benivolentiam ab ipsa re, de qua est tractaturus" (p. 223, ll. 27–28).

8–9 Cf. Caiazzo, "Un commento altomedievale," 127, ll. 26–27: "SYMMACHUS interpretatur conpugnans, sin 'con,' machia 'pugna' dicitur."

9–12 Cf. Caiazzo, "Un commento altomedievale," 127, ll. 23–26: "BOETIUS interpretatur adiutor, quod nomen non eius proprium fuit, sed Severinus. Nam ex accidente Boetius dictus est; plurimis enim prodesse satagebat, maxime pupillos et inopes suis fovens stipendiis."

12–15 Cf. Caiazzo, "Un commento altomedievale," 127, ll. 21–23: "BOETIUS consul fuit, eius dignitatem sui liberi post sunt adepti. SYMMACHUS quoque consul fuit, et suo tempore clarus, et fuit Boetii socer. Et hanc epistolam loco proemii et praefationis illi ponere placuit."

19–20 Cf. Caiazzo, "Un commento altomedievale," 127, ll. 33–34: "OFFICIA, scilicet amicitiae, seu largiendo in numerum; dicta officia ab efficiendo, quae pro amicitia ponuntur nonnunquam."

20–21 Thierry of Chartres, *Commentarius super libros De inventione* 1.1.1 (ed. Fredborg, p. 59): "... officium quod est congruus actus uniuscuiusque secundum morem

amicus amicum, cum eum laudando extollit. HAEC IPSE. Quod est dicere: quia liberale munus dandum est, ideo non opes attuli, sed munus honestissimum. Et captat auctor in hoc versu benivolentiam a propria persona, id est ab attributis personae. Ab officio captat iterum benivolentiam a persona auditoris Symmachi, id est ab attributis personae ipsius, scilicet ab habitu, ubi eum non esse cupidum designat. Commendat etiam rem ipsam. Opus etenim hoc ex graecis litteris extractum est; et est commendatio ab auctoritate. NON IGNAVA. Ignavus quasi ingenii igne vacuus. Ignavae dicuntur opes ab effectu, eo quod ignavum magis faciunt quam probum, et mentem aggravant. QUIBUS AD FACINUS. Quia cum homines sitiant habere aurum et argentum, pro eis quaerendis multa facinora committunt, et instruuntur eis ad homicidium, ad furtum et ad caetera vitia. CUM HABENDI SITIS. Hoc tractum est ab ydropicis, qui quanto plus bibent tanto plus sitiunt. Sic et avari quanto plus habent, tanto plus cupiunt. Unde est:

"Quanto plus potae, plus sitiuntur aquae."

INCANDUIT. Curialiter dicit, et est metaphora sumpta a ferro, quod quanto plus accedit ad ignem, tanto plus candet. Eodem modo et avarus quanto plus habet, tanto plus cupit habere. AD MERITUM. Et quia quidam habent aurum et argentum, qui non tantis ea sceleribus sed ex hereditate possident, et ideo viderentur non esse ignava sed aliquid gloriae et dignitatis seu meriti conferre, removet hoc, et dicit ea nullius meriti esse, quippe cum sancti et boni viri contemnant ea. CALCATA. Calcari dicuntur opes a virtuosis, et avaros calcare, quia pecuniae serviunt avari, pecunia vero virtuosis. SED EA. Non attuli opes, sed quae meliora et digniora videbantur.

patriae." See also: Marius Victorinus, *Explanationes in Ciceronis Rhetoricam* 1.5 (ed. Ippolito, p. 36): "Officium est unius cuiusque personae actus congruens secundum mores civitatis"; and William of Tyre, *Chronicon*, prol. (ed. Huygens, Mayer and Rösch, p. 11*): "... vere dicitur officium, congruus actus uniuscuiusque personae, secundum mores et instituta patriae." For more references to this definition of "officium," see William of Conches, *Glosae super Platonem* 10.12–14 (ed. Jeauneau, p. 18).

24–25 See above note to 6–8.

29 Cf. Caiazzo, "Un commento altomedievale," 127, ll. 37–38: "Ignavus ergo dicitur quasi ingenii igne vacuus." Cf. Thierry of Chartres, *Commentarius super Rhetoricam ad Herennium* 1.5.8 (ed. Fredborg, p. 230, ll. 52–53): "IGNAVIAM quantum ad animum; ignavus dicitur quasi igne, id est agilitate animi, vacuus"; and 3.3.6 (p. 281, ll. 83–85): "Gratia exempli: quem vident iustum, ignavum dicunt; hoc ad animum pertinet (dicitur enim ignavus quasi igne, id est agilitate animi, vacuus)."

36 Ovid, *Fasti* I, 216: "Quo plus sunt potae, plus sitiuntur aquae." A family of manuscripts of the *Fasti* contains the variant *quanto*.

Ex opulentia. Plus enim abundabat Graecia scriptoribus et artium inventoribus quam Roma. Graeci itaque opulentiam dicuntur habere, quia multa habent quae nos non habemus; nos autem thesaurum, quia quae meliora videbantur in latinum sermonem translata sunt. Et ideo bibliotheca Romanorum "thesaurus" esse dicitur, quasi "theca auri" per excellentiam. Conveximus. Simul a multis quaesivimus, quia non solum Nichomachum imitatus est, sed et plures. In thesaurum. De graeco in latinum transferendo ditavimus linguam Ro<manorum> artibus quibus caruerunt. Ita enim. Littera sic iungitur: ad te conveximus hoc opus, et merito, quia in auctoritate habebitur, si tuo comprobatur iudicio. Et captat hic iterum benivolentiam a persona Symmachi, commendando eum in sapientia, tum ab ipsa re, dicendo hoc opus extractum esse ex doctrinis sapientum, id est ex doctrinalibus istis artibus, quae vere scientiae sunt et vere sapientes reddunt, quia maior sapientia habetur in quadruvio quam in trivio. Mei quoque. Sicut et operis eorum a quibus transtuli. Vides igitur: quia hoc opus tuo commisi iudicio. Tuum igitur iudicium tantum expectat, quia si tu solus approbes, et alii comprobabunt; necesse est, cum tu sapientissimus sis. Astipulatione: confirmatione sive illuminatione, quia ignis astipulatus, id est cui stipula apponitur, magis illuminat circumstantia. In quo. Si nolit propalari, nisi prius tuo confirmetur, non debes mirari. Ibi enim non quaelibet parva et vilia tractantur, sed ea solum quae a sapientibus inventa sunt. Suis quippe. Instrumenta operis appellat sapientis iudicium, quo ipsum opus perficitur, quemadmodum aliquid perficitur instrumento. Et reddit causam cur Symmacho libros suos auctorandos tradiderit, quia tunc res sapienter discussa bene tractatur, cum iudicium sibi simile, id est sapientis, promeretur. Expenditur: trutinatur, iudicatur, firmatur extra pensum, id est dubitationem, ponitur. Quamdiu enim res pendet, etiam in dubio esse dicitur. Cum cogitur subire. Quod est dicere: quilibet physicus tractatus tunc primum extra dubitationem ponitur, cum de eo sapiens, qui instrumentum physicae est, iudicat. Et bene dicit cogitur, quia quaedam est coactio, et non aliter videretur opus perfectum. Sed huic. Aliae artes ad sui iudicium plura exigunt, quibus muniantur, ut perfectae esse comprobentur, sed huic muneri, id est arithmeticae, quam tibi offero, non sunt tot necessaria ad perfectionem sui, quot aliis artibus. Neque enim

49–51 On "thesaurus quasi theca auri," see William of Conches, *Glosae super Boetium* 3.m11 (ed. Nauta, p. 190, ll. 30–31); see also Isidore of Seville, *Etymologiae* 16.18.6, and 18.9.3.

FERE ULLA. De mechanicis, id est adulterinis artibus, id dictum intellige, de quibus subponuntur exempla. Unde etiam dicit FERE, illud phylosophicum temperamentum, ut excipiat septem liberales artes, quae non desiderant ita aperta et grossa adiumenta in sui iudicio. Quoniam vero inter ipsas liberales artes quaedam egent aliis, ut verbi gratia dialectica et rethorica egent grammaticae, grammatica dialecticae, et in quadruvio musica, geometria, astronomia, tres istae egent arithmetica, et ipsa non eget illis, quia est fundamentum ipsarum, ideo ad commendationem arithmeticae "fere" dixit exceptive, quia non eget aliqua illarum trium de se prodeuntium. NAM IN EFFIGIANDIS. Inductione utitur ad ostendendum quod artes caeterae adiumenta desiderent aliarum. Primo inductum est de arte marmorea, quae sculptoriam desiderat artem. MARMOREAM appellat artem secandi marmora. Solebant enim Romani hyrcino sanguine et aqua – nescio cuiusmodi – secare marmora facillime. Secundo inductum est de arte pictoria, cui fabrilis ars est necessaria. Tercio de arte fabrili, qua fiunt instrumenta bellorum. ALIUS LABOR. Hoc lathomus exercet. NITOREM POLITI OPERIS in imagine nonnisi peritissimus artifex novit facere, qui graece "aliptes" dicitur, id est politor imaginis. "Imago" graece "icona" dicitur. Cum enim imago politur, tunc pictores extremam manum illi imponere dicuntur. EIUSDEM. Qui scit formare, nescit eas polire ut niteant. NITOR. Non ex colore superposito, sed ex singulari arte quam sciunt marmorarii, ut opus perpolitum fiat nitidum. AT PICTURAE. In arte faciendarum statuarum diversi habentur artifices, sed in arte pictoria habetur multiplex materia. Tabula quae pingitur graece "pinax" vocatur. FABRORUM: carpentariorum qui faciunt ligneas tabulas quae pinguntur. Faber enim primo necessarius est, qui grossum tabulae cum dolabro auferat, postea panniculus super quem colores ponuntur, postea colores et alia multa sunt necessaria illi arti. Et notandum quod his verbis attenuat artes mechanicas et innuit earum utilitatem, quae tam vilibus instrumentis tam vilium artificum egent, ut tabulis quas faciunt

80 Hugh of Saint Victor, *Didascalicon* 2.21 (ed. Buttimer, p. 39): "Hae mechanicae appellantur, id est adulterinae, quia de opere artificis agunt, quod a natura formam mutuatur." Cf. Martin of Laon(?), *Scholica graecarum glossarum* (ed. Laistner, 439): "Moechus est adulter alterius thorum furtim polluens. Inde a moecho dicitur mechanica ars, ingeniosa atque subtilissima et paene quomodo facta vel administrata sit invisibilis in tantum, ut etiam visum conspicientium quodam modo furetur, dum non facile penetratur eius ingeniositas."

95–98 Cf. Caiazzo, "Un commento altomedievale," 128, ll. 60–63: "NITOREM, id est politi operis imaginem, non nisi peritissimus artifex novit facere, qui graece 'aliptes' vocatur, id est politor imaginis. Imago graece 'ycona' dicitur. Cum enim ipsa imago politur, pictores extremam manum illi imponere dicunt."

fabri, et cera quam colligunt rustici, et colore a mercatoribus comparato. Fucus proprie dicitur niger color, abusive vero pro omni colore accipitur. Decerptae: carptim collectae a rusticis qui observant tempus et modum in carpenda cera. Textrinum appellatur locus in quo textrices operantur. Multiplicem materiam: quia haec omnia tractant opifices in ista arte. Nonne idem. In arte statuarum habentur diversi artifices, et in arte pictoria diversa materia. Idem: id est utrumque fit in arte quae habetur de bellicis instrumentis. Torax graece dicitur "pectus," inde eodem nomine vocatur lorica muniens pectus. Gemit: metaphora ab animali ad inanimale. Umbonem appellat clipeum, et ponit partem pro toto. Orbem clipeum dicit, eo quod antiqui solebant clipeis rotundis uti. Orbi: orbiculato clipeo quem ipse elaboravit. Et quia ad eius circumductionem nescit artem faciendi coria cruda, ideo mercatur ea ab alio artifice. At nostri: ad aliarum artium iudicium multa exquiruntur, sed aliter nostro evenit operi, cui tu solus impones iudicium. Hoc autem per manum significat; tractum est autem hoc a pictoribus vel sculptoribus qui tum extremam manum imaginibus imponere dicuntur, cum ipsis perfectionem tribuunt, ne aliquis scrupulus ibi remaneat. In quo. Si tu emendes, non stabit labor, ut iudices consentiant. Te enim concedente, omnes ilico concedere necesse est, cum te sapientissimum iudicent. Et notat antiquorum consuetudinem, qui solebant suorum operum plures constituere iudices, ut quod quibusdam male aliis bene dictum videretur. Hoc tamen facere solebant praecipue, si ex pluribus esset opus extractum, sed intentione hac, quia unus in uno, alius valeret in alio. Quamlibet. Adversatur ei qui diceret hoc opus non a solo Symmacho debere iudicari, cum a multorum Graecorum multis artibus sit translatum. Vere non est curandum de iudicio aliorum, quia quamvis hoc opus ex pluribus extractum sit, tuum tamen solum sufficit iudicium, licet Graecorum iudicio probatum sit.

2. Experiare igitur. Quia iudicaturus es opus meum, igitur animadvertere debes quae sunt iudicis, quanto labore et tempore fecerim, si illud facere potuerim, quomodo fecerim. Longis otiis. Ociosus a negotiis secularibus. Otium quandoque ponitur pro "studio," quia studentes debent

109 cera] caera MS **137–138** animadvertere] animadverte MS

116–117 Cf. Caiazzo, "Un commento altomedievale," 128, ll. 66–67: "Torax graece, latine dicitur 'pectus'; in eodem nomine vocatur lorica muniens pectus."

123–125 Cf. Caiazzo, "Un commento altomedievale," 128, ll. 71–72: "Tractum est a pictoribus, qui tunc extremam manum imaginibus imponere dicunt, cum ipsis perfectionem tribuunt."

vacare a curis et liberum animum habere. EXPERIARE etiam an velox et mobile, id est leve, et acutum ingenium meum potuerit ea colligere quae subtilia sunt, et libenter leviterque elabuntur ab hominum memoria in numeris, quia non sufficit homini bonum ingenium habere ad colligendam artem aliquam, nisi etiam sit exercitatus in illa. UTRUM IEIUNAE EXPERIARE etiam utrum oratio nostra sufficiat ad expediendam rei difficultatem. IEIUNA vero et macra dicitur oratio, quae sine ornatu verborum et sententiarum est. Huic autem ornatui non intendunt, qui difficilia expedire cupiunt. Ornatus enim ad delectandum valet potius quam ad difficile aliquid expediendum. IMPEDITA iterum dicuntur esse quae tractat quadruvium, quia sunt difficilia intellectui. QUA IN RE. In tuo iudicio aliorum iudicia lucrabor, et omnes tuum sequentur iudicium, quia in graecis litteris peritus es, ex quibus hoc opus extractum est. PRAESCRIBERE est absque exceptione damnare. Vel PRAESCRIBERE: sicut et praeiudicare est ante iudicare, vel ante iudicium punire. Unde et praeiudicium dicitur absque cognitione.

3. AT NON ALTERIUS. Modum tractandi hic ostendit, in quo captat benivolentiam a persona sui propria. Sententia sua est haec: etsi hoc opus transtulerim, non hanc tamen legem translationis secutus sum, ut verbum ex verbo transferrem. Lex translationis triplex est: arta, artior, ARTISSIMA. Arta est cum ille qui transfert, sensum alterius sequitur, non ordinem nec verba. Artior est cum sensum sequitur et ordinem, etsi non verba sequatur. ARTISSIMA vero est cum sensum et ordinem et verba sequitur. ALIENO ITINERI. Metaphora est, quia quemadmodum aliquis eadem incedit via, non tamen per eadem vestigia, sic et auctor iste de arithmetica tractavit de qua et caeteri, quamvis non per eadem verba. FORMULIS. Formulas vocat figuras quas in sequentibus ponit, et descriptiones dispositionesque numerorum quae in graeco non habentur, sed ideo ista interposuit ut ars evidentior fieret. Et in hoc eodem versu reddit nos dociles et in parte benivolos. ALIQUANDO dicit quia tarde a Romanis haec ars perfecta est. SOBRIUS LECTOR, id est prudens qui recte potest iudicare de arduis rebus; FACILE intelliget quanto labore hoc opus transtulerim.

154 praescribere] scribere MS **162–163** itineri] tamen MS. itineri *ed. Guillaumin, ed. Oosthout and Schilling, Chartres 498*

156–157 See above note to 6–8.

165–170 Cf. Caiazzo, "Un commento altomedievale," 129, ll. 84–88: "FORMULAS vocat figuras, quas in sequentibus ponit, et descriptiones dispositionesque numerorum, quae in graeco non habentur, sed ideo iste interposuit, ut ars evidentior fieret. ALIQUANDO adiecit, quia tarde a Romanis haec ars percepta est. SOBRIUS LECTOR, id est prudens, qui recte potest iudicare de arduis rebus."

4. CUM IGITUR. Quia tractavi primum de arithmetica, et quia ipsa prima est et quasi mater caeterarum, ideo tu, qui sapientissimus es, tanto dignus munere videbaris. INERRATO: nomen est neutrum, non adverbium. SUSPECTA SECURITAS. Licet securus essem veniae apud te, quia socer meus es, tamen suspitionem habui ingratum tibi fore opus meum, si non esset perfectum. ARBITRABAR viro tantae auctoritatis nihil esse offerendum, quod maximo non esset studio perfectum. ELABORATUM INGENIO, PERFECTUM STUDIO. Bene coniunxit haec duo. Ex ingenio enim artificium, ex studio vero perfectio. DIGNUM POSTREMO. Tale volui opus meum esse, quod et te deceret et me, qui multum temporis et laboris apposui. Ipse enim, ut superius dixit, longo tempore studuerat, quod ei posset esse obprobrio, nisi operis sequeretur perfectio. NON IGITUR. Quia hoc opus subposui tuo iudicio, non igitur dubito quin emendes, cum sis sapientissimus. HIANTIA. Hiare est os aperire, et hiare dicitur res quae, quasi aperto ore stans, ostendit se aliquid ad sui perfectionem desiderare. QUAE RES: id est quia sciebam te emendaturum hoc opus, haec res ad scribendum de hac arte me IMPULIT. PIGRAM dicit quia diu deliberaverat utrum tanti operis translationem aggrederetur. NIMIOS. Bene dixi PIGRAM, quia ex translatione aliam utilitatem non desidero, nisi quod tibi placeat. NOVI QUIPPE. Non dubito quin opus nostrum tibi placeat, quia magis nostra quam caeterorum bona diligimus. Omnia vero mea pro tuis reputanda sunt, et tua pro meis. AUREOS CULMOS flavos ET MATUROS PALMITES, id est botros. Quoniam sapiens et plurium opera iudicare soles, ergo recte et convenienter tibi hoc nostrum opus offerimus, tamquam inventori artium et iudici, sicut ille convenienter faceret qui CERERI CULMOS et BACHO PALMITES offerret, utpote talium rerum numinibus et inventoribus. RUDIMENTA dicit non quod hanc artem primam composuerit – prius enim et de aliis scripsit artibus –, sed quia nuper hanc ediderat artem, ideo RUDIMENTA dicit secundum translationem de graeco in latinum. PRIMITIAS dicit de quadruvio; prius enim fecerat hunc librum quam vel musicam vel geometriam vel astronomiam.

172 Quia] de numeris *add.* MS

197–200 Cf. Caiazzo, "Un commento altomedievale," 129, ll. 102–105: "RUDIMENTA NOSTRI OPERIS: dicit quod hanc artem primam composuerit. Prius enim de aliis scripsit artibus, sed tamen nuper hanc edidit artem; ideo rudimenta dicit, secundum translationem de graeco in latinum." In the phrase, "Dicit non quod hanc artem," the commentary of Thierry of Chartres contains the same separative and conjunctive variant found in Reims, Bibliothèque municipale, MS 975 (that is, the twelfth-century copy of the manuscript of Chartres/Leiden; see above, pp. 73–77).

LIBER PRIMUS

CAPUT I

PROEMIUM. "Imen" Graeci dicunt "viam," "pro" "ante" vel "prae." Unde dicitur "proemium" vel "proimium" "praefatio" vel "praeviatio," quae fit antequam dicatur id quod susceptum est ad dicendum. Est autem proprie proemium tactus chordarum, qui et praeviatio potest dici, quae praecedit cytharizationem, per quam praeviationem ingreditur musicus ad certum et dispositum sonum quasi per viam. Et sicut illa praeviatio proemium vocatur, ita sententia, quae praecedit narrationem, prohemium appellatur.

I, 1. INTER OMNES. Quidam dicunt Boetium hunc prologum fecisse; quidam dicunt eum a Graecis transtulisse. Respondet autem in hoc prologo quibusdam qui dixerunt superfluam esse hanc artem, utpote sine qua omnes numeros suos computare possent et scire, commendans artem istam et ostendens eam multum esse utilem et necessariam, sine qua nemo sapiens dici poterit, quia sine ea nec tres illae, musica scilicet, geometria et astronomia, existerent. Haec enim posita non quidem ponit reliquas, sed haec destructa destruit alias. Et per hoc patet eam antiquiorem et digniorem esse caeteris, sicut ostendet in prologo isto. Est autem prologus iste "ante rem," cum praemissus fuerit "praeter rem." In isto enim praelibatur materia. PRISCAE AUC<TORITATIS>: priscos et auctorabiles VIROS. QUI PYTH<AGORA>. Pythagoras primus ingenium suum erexit ab istis visibilibus et sensibilibus rebus ad invisibilia et non comprehendenda sensibus corporeis, sed sola ratione et intellectu perceptibilia, scilicet ad universalia, id est X praedicamenta, quae verum esse habent non ab alio sed ex se, quae immutabilia sunt et invariabilia quantum in se et in sua natura, sed tactu

1 Liber primus] de libro I° MS

---

3 The title of *De arithmetica* is I, 1: "Proemium in quo divisio mathematicae."

3–9 Cf. Caiazzo, "Un commento altomedievale," 130, ll. 106–112: "PROEMIUM. Imen dicunt Graeci 'viam,' pro 'ante' vel 'prae'; unde proemium vel proimium 'praefatio' vel 'praeviatio,' quae fit antequam dicatur id quod susceptum est ad dicendum. Est autem proprie proemium tractus cordarum, quae et praeparatio potest dici, quae praecedit citharizationem, per quam praeviationem ingreditur musicus ad certum et dispositum sonum quasi per viam. Et sicut illa praeviatio proemium vocatur, ita sententia, quae praecedit narrationem, proemium vocatur."

18–20 Cf. Thierry of Chartres, *Commentarius super Rhetoricam ad Herennium* 4.1.1 (ed. Fredborg, p. 313). See praefatio 2–5, and the remarks above, p. 43.

variabilis vertuntur in variabilem inconstantiam. Haec autem generalissima sunt X praedicamentorum: substantia, qualitas et caetera. Quae quidem causae sunt rerum omnium, tam subsistentium quam etiam accidentium; et hoc ipso quod sunt causae rerum, sunt invariabiles, nec intenduntur nec remittuntur, sed effectus eorum in corporibus variatur circa corpora, et circa illa intenditur et remittitur. Verbi gratia: albedo illud universale nec intenditur nec remittitur, sed effectus albedinis intenditur et remittitur circa corpora, quorum quaedam efficit alba, quaedam plus alba. Has causas rerum primo investigavit Pythagoras, et quia in eorumdem inquisitione et tractatu plures phylosophi eum imitabantur, ideo vocat eum DUCEM. Ipse etiam primus arithmeticam docuit. Et cum alii doctores se sapientes appellari voluissent, iste primus humilitatis causa phylosophum, id est amatorem sapientiae, appellari se constituit. Quia ergo alii doctores sapientes se appellari censebant, ideo legimus fuisse VII sapientes ante Pythagoram. PURIORE MENTIS RATIONE. Ratione discernuntur eadem quae et sensu comprehenduntur et imaginatione, sed alio modo quam vel sensus vel imaginatio comprehendant. Et quia ratio discernit de istis sensibilibus quae incerta et non pura sunt, ideo dicit illam rationem quae considerat universalia puriorem esse. Vel vult intellegere quod duae sunt vires animae, superior et inferior, et superiorem hic appellat puriorem. HAUT QUEMQUAM: id est nullum posse evadere a libris poetarum. Vel notat difficultatem artis dicitque nullum posse pervenire ad perfectionem, nisi illum qui instructus erit in quadruvio. TALIS PRUDENTIAE. Talis et tam nobilis prudentia quae nobilitat animos discentium, quam etiam quondam soli nobiles apprehendebant. VESTIGATUR quatuor istis disciplinis, quasi IIIIor viis. QUADRUVIUM vocat quadripertitam mathematicam. INTUENTIS: intellegentis naturam istarum disciplinarum. "Sollon" dicunt Graeci "multum"; hinc SOLLERTIA peritia dicitur multarum rerum. EVADERE: bene dicit. Animus enim quodammodo

45 Haut] haud *ed. Oosthout and Schilling, Chartres 498*

35–39 Cf. Caiazzo, "Un commento altomedievale," 130, ll. 119–123: "Et ipse primus arithmeticam docuit, et cum alii doctores se sapientes appellari voluissent, iste primus, humilitatis causa, philosophum, id est amatorem sapientiae, appellari se constituit. Quia ergo alii doctores sapientes appellari se censebant, ideo legimus fuisse VII sapientes ante Pytagoram." Cf. Thierry of Chartres, *Commentum super Boethii librum De Trinitate* 2.7 (ed. Häring, p. 70); see above, pp. 28–29.

50–53 Cf. Caiazzo, "Un commento altomedievale," 130, ll. 123–126: "QUADRUVIUM vocat quadripertitam mathematicam. Sollon dicunt Graeci 'multum,' hinc SOLLERTIA dicitur peritia multarum artium. INTUENTIS, id est intelligentis, scilicet naturam istarum disciplinarum."

corporeis sensibus allicitur, qui postquam ad perfectionem cognitionemque veritatis pervenerit, tunc quasi carcere evasisse dicitur. TALIS PRUDENTIAE. Commendatio est ipsius scientiae, id est quadruvii. Quadruvium enim dicuntur hae quatuor disciplinae, quia sicut IIIIor diversis viis potest ad eundem locum pervenire quis, ita his IIIIor disciplinis nos ad cognitionem veritatis et virtutis. QUASI: quia non revera quadruvium dicitur, sed per similitudinem. EST ENIM SAPIENTIA. Probat argumento ac diffinitione quod sine his nullomodo potest perveniri ad cognitionem veritatis. SAPIENTIA EST COMPREHENSIO VERITATIS: id est vera comprehensio rerum. Et quia non omnia quae dicuntur esse, vere sunt, subiungit: "id est rerum quae ex se et ex sui natura non mutantur." Nota quod "immutabile esse" est nullomodo unquam posse mutari. SORTIRI IMMUTABILEM SUBSTANTIAM est ita manere aliqua ut dum ipsa permanent, nullomodo possint mutari. Quod igitur qualitas mutari videtur vel quantitas, mutatio quidem est substantiae subiectae, non qualitatis vel quantitatis. Nulla enim qualitatis mutatio, nisi de esse ad non esse, et econverso. Mutabilium itaque non est certa comprehensio, sed eorum QUAE SORTIUNTUR, id est per sortem accipiunt IMMUTABILEM SUBSTANTIAM. Sors enim lites dirimit; et quod videtur esse multorum, fit unius possessio. In illis itaque QUAE SORTIUNTUR IMMUTABILEM SUBSTANTIAM, certa comprehensio est de omnibus, quia comprehendendo illa quae adiacent rebus, comprehenduntur etiam res ipsae. Si opponatur quod sapientia non est certa comprehensio omnium rerum, quia non comprehendit materiam (quia potest mutari), dicimus nihil valere, quia cum scitur quid sit res, et qualis et quanta et caetera, manifestatur materia, et per hoc etiam est mutatio. Universa vero comprehensio veritatis in duobus consistit, in magnitudine et multitudine.

I, 2. HAEC AUTEM tractat quadruvium; quae quicumque praetermittunt, praetermittunt et sapientiam, quia quaecumque comprehenduntur sive sint QUALITATES sive QUANTITATES sive quaelibet alia huiusmodi praedicamentorum, aut in MAGNITUDINE aut in MULTITUDINE comprehenduntur; id est aut discernendo ea ab aliis, quod refertur ad multitudinem, id est ad

59 veritatis] veri *lect. dub.* MS

61–62 Cf. Thierry of Chartres, *Commentum super Boethii librum De Trinitate* 2.7 (ed. Häring, p. 70); see above, p. 29.

70–72 Cf. Caiazzo, "Un commento altomedievale," 131, ll. 133–134: "SORTIUNTUR, id est per sortem accipiunt. Sors lites dirimit, et quod videtur esse multorum, fit unius possessio."

numerum, aut iterum comprehenduntur in se, id est in magnitudine earum. Magnitudo autem rei est integritas ipsius, ita ut ne quid ultra nec citra sit. Rerum autem quae sunt IMMUTABILES SUBSTANTIAE et quarum esse est immutabile, sicut sunt X praedicamentorum X generalissima, proprium et naturale est per se subsistere, non ab alio, sicut effectus eorum faciunt, qui non per se sed ab ipsis tamquam a causis eorum proficiscuntur. Verbi gratia: haec albedo Socratis et haec albedo Platonis mutabilis est, quia intenditur et remittitur, nec habet esse suum verum et a se sed ab alio, a causa videlicet, et ab albedine, ab universali illo quod ex se subsistit, nec intenditur nec remittitur. ESSE AUTEM dicuntur quorum esse immutabile est. Nullus tamen intelligat qualitates, quantitates esse immutabiles, sed permanentes non posse mutari. Unde recte dicuntur IMMUTABILEM SORTIRI SUBSTANTIAM, quia nec qualitas potest mutari in qualitatem, nec aliquod accidens in aliud, nec augentur, nec minuuntur, nec permutantur. HAEC AUTEM SUNT. Decem kathegorias Aristotelis Boetius in hoc loco comprehendit, licet non eodem ordine et eadem litteratura. Est autem prima usia, ad quam reliquae VIIII pendent: qualitas et caetera. QUALITATES. Per QUALITATES et FORMAS unum praedicamentum intellige qualitatis, id est albedinem et nigredinem, et istarum media, pallidum et caeruleum. Per QUANTITATES et MAGNITUDINES et PARVITATES unum praedicamentum quantitatis accipe in plures et pauciores numeros, vel in mensuris diversis, maioribus scilicet minoribusque, ut est bipedalis, cubitalis, et caetera. Per AEQUALITATES et HABITUDINES, praedicamentum relationis. Haec enim habent se ad aliquid, ut aequale aequali aequale, spiritus spiritui, homo homini, corpus corpori, numerus numero. Per ACTUS, praedicamentum quod est agere. Per DISPOSITIONES, praedicamentum situm esse; per LOCA, praedicamentum ubi; per

96–101 Cf. Caiazzo, "Un commento altomedievale," 131, ll. 136–140: "Unde recte dicuntur 'inmutabilem sortiri substantiam,' quia nec qualitas potest mutari in quantitatem, nec aliquod accidens in aliud. Decem autem cathegorias Aristotelis Boetius hoc in loco comprehendit, licet non eodem ordine et eadem litteratura. Est autem prima usia ad quam reliquae VIIII pendent: QUALITATES et cetera."

98–101 Cf. Thierry of Chartres, *Commentum super Boethii librum De Trinitate* 2.7 (ed. Häring, p. 70); see above, p. 29.

103–109 Cf. Caiazzo, "Un commento altomedievale," 131, ll. 141–145: "QUANTITATES sunt in plures et pauciores numeros, vel in mensuris diversis, maioribus scilicet minoribusque, ut est bipedalis, tripedalis, cubitalis, ad quam quantitatem magnitudines pertinent et parvitates. AEQUALITATES sunt cum aequalis est spiritus spiritui, homo homini, corpus corpori, numerus numero, ut sunt X ad X, V ad V."

107–109 Boethius, *De Trinitate* 6: "Nam omne aequale aequali aequale est et simile simili simile est et idem ei quod est idem idem est."

TEMPORA, praedicamentum quando. FORMAE sunt formationes rerum; et formae in rebus corporalibus, figurae vel species in rebus incorporalibus. MAGNITUDINES sunt in rebus corporalibus et incorporalibus, sicut et PARVITATES, ut in unoquoque corpore et numero, quae et ad aliquid dicuntur. Magnitudo enim ad comparationem parvitatis dicitur. HABITUDINES sunt proportiones, cum unum refertur ad aliud, ut simplum ad duplum; sive etiam habitudines sunt species et similitudines hominum aliarumque creaturarum, dictae ab habendo, quod habeant semper quandam comparationem. ACTUS sunt operationes quae sunt in rebus se semper moventibus, sicut nos et caelum seu quaeque animata semper sumus in motu. DISPOSITIONES sunt ordinationes et diffinitiones, per quas ostenditur quod primum quodve secundum sit; seu etiam dispositiones naturales sunt rerum consequentiae, id est ordines, ut ignis propriam obtinet dispositionem, ut superna petat. LOCA: circumscriptiones corporum quibus et propria sunt. TEMPORA sunt in quibus corpora transeunt cuncta, in quibus etiam homines et angeli esse ceperunt, sed non esse in eisdem desinent. Ista vere esse dicuntur, quia uniformiter semper manent; quamvis et videantur cum corporibus iunguntur perire, non tamen pereunt. Si enim album sit corpus vel nigrum, perire potest; non tamen ideo albedo aut nigredo perit, quia incorporalia sunt et immutabilia. Ergo quale perit, id est corpus quod qualitatis participatur, dum fit album vel nigrum. Qualitas vero in suo genere consonanter et perpetuo manet, sic et caetera. ET QUICQUID ADUNATUM CORPORIBUS INVENITUR: scilicet participatione non generalitate, ut longitudo, paternitas, filiolitas. Ideo dicit: CORPORIBUS ADUNATUM QUODAMMODO, quia alii dicunt quod circa corpus sint, alii non circa corpus sed intus inesse asserunt, sicut habere et pati, quae sunt duo praedicamenta.

132 quicquid] *Chartres 498* quidquid *ed. Oosthout and Schilling*

111–136 Cf. Caiazzo, "Un commento altomedievale," 131–132, ll. 145–167: "FORMAE sunt formationes rerum et formae in rebus corporalibus, figurae vel species in rebus incorporalibus. HABITUDINES sunt proportiones, cum unum refertur ad aliud, ut simplum ad duplum. Sive etiam habitudines sunt species et similitudines hominum aliarumque creaturarum, dictae ab habendo quod habeant se per quandam comparationem. MAGNITUDINES sunt in rebus corporalibus et incorporalibus sicut et parvitates, ut in unoquoque corpore et numero, quae ad aliquid dicuntur; magnitudo enim ad comparationem parvitatis dicitur. ACTUS sunt operationes quae sunt in rebus semper, sed moventur ibi, sicut nos et caelum seu quoque animata semper sumus in motu. DISPOSITIONES sunt ordines et diffinitiones per quas ostenditur quid primum, quidve secundum sit; seu etiam dispositiones naturales rerum consequentiae et ordines, ut ignis propriam habet dispositionem, ut superna petat. LOCA: circumscriptiones corporum proprie sunt. TEMPORA sunt in quibus corpora transeunt cuncta, in quibus etiam homines

Sic enumeravimus VIIII praedicamenta; decimum enim ipse non commemorat, neque mentionem eius facere necessarium puto, utpote quod est fundamentum aliorum. Novem enim ista in substantiis habent existere, per effectum dico. Ipsae quoque causae referuntur ad substantiam, tamquam ad maius et principalius, et genera cuius sunt praedicabilia, et de quo habent praedicari. Si enim non esset substantia, nihil diceretur tale vel quantum, et caetera. QUODAMMODO, id est per effectum ADUNANTUR CORPORIBUS ipsae causae. Quaedam enim ex his sicut agere et pati et ad aliquid et caetera, non ita principaliter nec affiniter habent se ad substantiam, sicut qualitates et quantitates, nec etiam eorum praedicatione tam frequenter utimur sicut qualis et quanti, et ipsius substantiae. VARIABILIS, id est corporis quod semper mutatur et numquam stat in eodem statu, sed est incertum, nec potest veraciter comprehendi aliquo sensu nec aliqua alia cognitione, quia si iam consideravimus tale vel tale esse aliquod corpus, et secundum hunc statum vellemus iudicare, non possemus, quia in eodem momento esset mutatum. Et ideo vocatur eius cognitio "opinio," quam quidem praecedit sensus aliquis, sed ille est irrationabilis quia fallitur. Quod Plato non nesciens, ait archetipum mundum intellectu percipi, sensibilem vero mundum ita incertum esse, ut eius perceptionem neque sensum neque imaginationem nec rationem nec intellectum appellet, sed opinionem cum irrationabili sensu, dicitque de eodem in eadem lectione sensibilia esse illa de quibus opinio aliquo sensu commota praesumit, id est praesumptuose iudicat, quia fallitur illa opinio quocumque sensu praecedente. Quando enim opinamur lupum esse, quod canis est, ibi opinio mixta est sensui, quia opinamur de re visui

et angeli esse coeperunt, sed non esse in eisdem desinent. Ista vere esse dicuntur, quia uniformiter semper manent, quamvis et videantur cum corporibus perire, tamen non pereunt. Si enim album sit corpus vel nigrum, perire potest, non tamen ideo albedo aut nigredo perit, quia incorporalia sunt et inmutabilia; ergo quale perit, id est corpus, quod qualitatis participatur, dum sit album vel nigrum. Qualitas vero in suo genere qualitatis constanter et perpetuo manet, sic et cetera: ET QUICQUID ADUNATUM QUODAMMODO CORPORIBUS INVENITUR, scilicet participatione non generalitate, ut longitudo, paternitas, filiolitas. Ideo dicit 'corporibus adunatum quodammodo,' quia alii dicunt quod circa corpus sunt, alii non circa corpus sed intus ea asserunt." On the *filiolitas*, see above, p. 73.

153–155 Plato, *Timaeus* 28A (ed. Waszink, p. 20): "… alterum intellectu perceptibile ductu et investigatione rationis, semper idem, porro alterum opinione cum inrationabili sensu opinabile proptereaque incertum, nascens et occidens neque umquam in existendi condicione constanti et rata perseverans."

156 *Ibid.*

157–158 Plato, *Timaeus* 28C (ed. Waszink, p. 21): "… sensilia porro ea quae opinio sensu aliquo commota praesumit eaque omnia facta sunt habentque ex aliqua generatione substantiam."

subdita. De hoc etiam quod odoramus, tangimus, audimus, opinamur et iudicamus quid sit; et illam opinionem quae sic fallitur, praecedit sensus quicumque. HAEC IGITUR: quia ex se immutabilia, igitur ex hoc proprie dicuntur esse. HORUM IGITUR. Quia sapientia est comprehensio veritatis eorum quae sunt, haec autem sunt ea quae sunt, igitur sapientia est comprehensio istorum. Distat inter SAPIENTIAM et SCIENTIAM. Sapientia est rerum spiritualium, scientia corporalium. Aliter: sapientia est rerum intelligibilium, scientia sensibilium. Cum dicit: ESSE SAPIENTIAM COMPREHENSIONEM VERITATIS EARUM RERUM QUAE SUNT, excluduntur illi qui falsis argumentis comprehendunt. Excluduntur etiam ab hac diffinitione cuncta visibilia ab ortu ad occasum currentia, quae dum immutabiliter non subsistunt, quodammodo non sunt. Nam fuit quando non fuerunt, et erit quando non erunt; ergo non vere sunt.

I, 3. ESSENTIAE: non dico cuiuslibet, sed eius quae est quantitas, quia indefinita equipollet particulari. Ostensis autem illis X, sine quorum scientia nemo sapiens dicitur, accedit ad materiam praelibandam, et ad artem istam commendandam. Et quia dicturus est quod numerus sit quantitas, assumit unum ex his X praedicamentis, scilicet quantitatem, eamque dividendo in partes et ipsas partes subdividendo, convenienter incidit numerum qui est quantitas discreta. Non dividit ea quae sunt, ut quidam dicunt, sed duas partes eorum ostendit quibus omnia, ut supra dixi, comprehendi possunt, multitudinem scilicet et magnitudinem. Et hoc facit, ut ostendat quam illarum tractet unaquaeque istarum IIII disciplinarum. Et sic bene probabit propositum, quia per quadruvium tractat haec sine quibus non potest haberi perfecta de rebus cognitio, quia qui negligit quadruvium, negligit et sapientiam. Et nota quod omnis comprehensio in his duobus notatur, secundum multitudinem et magnitudinem, id est in numero et mensura. Discernitur enim res a re numero, id est discretione; discernitur etiam magnitudine, id est mensura, cum consideratur in se, an simplex sit an composita. Et ideo

188 enim] *iter.* MS

166–170 Cf. Caiazzo, "Un commento altomedievale," 130, ll. 126–129: "Distat inter sapientiam et scientiam: sapientia est rerum spiritalium, scientia corporalium; aliter: sapientia est rerum intellegibilium, scientia sensibilium. Cum dicit tunc: SAPIENTIAM COMPREHENSIONEM VERITATIS, excludantur illi qui falsis argumentis falsa comprehendunt."

185–186 It is William of Conches who, in his *Glosses on the Consolation of Philosophy*, distinguishes *eloquentia* from *sapientia*, associating the first with the *trivium* and the second with the *quadrivium*, a distinction that may have been inspired by Cicero's *De inventione*. The identification of *sapientia* with the *quadrivium* is based on the prologue to Boethius's *De arithmetica* itself.

dividit essentiam. Et nota quod numerus pro discretione, magnitudo pro mensura ponitur et pondere, cum deus dicatur omnia fecisse in pondere, numero et mensura. Haec itaque tria sub illis duobus comprehenduntur. Et sicut dictum est, in omnibus essentiis sunt ista duo, scilicet numerus et mensura. NEC ULLIS: fines quidem habet, sed quid intendat, melius in disiuncta videbis, quia in choro tritici multa grana sunt, et unumquodque suis continetur finibus; sic autem non est in linea nec superficie, vel alia continua quantitate. UT EST ARBOR. Cum deberet exemplificare de linea et superficie et aliis continuis quantitatibus, exemplificat de his quae suscipiunt illas quantitates et in quibus sunt, sicut in corporibus quibuslibet. QUAE PROPRIE: quae corpora proprie appellantur magnitudines, quia sunt magna. Vel QUAE continuae quantitates sunt magnitudines, quia secundum eas dicuntur aliqua quanta, scilicet magna vel parva, vel alicuius alterius quantitatis. UT GREX. Iterum hic deberet exemplificare de discreta quantitate, scilicet de numero vel oratione, sed transfert se ad corporea. ALIA VERO. Haec sunt continua, sed alia disterminata suis partibus, quorum partes propriis terminis inter se divisa sunt. Et haec vocantur multitudines, quia proprie per multitudinem discernendo unumquodque eorum AB ALTERO quid sit, per numerum comprehenduntur.

I, 4. RURSUS MULTITUDINIS. Dividit multitudinem in duo, ut ostendat quam earum tractet arithmetica et quam musica. Quae divisio talis est: ALIA PER SE, alia relata AD ALIUD. Quae autem per se, quae vero relata sit, ipse exponit. Posterius huius divisionis membrum prius subdividit. QUOD NISI RELATUM SIT: non quod id quod duplum EST NON POSSIT esse sine eo cuius est duplum, sed quia nisi sit cuius dicatur duplum, non ipsum poterit esse duplum. MAGNITUDINIS. Horum duum generum magnitudinis et multitudinis ostendit quae earum cuius artis sit materia. NEC ULLIS. Hoc ideo dicit, quia possunt moveri et post tempus cessare, ideo addit ea numquam a motu quiescere.

HORUM ERGO. Hic incipit assignare materias disciplinarum IIII matheseos, quae cui subiciatur. Quia multitudinis duae sunt partes, et item magnitudinis duae, ERGO HORUM quatuor membrorum quatuor varietates et naturas speculantur IIII istae artes. ARITHMETICA INTEGRITAS: quia vis et proprietas numeri qui per se est, integre discitur in arithmetica. Partes itaque essentiae sunt duae, quarum una multitudo, altera dicitur magnitudo. Rursus multitudinis altera species est, illorum quae per se constant, nullius indiga supplementi, sicut sunt omnes numeri naturales; cuius multitudinis per se constantis scientiam arithmetica profitetur. Altera vero eorum quae per se non constant, sed ad aliquid referuntur, ut duplum ad simplum et

191–192 Sap. 11:20.

quicquid ad aliud relatum non tamen per se constat. Huius multitudinis peritiam musica complectitur disciplina. Porro magnitudinis geminae similiter sunt species, quarum una motu carentia et in sui firmitate persistentia continet, sicut est terra; cuius immobilis magnitudinis noticiam geometrica vendicat sollertia. Altera vero huius magnitudinis species mobilia quaeque suscipit, sicut est caelum, quod semper mobili volubilitate rotatur, quae astronomicis cessit speculationibus. MUSICI MODULAMINIS, id est musica, quae bene temperatur ad faciendas modulationes symphoniarum.

I, 5. QUIBUS IIIIOR. Haec tractat quadruvium; his autem omnia comprehenduntur. Qui ergo haec negligit, negligit sapientiam, quae sine his non potest haberi, cum in his omnibus comprehensio sit veritatis. NON POSSIT habere scientiam eorum quae vere sunt et verum esse habent. SINE HAC SPECULATIONE: sine his viis quisquis speculatur ea quae veraciter sunt non dicetur nec erit sapiens. Et hoc probat per repetitionem diffinitionis supradictae. EST ENIM. Septem sunt artes liberales, quarum tres in recte loquendo et recte disputando constant – grammatica videlicet, dialectica et rethorica –, quatuor reliquae matheseos disciplinae numero constant, quibus veritatis et sapientiae ratio colligitur et comprehenditur. Unde et INTEGRA COMPREHENSIO dicitur. Sapientia enim proprie in his IIIIor artibus versatur. SIQUIDEM PHILOSOPHIA. Subponit aliud argumentum a nominis ethimologia. Si quis spernit haec, spernit philosophiam. PHILOSOPHIA EST AMOR SAPIENTIAE. Ergo qui spernit haec, spernit amorem sapientiae, non igitur est philosophus.

I, 6. ILLUD QUOQUE. Commendavit huc usque quadruvium, in hoc quod sine quadruvio nullus esse potest phylosophus; et iterum commendat in hoc quod per illud vis infinitorum per quaedam finita docetur. Cum enim per multiplicationem multitudo in infinitum procedat, magnitudo vero in infinitum per divisionem decrescat, utramque infinitatem doctrina quadruvii repudians, per quaedam finita, horum vim et proprietatem demonstrat. AB UNO TERMINO. Ab unitate incipientes, possumus computando procedere in infinitum, sive in ea specie multitudinis quae per se est, ut I, II, III, IIII et caetera, sive in ea specie multitudinis quae est ad aliud, ut duplum, sesqualterum et caetera. MODUM IN DIVISIONE NON RECIPIT: tam in mobili quam in immobili magnitudine, quia deficientes in integris sectionibus, imus ad minutias et ab illis ad intelligibiles. INFINITISSIMAS dico partium denominatione, non actu. Nullum enim totum est, quod non possit dividi per duo media, et quod iterum non habeat terciam partem, quartam, quintam et caetera. NICHIL POTEST: a nobis, quia deo finitus est et omnis

237–239 Cf. Thierry of Chartres, *Commentum super Boethii librum De Trinitate* 2.2 (ed. Häring, p. 68); see also above, p. 30.

numerus et divisio. SED HINC: id est ab infinitate sumpsit id quod finitum est, ut ibi indagaret veritatem et exerceret sollertiam. "Indago, indaginis," inde "indago-indagas," id est inquiro, quia in indagine quaeruntur ferae. Ponitur etiam indago nomen pro inquisitione, id est indagatione. FINITAE TERMINUM QUANTITATIS, id est quantitatem quae terminatur aliquo fine.

I, 7. HOC IGITUR ILLUD QUADRUVIUM. Quandoquidem hoc quadruvio praetermisso perditur philosophia, igitur hoc assumpto acquiritur. Et quia haec IIII disciplinae illae tractant, sine quibus res comprehendi non possunt, igitur sunt IIII viae quibus ad cognitionem veritatis animus studentis ascendit. QUO HIS VIANDUM. Determinat quibus viandum sit, quia non quibuslibet sed his quorum animus factus est excellentior per doctrinam et studium. QUIBUS EXCELLENTIOR. ANIMUS enim per se tantum spiritualia intuetur, et quia in corpore manet, non valet aliquid discere per se nisi per corporales sensus. Sed dum per sensus exteriores doctrina a magistro audita ad interiorem animi intellectum perducitur, quae ipsa ratio probat, haec discit et intuetur. NOBISCUM PROCREATIS. NOBISCUM in una natura viventibus, illis scilicet V, qui cum corpore creantur ad percipienda corporea, sicut intellectus creatur cum anima et in anima ad percipienda incorporea. ANIMUS perducitur A SENSIBUS, quia sensus sunt ministerium et via animo. Illis enim deficientibus nec valentibus comprehendere superiora, ipsi sensus mittunt animum ad illa tamquam vicarium. Et bene dicit: A SENSIBUS. Philosophi animam enim dicunt esse antiquiorem corpore, in coniunctione vero corporis et animae nasci sensus, cum sensus nihil aliud sit quam passio ex utriusque coniunctione nata. SUNT ENIM. Et vere his est viandum, cum ipsae sint gradus et dimensiones certae, ultra quas vel citra non sit fas progredi, quae quodam ordine et gradatim legendae sunt et discendae. UT ANIMI OCULUM, id est animae, quae per animum tamquam per oculum suum comprehendit intelligibilia et generalissima. Animae itaque oculus est intellectus, qui ideo plus CORPORALI OCULO SALVARI debet ab

269 The scribe add the word *hage* above the line, which is a vernacular translation of Latin *indago*. **288** Philosophi animam] philosophiam MS **293** Ut] ad MS

278–282 Cf. Caiazzo, "Un commento altomedievale," 132, ll. 172–175: "Animus enim per se spiritalia tantum intuetur. Et quia in corpore manet, non valet aliquid discere per se, nisi per corporales sensus; sed dum per sensus exteriores, doctrina, a magistro audita, ad intentionem animi per intellectum ducitur, quam ipsa ratio probat, discit et intuetur."

294–302 Cf. Caiazzo, "Un commento altomedievale," 132–133, ll. 176–181: "OCULUS ANIMI est intellectus, qui ideo plus CORPORALI OCULO SALVARI debet ab omni vitiorum pruritu, quia illo solo intellectualiter scimus veritatem. Et corporalibus oculis cum muscis et belvis in commune utimur; intellectum vero cum solis angelis possi-

omni vitiorum pruritu, quia illo solo intellectualiter scimus veritatem, et corporalibus cum muscis et belvis in commune utimur, intellectum vero cum solis angelis possidemus. Oculus itaque animae est intellectus, visus eius est ratio, id est intellectus divinitatis vel filii dei, quae in evangelio appellatur vita; aspectus eius ratiocinatio, id est sermo cum ratione. Aspectus autem est subita visio, cum quis aciem oculorum subito aperit. Visus vero est dum aspectus fit continuus. QUI, UT AIT. Nos quidem laboramus pro salute et conservatione oculi nostri corporalis, sed PLATO AIT oculum animae digniorem esse SALVARI, ut salvetur et custodiatur ab impedimento sensuum, ET CONSTITUI, id est stabiliri in consideratione superiorum a qua retrahitur sensibus, quando vel videt vel audit vel olfactat vel gustat vel tangit aliquid delectabile et dulce. Pro hoc impedimento sensuum vitando, quidam ex phylosophis oculos sibi eruit. EO SOLO: et non inferiori illa vi animae, scilicet sensualitate, neque aliquo sensu nec imaginatione, vix etiam ratione, quia ubi ipsa deficit, deserit intellectum veritas, id est ea quae verum esse habent. VESTIGARI. Bene dicit "vestigari" et post INSPICI; primitus enim res investigatur, id est inquiritur, postea inspicitur, id est perfecte comprehenditur. HUNC INQUAM. "Inquam" dicit propter nimiam interpositionem. DEMERSUM: subiugatum, ORBATUM: cecatum corporeis sensibus. Referunt enim physici quod quanto gravior conspersio est corporis, tanto tardior et hebetior animae natura; et quanto levior, tanto fieri acutiorem intellectum animi ad veritatem investigandam. RURSUS ILLUMINENT. Quia animae antequam corpora ingrederentur, secundum philosophos, omnia praesciebant. Cum itaque dicit: HAE DISCIPLINAE RURSUM ILLUMINENT, tangit opinionem phylosophorum qui dicunt animam naturaliter habere in

297 corporalibus] corporaliter MS **299** evangelio] evangeli MS **300** ratiocinatio] ratiotinatio MS **303** conservatione] observatione MS **305** et constitui] constituique *ed. Oosthout and Schilling* **319** rursum] rursus *ed. Oosthout and Schilling; some of the manuscripts contain the variant* rursum (*cf. ad appar.*), *including Chartres 498.*

---

demus. OCULUS ANIMI est intellectus. Visus eius est ratio, aspectus eiusdem ratiocinatio esse dicitur. Aspectus est subita visio, cum quis aciem oculorum subito aperit. Visio vero est dum aspectus fit continuus."

299–300 Ioh. 14.6: "Ego sum via, veritas et vita."

315–317 Cf. Caiazzo, "Un commento altomedievale," 133, ll. 183–185: "Beatus enim refert Isidorus quod, 'quanto gravior est conspersio corporis, tanto tardior et hebetior fit animae natura; et quanto levior, tanto fieri acutiorem intellectum animi ad veritatem investigandam.'" I was unable to locate this citation in Isidore of Seville.

319–323 Cf. Caiazzo, "Un commento altomedievale," 133, ll. 185–188: "Cum dicit: HAEC DISCIPLINAE RURSUS ILLUMINENT, tangit id quod sepius narrant philosophi;

se omnes artes, sed mole corporis praegravata non potest vim suam explere; at postquam accedente doctore excitatur et sensibus corporis admonetur, non discit, sed ea quae oblita fuerat, quodammodo immemor recordatur. Unde et Boetius in libro *Phylosophicae consolationis*:

"Quodsi Platonis – inquit Musa personat verum,
quod quisque discit, immemor recordatur."

Cui sententiae etiam beatus Augustinus assensit, sed postea in libro *Retractationum* se ipsum reprehendit, dicens neque velle aut habere posse animam bonum, nisi divina largiente gratia.

I, 8. QUAE IGITUR. Commendavit huc usque quadruvium totum, modo commendat ipsam arithmeticam specialiter, quia de ea intendit, ostendendo ipsam sine aliis sciri posse, alias vero sine ea minime, cum eius materia materiis earum praeiaceat. PORTIONEM, id est similitudinem. Sicut enim matris lacte nutriuntur filii, sic huius doctrina et introductione ad caeterarum praeducimur cognitionem. HAEC ENIM CUNCTIS: vere principium, quia EST PRIOR. "Prior" dico non solum quod ad eius exemplar omnia facta sunt, sed etiam quod arithmetica destructa, destruuntur reliquae, sed non econverso. ILLE CONDITOR: qui scilicet numquam debet excidere ab animo, demonstrative dixit. SUAE RATIOCINATIONIS. Inter rationem et ratiocinationem

339 ratiocinationis] ratiotinationis MS   ratiocinationem] ratiotinationem MS

---

omnes scilicet artes naturaliter omnes homines scire, sed torpere eas in cunctis ob molem corporis, quasi gravi sopore sint praessae, quoad studio discendi et doctrina magistri excitentur."

324–326 Boethius, *Consolatio Philosophiae* 3.m11.

327–329 Augustine, *Retractationum libri duo* 1.8.2 (CCL 57: 22): "In quo libro illud quod dixi *omnes artes animam secum attulisse mihi videri, nec aliud quidquam esse id quod dicitur discere quam reminisci et recordari,* non sic accipiendum est, quasi ex hoc adprobetur animam vel hic in alio corpore vel alibi sive in corpore sive extra corpus aliquando vixisse, et ea quae interrogata respondet, cum hic non didicerit, in alia vita ante didicisse. Fieri enim potest, sicut iam in hoc opere supra diximus, ut hoc ideo possit, quia natura intellegibilis est et conectitur non solum intellegibilibus, verum etiam immutabilibus rebus, eo ordine facta, ut cum se ad eas res movet quibus conexa est vel ad se ipsam, in quantum eas videt, in tantum de his vera respondeat. Nec sane omnes artes eo modo secum attulit ac secum habet; nam de artibus quae ad sensus corporis pertinent, sicut multa medicinae, sicut astrologiae omnia, nisi quod hic didicerit, non potest dicere. Ea vero quae sola intellegentia capit propter id quod dixi, cum vel a se ipsa vel ab alio fuerit bene interrogata, et recordata respondet."

338–347 Cf. Caiazzo, "Un commento altomedievale," 133, ll. 189–194: "Ille CONDITOR, qui scilicet nunquam debet excidere ab animo, demonstrative dixit. EXEMPLAR est forma quaevis, quae ad imitandum proponitur. Exemplum autem illud, quod sequimur,

hoc interest, quia ratio est dum adhuc in mente versatur, ratiocinatio vero dum ad effectum perducitur. EXEMPLAR est quaevis forma quae ad imitandum proponitur; exemplum autem illud quod sequimur, quamvis interdum confunderant auctores. Ad exemplar numeri dicitur deus res creasse, quia omnia ex discretione, quae est ex numeris, habent esse. Et nota quod deus dicitur habuisse arithmeticam, id est numerum, exemplar suae ratiocinationis, id est dispositionis, quia secundum numerum, id est secundum rerum discretionem in mente provisa, CUNCTA disposuit. "Cuncta" dico quae concordant per numerum, id est iuxta proportionem assignantur, id est quae sunt assignatus ordo in mente divina. Ex his enim omnis rerum ordo procedit. ASSIGNATI: tractum est a signo, quod ovibus quodam cauterio ob recognitionem imprimitur. SED ob HOC QUOQUE, PRIORA NATURA dicuntur quae destructa, POSTERIORA destruunt, nec illata inferunt. POSTERIORA vero quae posita ponunt priora, nec destructa destruunt; et hoc vult littera sequens, ubi tanguntur duae species dialecticae, diareticae et analiticae. Diaretice est quae a generalissimo in specialissimum dividitur, ut ab animali ad unum solum. Analitice est quae a specialissimo in generalissimum resolvitur, ut: quid est Cicero? Animal. UT ANIMAL PRIUS EST HOMINE: non in praedicatione, nec in tempore vel inferentia (non enim sequitur: "si est animal est homo"), sed in continentia et dignitate, quia ad hoc ut aliquid sit homo, oportet illud esse animal. NAM SI TOLLAS. Per

340 ratiocinatio] ratiotinatio MS **345–346** ratiocinationis] ratiotinationis MS

---

quamvis interdum confundant auctores. Ergo exemplar dicitur deus arithmeticam habuisse, cum scilicet sex diebus cuncta creavit, ut inde quodammodo videatur sumpsisse similitudinem, quia cuncta in numero deus condidit."

350–351 Cf. Caiazzo, "Un commento altomedievale," 133, ll. 196–197: "ASSIGNATI ORDINIS, id est ascripti; tractum a signo, quod ovibus quodam cauterio ob recognitionem inprimitur."

354–357 This division of the dialectic is found in Iohannes Scotus, *Expositiones in Hierarchiam coelestem* 7.578–603 (ed. Barbet, pp. 106–107).

354–360 Cf. Caiazzo, "Un commento altomedievale," 133–134, ll. 200–209: "Notandum ubi dicitur: SED HOC QUOQUE PRIOR ARITHMETICA DECLARATUR et cetera, ibi tanguntur duae species dialecticae: diaretike et analitike. Diaretike, quae a generalissimo in specialissimum dividitur, ut ab animali ad unum solum hominem. Analitike, quae a specialissimo in generalissimum solvitur, ut: quid est Cicero? Animal. Generale enim seipsum tantum ostendit, ut animal; speciale vero se et genus suum, ut cum dico 'homo,' 'animal' intelligo. Omne quod prius est, et quod praecedit, tribus modis praecedit: natura, dignitate, aeternitate. Natura vel tempore: sicut Esau praecessit Iacob nascendo, sicut flos praecedit fructum. Dignitate: sicut fructus praecedit florem. Aeternitate: quia deus cuncta praecedit." On this passage, see the remarks above, p. 77.

negationem de aliqua re quod sit animal, de eadem sequitur quod non sit homo. Si autem negaveris de ea quod sit homo, non ideo sequitur de ea quod non sit animal. UT IN EODEM HOMINE, id est ut in dicto posito eodem exemplo utar, quo prius scilicet de homine. IDEM EST ENIM HOMO QUOD ANIMAL, id est eadem res quae est homo, est animal, sed non convertitur, ideoque subiungit: SI ANIMAL DIXERIS, id est si de aliquo affirmaveris quod sit animal, NON SPECIEM SIMUL HOMINIS INTULISTI, id est non ideo dixisti eum esse de illa specie animalis, quae est homo. NON EST ENIM IDEM ANIMAL QUOD HOMO: quia animal praedicatur de omni anima vigente, id est non omnis res quae est animal est homo.

I, 9. HOC IDEM. Coaptat similitudinem quod, quemadmodum animal – quod est genus hominis – positum non ponit hominem, sed destructum destruit, ita arithmetica posita non ponit geometriam, sed ea destructa geometria destruitur, cum sine ea esse non possit. HOC IDEM, scilicet quod sublatis prioribus, intereunt posteriora, sed non convertitur. QUAE OMNIA. Denominativa numerorum sunt quae denominantur ab eis, ut triangulum a tribus, quadrangulum a IIIIor, et sic de reliquis. AT VERO. Destructa arithmetica, destruitur geometria, sed non convertitur, quia etsi non sit quadratum, poterit tamen esse quaternarius, a quo denominatur quadratum. GEOMETRICAM FORMAM: vel trigonam vel tetragonam vel pentagonam.

I, 10. MUSICA VERO. Ex supradictis potest probari quod arithmetica sit prior quam geometria. Nunc iterum probat eam priorem esse musica, et hoc secundum materiam sicut prius, quia materia arithmeticae praecedit naturaliter materiam musicae, cum numerus per se prior sit relato. Arithmetica prior est tribus artibus, sicut et genus prius est specie, quia plura continet in se. Prior est musica, quia ex numero proportio numeri. Prior est geometria, quia de proportionibus tractat, quales circa mensuras geometria considerat. Geometria prior est astronomia, quia geometria qualitates prius pertractat, quales astronomia circa motum considerat. DYATESSERON. Sicut triangulum et quadrangulum a numeris denominatur, sic diatesseron et dyapente.

381 Musica] The scribe added *ablativus* above the line. **386** musica] The scribe added *ablativus* above the line. **388** Geometria] The scribe added *nominativus* above the line. astronomia] The scribe added *ablativus* above the line.

376–377 Cf. Thierry of Chartres, *Glosa super Boethii librum De Trinitate* 1.17 (ed. Häring, p. 263): "Si aliquid denominative predicatur de aliquo ipsum non est id quod dicitur ut quia bonus est Omerus ipse non est bonitas et quia sapiens est Cicero ipse non est sapientia."

Dyatesseron est de quatuor sonis, vel chordis modulatio. Dia enim quando per "y" scribitur "duo," quando per "iotam" "de" vel "ex" significat. Unde dyatesseron de quatuor sonis dicitur, quae modulatio fit in musica, cum maior sonus minorem tercia parte superat. ET IDEM IN HAC. Quia si tollas musicas istas symphonias, numeri tamen remanebunt, et non convertitur, quia si non sunt numeri, nec remanebunt symphoniae. IPSORUM QUOQUE. Duplum et triplum in nullis, nisi secundum numeros invenitur. Bene dico in numeris, quia dyapason ex duplo, dyatesseron ex sesquitercio, et sic de aliis. "Dia" "de," "pason" "totum" dicitur, inde diapason "de toto." In arithmetica vero epitrita vel sesquitercia dicitur proportio, ut IIII ad tres comparati. Veniunt enim ex arithmetica proportiones musicae. DIAPENTE: de quinque. In quinque enim chordis ille fit sonus in musica, ubi sonus sonum medietate superat, quod in arithmetica sesqualtera vel hemiolia proportio dicitur. HEMIOLION compositum est ex "hemis" et "olon," id est "semis" et "totum," ut III ad II, ubi scilicet maior numerus habet totum minorem et eius medietatem. DIAPASON: de omnibus. Omnes enim minores in se colligit proportiones. Vocatur autem dupla in arithmetica, diapason in musica. EPOGDOUS dicitur quasi "super octo" vel "superoctavus," videlicet cum chorda chordam superat in nona parte sive in acumine seu in gravitate. Epogdous dicitur et sesquioctavus, ut sunt VIIII ad VIII. TONUS dicitur in musica vox perfecta.

393 dyatesseron] dyathesseron MS **404** Hemiolion] emiolion MS hemis] emis MS

391–394 Cf. Caiazzo, "Un commento altomedievale," 134, ll. 211–214: "DIATESSARON, id est de quattuor sonis dicitur, vel cordis modulatio. Dia enim, quando per 'y' scribitur 'duo,' quando per 'iota' 'de' vel 'ex' significat; unde diatessaron de quattuor sonis dicitur, quae modulatio fit in musica, cum maior sonus minorem tertia parte praecedit."

399–411 Cf. Caiazzo, "Un commento altomedievale," 134, ll. 214–225: "In arithmetica vero epitrita vel sesquitertia dicitur proportio, ut quattuor ad tres comparati. Veniunt enim ex arithmetica proportiones musicae. DIAPENTE: de quinque. In quinque enim cordis fit ille sonus in musica, ubi sonus sonum superat medietate, quod in arithmetica sesqualtera et sescupla atque hemiolia proportio dicitur. Hemiolion compositum est ex 'hemis' et 'olon,' id est 'semis' et 'totum,' ut III ad II; maior enim numerus minorem totum numerum habet in se, et insuper eius medietatem. DIAPASON interpretatur 'de omnibus,' et est ex diapente et diatessaron; omnes enim minores colligit in se proportiones. Vocatur autem dupla in arithmetica, et diapason in musica. EPOGDOUS dicitur quasi super octo vel superoctavus; ubicumque enim corda cordam superat in nona parte, sive in acumine sive in gravitate, epogdous dicitur, et in arithmetica sesquioctavus, ut sunt VIIII ad VIII. TONUS dicitur in musica vox perfecta."

I, 11. Sperica et astronomia unum est. Sperica vocatur propter rotunditates circulorum; astronomica a lege astrorum. Astronomia non prohibetur disci, quia nil habet in se quod rei veritas repudiet. Astronomica vero habet in se quaedam quae prohibentur disci, quia mathematica ac constellatio, secundum quod dicunt mathematici genituram hominum constare, in illa subsistunt. In astronomia vero sunt circuli VII planetarum. "Planes" graece, item "errans." Et planetae dicuntur stellae errantes, quae aliquando cum sole, aliquando contra solem gradiuntur, modo cum mundo, modo contra mundum feruntur, modo sunt stationariae propter solem et lunam. Speram vocat ambitum totius mundi, vel orbem XII signorum. Centrum dicunt Graeci punctum circa quem volvitur circulus. Centrum etiam dicunt medietatem caeli. Paralleli circuli sunt <V>, sicuti sunt V zonae, quarum duae nimio frigore squalent, a septentrionali parte <et a australi parte>, et sunt inhabitabiles, duae temperatae, una torrida, quae in medio est, per quam sol currit, ideoque inhabitabilis. Est et orizon circulus qui dicitur limitans, eo quod caelum ibi dicatur a terra seiungi. Axem dicit lineam intellectualem, quae a polo usque ad polum, id est a septentrionali plaga usque ad nothium vel austrinum polum tendit. Duo enim poli sunt,

423 Paralleli] parallelique *ed. Oosthout and Schilling, Chartres 498* **426** medio] austro MS

412–433 Cf. Caiazzo, "Un commento altomedievale," 135, ll. 226–245: "Spherica et astronomia unum est. Inter astronomiam et astronomicam beatus Isidorus differentiam facit, et dicit quia astronomia non prohibetur disci, quia nihil habet in se quod rei veritas repudiat. Astronomica autem habet quaedam quae prohibentur disci, quia mathematica ac constellatio, secundum quam dicebant mathematici genituram hominum constare, in illa subsistit. In astronomia enim circuli sunt scilicet VII planetarum. 'Planes' graece, latine dicitur 'errans'; hinc planetae dicuntur stellae errantes quae, aliquando cum Sole aliquando contra Solem gradiuntur, modo cum mundo modo contra mundum feruntur, modo sunt stationariae praeter Solem et Lunam. spheram vocat ambitum totius mundi, vel orbem syderum. Centrum Graeci dicunt punctum, circa quem volvitur circulus. Centrum autem dicitur medietatem caeli. Paralleli circuli sunt sicuti sunt V zonae, quarum duae nimio frigore squalent: a septentrionali parte una, altera ab australi, et sunt inhabitabiles, duae temperatae, una torrens est, quae est in austro, per quam Sol currit, ideoque inhabitabilis. Est etiam orizon circulus, qui interpretatur limitans, eo quod ibi videtur caelum a terra seiungi. Axem vocat lineam intellectualem, quae a polo usque ad polum, id est a septentrionali plaga usque ad notium vel austrinum polum tendit. Duo enim poli sunt: septentrionalis et australis; tali enim modo caelum in transversum positum est. Et habet septentrio stellam, quae dicitur polus; auster quoque similiter, sed non apparet, quia infra terram est. In his duabus stellis totum vertitur firmamentum, et ipse polus centrum vocatur." Cf. Isidore of Seville, *Etymologiae* 3.27.1.

septentrionalis et australis. Tali enim modo caelum in transversum positum est. Et habet septentrio stellam quae dicitur polus; auster alterum habet polum, sed non apparet, quia infra terram est. In his duabus stellis totum vertitur firmamentum, et ipse polus centrum vocatur.

Hucusque autem ostendit arithmeticam esse priorem geometria et musica, modo ostendit eam priorem esse astronomia etiam secundum materiam. ASTRONOMIAM VERO SPERICAM: quia principaliter de speris agit. Vel aliter, duae sunt species astronomiae: sperica et gnomonica. Sperica est quae agit de motu planetarum, gnomonica vero computativa, a gnomone stilo horologii dicitur, in qua de planetis, quantum unusquisque in cursu suo moretur, agitur. Haec vero ex nimia dignitate nomen imponit generi. QUARE. Alia causa quare iterum geometria praecedat astronomiam, haec scilicet: quia geometria tractat de immobilibus, astronomia vero de mobilibus. Principium autem motus immobile est. Hoc quidem dicimus secundum Platonis sententiam, quem Boetius in hoc prologo imitatur. Secundum Aristotilem autem principium motus non est quietum sed vere movetur, sicut in fine Macrobii docemur, qui etiam fuit Platonicus et oppositus Aristoteli.

MUSICAE VIM. Tres sunt soni musici: cromaticus, diatonicus, armonicus, quorum primus est levissimus, alter gravissimus et durissimus, tercius vero suavissimus.

CAPUT II

II, 1. OMNIA. In hoc capitulo intendit ostendere dignitatem numeri, et convenienter. Cum enim in praecedenti prologo praemiserit dignitatem arithmeticae, merito subponit dignitatem numeri, de quo tractat arithmetica. Est autem haec dignitas numeri: quod ad eius similitudinem cuncta facta sunt, et quod ipse est invariabilis, et caetera quae in littera

443 mobilibus] mobililibus MS

437–440 Cf. *Commentary on Boethius's De arithmetica*, Bern, Burgerbibliothek, Cod. 633, fol. 25ra: "Vel possumus dicere quod due partes astronomie sunt: gnomonica et sperica. Gnomonica est de motu per quem futura cognoscimus, et hanc per excellentiam vocat hic astronomiam. Sperica vero est de situ et magnitudine sperarum."

445–447 Macrobius, *Commentarii in Somnium Scipionis* 2.15.3–5.

448–450 See below *ad Arithmeticam* II, II, 1.

assignabimus. Et notandum quia agit hic promiscue de numero: modo de hoc universali quod est numerus, modo de numero per quem numeramus. Et haec diversa sunt, quia universale solo intellectu concipitur. Verbi gratia: dualitas et trinitas ut ita loquitur, secundum quam isti duo dicuntur duo, et isti tres dicuntur tres, solo intellectu concipitur esse. Numerus autem realis, per quem numeramus, efficitur ex his universalibus, et sensu corporeo percipitur. Virtus autem numerorum diversa est, quia alia est virtus binarii, alia ternarii, et sic usque ad X. Nam quod supra X est, numerorum replicatio est. Virtus binarii est quod confert rebus alteritatem, ternarii virtus quod facit unam medietatem, quaternarii quod duas. Artes itaque quadruvii dicuntur certae dimensiones, id est determinatae, ultra quas, si quis tendat, decipitur, si infra, errat. Sensus comprehendit corpora, imaginatio formas, ratio universalia, intellectus divina. Ista igitur possunt dici gradus ad certiorem intelligentiam, sicut et quadruvium. Finito autem prologo, de numero hoc ordine praemittit quaedam elementa ad probanda theoremata quae sequuntur, quod facit more mathematicorum. Elementa considerantur in his tribus: an sit, quid sit, quale sit. Diffinitionibus probatur quid sit, divisionibus quale sit. Et dicuntur haec elementa quasi initia, quibus incognitis incognita erunt sequentia. Theoremata vero vocamus speculationes in numero et in proportione et in motu et in mensura. In hoc itaque capitulo agit de substantia numeri, scilicet unde numerus habeat esse. Procedit autem sic: primum ostendit an sit numerus, deinde quid sit, postea qualis sit.

OMNIA QUAECUMQUE. In hoc primo capitulo ostendit numerum esse aliquid, in secundo definit, in reliquis eius qualitatem, id est virtutem, ostendit, et sic tractatum propositum finit. Probat igitur numerum esse hoc argumento, quia omnia coniuncta sunt proportione numerorum. A PRIMAEVA RERUM NATURA, id est a deo faciente nasci res, qui omnium rerum principium est. Vel PRIMAEVA, id est prima omnium creatura. Vel PRIMAEVAM NATURAM dicit illam informem materiam quae fuit initium quatuor elementorum. RATIONE NUMERORUM. Ratio est intuitus vel aspectus. RATIONE, id est similitudine numerorum rationabiliter facta sunt. Vel RATIONE, id est proportione quae procedit ex numeris vel per numeros, ut per

457–458 Boethius, *De Trinitate* 3: "Numerus enim duplex est, unus quidem quo numeramus, alter vero qui in rebus numerabilibus constat."

473 Cf. Quintilian, *Institutio oratoria* 3.6.80: "... tria esse, quae in omni disputatione quaeruntur: an sit, quid sit, quale sit; quod ipsa nobis etiam natura praescribit"; see also 5, 10, 53. These three questions are also evoked by Claudian Mamertus, Martianus Capella, and Isidore of Seville.

484–486 Cf. Caiazzo, "Un commento altomedievale," 136, ll. 254–255: "PRIMAEVAM NATURAM vocat illam informem materiam, quae fuit initium quattuor elementorum. RATIO est mentis intuitus vel aspectus."

binarium alteritatis, per ternarium primum vinculum, et deinceps. PRINCIPALE EXEMPLAR. PRINCIPALE dicitur fuisse NUMERUS IN ANIMO CONDITORIS EXEMPLAR, id est in dispositione eius, quia diffinitum fuit ab eo quasi per numerum quot faceret creaturas.

HINC ENIM. De generali descendit ad singulare, et utitur inductione probans omnia fieri numerorum ratione. HINC TEMPORUM VICES: quia elementa in certo sunt numero, ut concordia ipsorum ex certis proportionibus est. Notandum est IIIIor tempora esse ex IIIIor elementis, et etiam IIIIor humores, hoc modo. Quatuor elementa sunt mundi – ignis, aer, aqua, terra – cum totidem aetatibus hominum concordantia – pueritia, adolescentia, iuventute, senectute – et cum IIIIor anni temporibus – vere, aestate, autumno, hyeme – cumque paribus etiam convenientia humoribus – sanguine, colera rubea, melancolia, flegmate. Quae omnia hoc modo concordant: ignis, aestas, adolescentia, colera rubea; terra, autumnus, iuventus, melancolia; aqua, hyemps, senectus, flegma; aer, ver, puericia, sanguis. Ignis terrae siccitate coniungitur, terra aquae frigiditate, aqua aeri humiditate.

501 flegmate] flecmate MS   **503** flegma] flecma MS

490–492 Cf. Caiazzo, "Un commento altomedievale," 136, ll. 255–257: "PRINCIPALE EXEMPLAR dicitur fuisse numerus IN ANIMO CONDITORIS, id est in dispositione eius, quia definitum erat ab eo, quasi per numerum quot faceret creaturas."

496–504 See the "Remigian" commentary on the *Consolatio Philosophiae*, appendix (ed. Edmund T. Silk, *Saeculi noni auctoris in Boetii Consolationem Philosophiae commentarius* [Rome, 1935], pp. 336–337): "Sicut enim mundus quatuor elementis et quatuor temporibus constat, ita et homo quatuor humoribus et quatuor temporibus. Videamus ergo mundi et hominis concordiam. Quatuor sunt elementa: aer, ignis, aqua, terra. Aer calidus et humidus est, ver calidum et humidum. Similiter et sanguis est humidus in puero et calidus. Pueritia enim calida et humida est. Ignis calidus est et siccus; aestas calida et sicca. Cholera rubea, quae abundat in adulescente, calida et sicca est. Adulescentia etiam calida et sicca est. Terra frigida et sicca; autumnus frigidus et siccus. Melancholia, id est cholera nigra, quae est in iuvenibus, frigida et sicca. Iuventus: frigida et sicca. Aqua: frigida et humida est, hiems frigida et humida. Flegma, quae abundat in senibus, frigida et humida." The attribution to Remi of Auxerre has been questioned by Fabio Troncarelli: "Per una ricerca sui commenti altomedievali al *De Consolatione* di Boezio," in *Miscellanea in Memoria di Giorgio Cencetti* (Turin, 1973), pp. 377–378, and *Tradizioni perdute: La* Consolatio Philosophiae *nell'alto medioevo* (Padua, 1981), pp. 144–149. See also Joseph S. Wittig, "King Alfred's Boethius and Its Latin Sources: A Reconsideration," *Anglo-Saxon England* 11 (1983), 157–198; Colette Jeudy, "L'œuvre de Remi d'Auxerre: Etat de la question," in *L'Ecole carolingienne d'Auxerre: De Murethach à Remi, 830–908*, ed. Dominique Iogna-Prat, Colette Jeudy, and Guy Lobrichon (Paris, 1991), pp. 373–397; Colette Jeudy, "*Remigii Autissiodorensis opera (Clavis)*," ibid. pp. 457–500; Joseph S. Wittig, "The 'Remigian' Glosses on Boethius's *Consolatio Philosophiae* in Context," in *Sources of Wisdom: Old English and Early Medieval Latin Studies in Honour of Thomas*

Hinc motus. Omnis sonus motum habere dicitur, et ubicumque motus est, sonus subsequitur, quia sonus ex motu fit. Unde cum dicit: motus ipse celebratur astrorum, tangit musicam caelestem, quae, quia motum habet, habet et sonum. Semper enim volubili lapsu vergitur, de quo formarum modulatio nota est, cuius medium dicitur obtinere sol. Est enim medius in armonia caelesti, sicut mese in chordis. Nam tantus illic sonus est tantaeque dulcedinis, ut dicit beatus Ysidorus, ut si humanae aures valerent illum audire, ab oriente in occidentem transirent sequentes motum sperae caelestis. Motus solis quadrifarius est: ab ortu ad meridiem, inde ad occasum, ab occasu ad meridiem sub terra, inde rursus ad ortum. Caelique conversio: in quot horis volvitur, scilicet in XXIIII.

D. Hill, ed. Charles D. Wright, Frederick Biggs, and Thomas N. Hall (Toronto, 2007), pp. 168–200. A figure, entitled "De aetatibus hominis" and showing the elements, the humours, the seasons, and the ages of man is found in Chartres, Bibliothèque municipale, MS 62, fol. 37v (tenth century). The manuscript was destroyed in 1944 but there is a microfilm made before the Second World War preserved at the Institut de recherche et d'histoire des textes, Paris; see Ernest Wickersheimer, *Les manuscrits latins de médecine du Haut Moyen Age dans les bibliothèques de France* (Paris 1966), plate 1.

505–513 Cf. Caiazzo, "Un commento altomedievale," 135–136, ll. 245–253: "Omnis sonus motum habet, et ubicumque est motus, sonus subsequitur, qui semper ex motu fit. Unde dicens: motus ipse celebratur astrorum, tangit musicam caelestem, quae, quia motum habet et sonum, semper enim volubili lapsu vergitur, de quo maxima modulatio procedit, cuius medium dicitur obtinere Sol; est enim medius in armonia caelesti, sicut mese in chordis. 'Nam tantus illic sonus est, tantaeque dulcedinis - ut beatus dicit Isidorus -, ut si humanae aures valerent illum audire, ab oriente in occidentem transirent, sequentes motum spherae caelestis.'" Cf. Remi of Auxerre, *Commentum in Martianum Capellam* 22.8 (ed. Lutz, 1: 107): "Eruditas a se. Apollo erudisse dicitur novem Musas quia ipse et princeps et maximus musicae caelestis et eum locum obtinet in planetis quem mese in chordis"; and 73.16 (ed. Lutz, 1: 198): "Eum enim locum obtinet sol in musica caelesti quem mese, id est media chorda, in musica artificiali." See also the glosses of Byrhtferth of Ramsey on Bede's *De temporum ratione* (PL 90: 211): "Medius sol. Medius Sol dicitur esse, quia in medio planetarum est, tres supra se, et tres infra se habens planetas. Dictus autem Sol eo quod exortus solus luceat, obscuratis omnibus sideribus. Et quia medius est in planetis, mese vocatur a chorda quae, similiter media in lyra, mese vocatur."

510–513 This citation is not found in Isidore; this same link to Isidore is found in the old commentary (see previous note) and in the glosses of Byrhtferth of Ramsey on Bede's *De temporum ratione* (PL 90: 442): "Nam tantus illic sonus est, tantaeque dulcedinis, ut beatus Isidorus ait, ut si humanae aures valerent illum audire, ab oriente in occidentem transirent, sequentes motum sphaerae caelestis."

513–515 Cf. Caiazzo, "Un commento altomedievale," 136, ll. 257–260: "Motus Solis quadrifarius est: ab ortu ad meridiem, inde ad occasum, ab occasu ad meridiem sub terra, inde rursus ad ortum. Caelique conversio, scilicet sumpta est in quot horis volvitur, scilicet in XXIIII."

II, 2. Quae cum ita sint. Infert a praemissis sic: quandoquidem omnia variabilia facta sunt ad similitudinem numerorum, necesse est eum esse variabilem. Et videtur hic assentire Aristoteli, qui probavit in fine Macrobii, licet sophistice, principium motus non moveri. Quod etiam Boetius paulo ante voluit ubi dixit: Omnis motus est post quietem. Eumque compositum: ad differentiam intellectualis qui simplex est, et quia est compositus, ideo etiam variabilis, propter tactum subiectae rei variabilis quam numerat. Ipse enim est effectus illius universalis.

Non ex diversis. Nihil ex diversis componi potest, sicut vinum et aqua non possunt componere vinum. Licet enim numerus compositus sit, tamen sicut iam dicet, ea quae substantiam numeri iungunt non sunt penitus diversa, quia ab una unitate procedunt, scilicet paritas et imparitas. Quamvis enim par et impar componant numerum, et sint diversa inter se, non tamen diversa sunt a numero. Quid enim. Probat illum intellectualem esse simplicem, et non compositum; quid enim componeret illud unde omnia sunt composita? Sed ex se. Mihi quidem videtur esse simplex, sed aliis videtur esse compositus, non ex diversis, sed ex se. Et hic respondet obiectioni eorum, ostendens eum nec etiam ex se compositum, quod magis rationi consentiret, nec ex diversis, quia penitus contraria nihil componunt, ut terra et ignis, et penitus similia nihil componunt.

II, 3. Nec ex his quae nulla. Hoc ideo dicit, non quod adversarius voluerit numerum esse compositum ex diversis, sed ideo quia, sicut est impossibile aliquid ex similibus constitui, ita etiam ex penitus diversis. Constat igitur. Ostensa natura et substantia universalis numeri, scilicet quod simplex est, subdit naturam et substantiam eius quo numeramus, et sic infert a praemissis. Quandoquidem ex similibus vel penitus dissimilibus nihil potest constare, ergo nec etiam iste numerus usualis quo numeramus. Et ideo cum ex aliquibus sit compositus, videamus quae ipsum constituant. Sed hoc facit paritas et imparitas subiectorum, quae habet numerus numerare. Ex hoc enim quod paria et imparia sunt subiecta, numerus habet existere, scilicet in talium computatione. Erunt ergo. Primo debet constare quod sint illa ex quibus constat numerus, antequam ipse dicatur ex

518 The scribe added *in fine Macrobii* above the line. **539** igitur] ergo *ed. Oosthout and Schilling, Chartres 498*

518–519 Cf. Macrobius, *Commentarii in Somnium Scipionis* 2.15.3–5.

520–521 Cf. Boethius, *De arithmetica* I, 1, 11.

539–540 Cf. Boethius, *De Trinitate* 3: "Numerus enim duplex est, unus quidem quo numeramus, alter vero qui in rebus numerabilibus constat." Cf. above ll. 457–458.

eis constare. PRIMA, id est initia et fundamenta atque principia, QUAE CONSTENT SEMPERQUE PERMANEANT. Numerus enim aeternus est et immutabilis. Ut enim beatus Augustinus dicit: “Ante potest fieri ut terra sursum, caelum deorsum vertatur, antequam tres non sint tres, et quatuor quatuor.” ET SUNT IPSA. Fundamenta numerorum DISSIMILIA sunt <ET> COMPONENDI POTENTIA, id est paria atque imparia, quae dissimilia quidem sunt, sed in unum possunt componi numerum.

Posito itaque et pro constanti habito quod sint, ostendit qualia sint. SUNT autem DISSIMILIA, et tamen constituentia, quod prius impossibile esse docuerat. HAEC vero POTENTIA divinitus eis conceditur, quod habent ex unitate generari. Et nota quod numerus dicitur constare, sive ad substantiam iungi, ex pari et impari. Hoc ideo dicitur, quia nullus est numerus qui non sit utrumque, et in quo non inveniatur tam paritas quam imparitas. Quod quidem patet in omnibus nisi in solo binario. Unde quidam volentes paritatem et imparitatem assignare binario, dicunt eum constare ex uno et altero: per unum intelligi volentes imparitatem sive identitatem, per alterum vero intelligi volentes paritatem sive variationem, cuius variationis principium est binarius. Alii dicunt omnes numeros constare ex pari et impari, excepto binario, et ita excipiunt binarium, quia de eo non facile patet. Ita autem constat binarius etiam ex impari et pari, sicut omnis numerus. Nam constat ex duabus unitatibus: altera est impar, altera par. Et hoc probo. Ternarius constat ex binario pari et unitate impari, et quaternarius constat ex ternario impari et unitate pari. Ecce vides quod unitas est par et impar, sed prior unitas est impar, secunda par, quia ab impari unitate incipit numerus, et secunda facit parem binarium. Componitur itaque binarius ex pari et impari, id est ex additione alteritatis super unitatem, quae alteritas ad parem, unitas vero refertur ad imparem. Ternarius similiter componitur ex uno et altero, scilicet ex binario et additione tercii, non ex unitatis additione. Si enim adiciatur unitas, et ponatur quod non sit tercia – quod in philosophia licet facere – non inde fit ternarius. Et ideo diximus componi ternarium ex

563 identitatem] identatem MS

548–554 Cf. Caiazzo, “Un commento altomedievale,” 136, ll. 267–272: “PRIMA, id est initia et fundamenta atque principia, QUAE CONSTENT SEMPERQUE PERMANEANT. Numerus enim aeternus est et inmutabilis, ut enim beatus dicit Augustinus: ‘Ante potest fieri, ut terra sursum caelum deorsum vertatur, antequam tres non sint tres et quattuor quattuor.’ IPSA fundamenta numerorum DISSIMILIA SUNT ET COMPONENDI POTENTIA, id est paria atque imparia, quae dissimilia sunt quidem, sed unum possunt componi numerum.”

550–551 Augustine, *locum non inveni.* See the previous note.

duobus praecedentibus et tercio, non ex unitate, etsi tercium sit unitas. Et binarius ad paritatem, ut supra dictum est, refertur, quia cardinalis dicitur numerus, ut unus, duo, tres; ordinalis qui ordinem significat, ut primus, secundus, tercius; partialis vero ut totum, medium et similia. Eodem modo de quaternario. Quaternarius componitur ex uno et altero et tercio et additione quarti, et ita ex pari et impari. Si vero opponatur: quaternarius componitur ex binario et binario, nihil valet haec obiectio, cum oppositio ista non sit ad substantiam sed ad quantitatem. Et nota, ut dictum est, quod alia est essentia numeri, alia quantitas, ut ternarii quantitas est ex tribus unitatibus, eius essentia ex duobus et tercio.

Caput III

III, 1. Et primo. Postquam ostendit unde constat numerus, convenienter assignat quid sit numerus. Quae quidem diversa sunt. Aliud est enim ostendere quid sit homo per diffinitionem congruam, scilicet "homo est animal rationale, mortale"; et aliud est ostendere unde constet, sicut ex quatuor elementis. Vel quantitatis acervus, id est quantitas coacervata et profusa ex unitatibus. In sensu est eadem haec diffinitio cum priore, sed quia non omne quod colligitur ex aliquo nascitur ab eodem, ideo plus videtur ponere consequens, quia ostendit numerum et nasci et colligi ex unitatibus.

Caput IIII

IIII, 1. Par numerus est qui sub eadem divisione, id est sub pari, ubi divisio fit in duo aequalia. In maxima: quantum ad spacia sive ad partes. Verbi gratia: octonarii non possunt esse maiores partes quam duo quaternarii. Et si quaeratur quanta sit illa divisio, vel quae sit quantitas illius divisionis, respondetur: dupla est vel dualis. Et haec est minima divisio in quantitate, id est minimae quantitatis. Maior enim est divisio, ubi aliquid dividitur in tria vel in ampliora, quam illa quae dividitur in duo; et econverso ubi minor est divisio, ibi maiores sunt partes. Agit autem hic de pari numero, et hoc ordine: diffinit primitus parem numerum, secundo loco ostendit unitatem esse principium numeri, tercio loco dividit eum per species suas, ad ultimum exequitur illas. In maximas partes par numerus

589 primo] primum *ed. Oosthout and Schilling, Chartres 498*

608–619 Cf. Caiazzo, "Un commento altomedievale," 136–137, ll. 274–284: "In maximas partes par numerus dividitur, cum in duas medietates partitur, nec est pars ulla

dividitur, cum in duas medietates partitur, nec est pars ulla quae non sit minor medietate. Cum ergo in has medietates maximae partes paris numeri divisae sunt, simul etiam quaedam parvissima sunt, id est partitio et divisio quae in duabus partibus facta est. Verbi gratia: XVI par numerus est. Hic ergo in maxima dividitur, scilicet secundum summam partium, id est VIII et VIII; in minima secundum numerum partium. Duae enim tantummodo partes sunt, quae in omni divisione minimae sunt, quia omnis numerus primo in duas partes, inde in tres (si terciam habet), deinde in quatuor partitur, quosque perveniatur ad eius unitatem. Similiter MAXIMA in SPATIO, id est magnitudine divisi numeri, PARVISSIMA QUANTITATE, scilicet divisionis, quia adhuc semel divisio facta est. SECUNDUM ISTORUM DUORUM GENERUM, id est multitudinis et magnitudinis, CONTRARIAS PASSIONES, quorum hic iam fecit mentionem, vel etiam superius in prologo, ubi mentionem utrorumque istorum generum faciens, dixit multitudinem ad infinita crescere, magnitudinem vero ad infinita decrescere. Cuius proprietatis contrarietatem haec duo genera in hac paris numeri divisione patiuntur, quia hic spatia sive magnitudo crescit, quantitas vero sive multitudo decrescit.

Passio est contrarius sensus; PASSIONES vocat motus, id est augmentum et detrimentum. Quanto enim magis numerus crescit, magnitudo decrescit. Ideo dicit CONTRARIAS, quia quo maior est summa, id est spacium, eo minor

611 etiam] et id MS **619–620** Secundum istorum duorum generum] secundum duorum istorum generum *ed. Oosthout and Schilling*, secundum duum istorum generum contrarias passiones *Chartres 498*

---

quae sit minor medietate. Cum ergo in duas medietates maximae partes paris numeri sunt, simul etiam parvissima quaedam sunt, id est particio et divisio, quae tantum in duabus partibus facta est. Verbi gratia: XVI par numerus est; hic ergo in maxima dividitur, scilicet secundum summam partium in VIII et VIII, et in minima, scilicet secundum numerum partium. Duae enim tantummodo partes sunt, quae in omni divisione minimae sunt, quia omnis numerus primo in duas partes, inde in tres, si tertiam habet, deinde autem in IIII partitur, quoad perveniatur ad eius unitatem similiter. MAXIMA in SPATIO est in magnitudine divisi numeri. PARVISSIMA QUANTITATE, scilicet divisionis, quia adhuc semel divisio facta est."

626–636 Cf. Caiazzo, "Un commento altomedievale," 137, ll. 285–293: "Passio est contrarius sensus; ideo dicit CONTRARIAS, quia quo maius est spatium, id est summa, eo minor est divisio, quae dicitur quantitas; et quo maior est quantitas, eo minus est spatium. Verbi gratia: si dividas VIII in quattuor et IIII, maximum quidem spatium est IIII, minima vero quantitas, id est divisio. Non enim nisi semel est divisum. Iterum IIII si dividas in II et II, spacium, id est summa supradicta, imminuta est, sed quantitas, id est divisio, aucta. Similiter, si duo in I et I partiaris, spatium minimum est, quantitas vero maxima. Non enim VIII minus spatium est quam unus, neque maior quantitas, id est divisio, quam dividere VIII octies."

est divisio, quae dicitur quantitas; et quo maior est quantitas, eo minus est spacium. Verbi gratia: si dividas VIII in IIII et IIII, maximum quidem spacium est IIII, minimum vero quantitas, id est divisio. Non enim nisi semel est divisum. Iterum IIII si dividas in II et II, spacium, id est summa supradicta, imminuta est, sed quantitas, id est divisio, aucta. Similiter si II in I et I partiaris, spatium minimum est, quantitas maxima. Non enim VIII minus spacium est quam I, neque maior quantitas, id est divisio, quam dividere VIII octies.

Caput V

V, 1. Quodammodo. Quia eadem ratione qua dicimus unitatem esse parem, scilicet quia est principium et origo parium, eadem ratione possumus eam dicere imparem.

Caput VII

VII, 1. Omnis quoque numerus. Ostensa diffinitione numeri et facta eius divisione in parem et imparem, data etiam diffinitione paris et imparis, prima vulgari, secunda Pythagorica, tercia secundum antiquissimos, ad ultimum secundum altrinsecas species, prius quam tractet de pari et impari, praemittit quamdam specialem dignitatem unitatis, quae omnium fons et origo numerorum est. Quilibet numerus est medietas duorum sibi proximorum, vel etiam IIIo et quarto loco altrinsecus circa se positorum. Sola vero unitas est medietas unius, scilicet binarii; et haec est eius principalis dignitas et propter principium sibi conveniens, sicut in exemplis apparet. Praemissis itaque elementis, subponit theorema; ex quo probato, probat unitatem esse principium numeri. Quod theorema sic probatur. Si aliquid cum alio aliquod totum componat integre, ipsum nec maius nec minus cum eodem potest illud componere, nec iterum cum eodem maiore vel minore potest illud componere. Sed omnis numerus est par<s> illius numeri qui ex circum se positorum aggregatione componitur. Quod autem sit pars ex hoc, quod est pars partis et pars totius. Cum igitur sit pars, componit ipsum. Sed hoc habemus ex regula quod cum neutro circum se positorum componit, quia nec cum maiore, nec iterum cum minore eo, nec iterum cum minore vel cum maiore eo. Igitur duplicatum componit, ergo est eius medietas. Et hoc est quod omnis numerus cum tercio a se componit parem numerum. Si enim medius illorum numeri, qui ex eis componitur, medietas est, quod iam probatum est, necesse est parem illum esse, cum medietatem habeat.

VII, 3. SOLA UNITAS. Omnis numerus hoc habet, quod est medietas circum se positorum. Unitas vero non habet hoc, sed ipsa componit numerum; ergo est principium numeri.

VII, 5. NATURALEM DISPOSITIONEM vocat I, II, III, et ita secundum naturalem calculationem ascendere.

VII, 1. SUPRADICTUS: quilibet. Ipse autem exemplificabit de quinario. SI COMPONANTUR. Et hoc ideo apponit, quia senarius et IIII nisi iungantur, non clare patet qualiter V sit medietas illius compositi ex ipsis iunctis. CUM IPSI QUOQUE. Hoc iterum ideo addit, quia constare debet eorum compositio. Illius enim compositi medietas est V, et non est medietas de III et VII, quia isti carent media. SUPER SECUNDO. Hoc dicit quantum ad dignitatem unitatis, quae, cum sit princeps omnium, sedet in arce et solio eius tamquam domina, et numeri aliqui quanto sibi sunt viciniores, tanto sunt altiores; et ideo de quaternario dicit quod sit superior V, et de VII quod sit inferior V.

VII, 6. QUAMVIS PROLIXAE. Quantumcumque enim ascendas calculando, omnes ab unitate generantur.

CAPUT VIII

VIII, 1. PARIS AUTEM NUMERI. Post diffinitionem paris numeri, subdit eius divisionem, quae triplex est. PARTICIPAT. Participo illa re et participo illius rei: bene utrumque dicimus.

CAPUT VIIII

VIIII, 1. PARITER PAR. Tractat de prima specie paris numeri, scilicet de pariter pari, sic: diffiniendo, causam vocabuli assignando, generationem ipsius ponendo, et quae sibi sint accidentia enumerando. Et sunt in hoc tractatu V theoremata: primum est quod omnis eius pars et nomine et quantitate pariter par invenitur; secundum, quod ex duplis solis nascuntur, et non ab aliis; tercium, quod omnis eius pars ab aliqua parte, quae intra ipsum continetur, denominatur; quartum, quod si minores summae sibi aggregentur, exequabuntur sequenti pariter pari minus uno; quintum, quod si una partium quae respondent sibi per aliam multiplicetur, efficitur totum.

HUIC NUMERO. Diffinitioni deberet statim causam vocabuli subponere; quod non facit, sed praemittit quoddam eius accidens, unde pariter par appellatur. Hoc autem est illud quod quaelibet pars eius denominatur

669 Thierry here returns to the beginning of chapter VII.

ab alio pari numero, et quota pars fuerit totius illa denominatio, tota erit multitudo sive quantitas denominatae partis, tot unitatum erit pars denominata. Verbi gratia: XVI est IIIIa pars de LXIIII. Unde ergo denominatur, id est nomen accipitur? A IIII, qui est par etiam numerus. Et haec denominatio, scilicet quaternarius, quota pars est totius, scilicet LXIIII? XVIa. XVI ergo unitates continebit pars denominata, quae est XVI; et ita par numerus est denominatio, et par numerus est quantitas denominati. CUM NOMINE: quo vocatur et denominatur; vel ideo haec duo dixit, quia numeri binarius, ternarius et caetera vocabuntur his nominibus secundum hoc quod sunt res incorporeae, id est universalia, et eisdem nominibus habent nominare subiecta. Et nota hic "nomen" quodammodo ad corporales, "vocabulum" ad incorporales res pertinere, atque ideo utrumque hic ponit, quia et numeri corporibus iunguntur, unde et nominantur, et incorporales sunt natura unde vocantur. SED IDEO. Aliis videtur quod ideo dicatur pariter par numerus, quia dividitur in duo paria, iterum et iterum, sicut praemissum est: SED MIHI IDEO VIDETUR. POST DICEMUS. Ostensa generatione quam praemittit "ideo," quia facilius patebit paritas nominis et quantitatis.

VIIII, 6. ILLUD AUTEM. Generatio videnda fuit, etiam istud videri DIGNUM EST QUOD OMNIS EIUS PARS, id est dispositionis, AB UNA PARTE QUAE INTRA IPSUM EST, scilicet numerum, denominatur. Verbi gratia: sit summa pariter paris numeri CXXVIII, media eius pars LXIIII, et haec pars

706 The scribe added *id est universalia* above the line. **715–716** ab una parte quae intra ipsum est] ab una parte quaque quae intra ipsum numerum est *Chartres 498*

716–733 Cf. Caiazzo, "Un commento altomedievale," 138, ll. 319–336: "Verbi gratia: sit summa pariter paris numeri CXXVIII, media eius pars LXIIII sunt; et haec pars denominata est 'ab una parte quae intra ipsum numerum est,' id est a binario. Bis enim LXIIII colligunt CXXVIII. Adverbia namque numeri denominationem ostendunt, idemque binarius pars est LXIIIIma, sic et de caeteris partibus. Quarta pars supradictae summae XXXII sunt, et haec quoque pars denominatur per adverbium ab una parte eiusdem numeri propositi, id est a quaternario, ut quater XXXII, CXXVIIII; et idem quaternarius XXXIIma pars est eiusdem numeri. TANTAMQUE SUMMAM, id est collectionem numeri, QUANTITATIS, id est multitudinis, INCLUDIT, QUOTA PARS EST ALITER NUMERUS PARITER PARIS ILLIUS, id est numeri, QUI eandem partem CONTINET QUANTITATIS. Verbi gratia: si quarta pars fuerit cuiuscumque pariter paris numeri, ipsa quarta pars per quaternarium, ex quo denominatur, totam summam includit, ut est hoc. Pono LXIIII, eius quarta pars XVI, quae quarta pars per quaternarium multiplicata, supradictam summam colligit, id est LXIIII. Similiter, caeterae partes per ipsas, ex quibus denominantur, multiplicatae supradictam summam concludunt. Et hoc omnino dicit, quotacumque fuerit pars totius multitudinis, id est sive secunda sive quarta sive alia quaelibet per ipsam, ex qua denominata est, multiplicata totam numeri summam comprehendit."

denominata est ab una parte, quae est intra ipsum numerum, id est a binario. Bis enim LXIIII colligunt CXXVIII; adverbia namque numeri denominationem ostendunt, isdemque binarius pars est LXIIII. Sic et de caeteris partibus. Quarta pars supradictae summae XXXII sunt. Et haec quoque pars denominatur per adverbium ab una parte eiusdem numeri proposita, id est a quaternario, ut quater XXXII, CXXVIII. Et idem quaternarius XXXaIIa pars et eiusdem numeri tantamque summam includit quantitatis, quota pars est alter numerus pariter paris illius, qui eum continet quantitatis. Verbi gratia: si quarta fuerit pars cuiuscumque pariter paris numeri, ipsa IIIIta pars per quaternarium, de quo denominatur, totam summam includit, ut est hoc. Pone LXIIII, eius quarta pars XVI, quae quarta pars, per quaternarium multiplicata, supradictam summam colligit, id est LXIIII. Similiter: caeterae partes, per ipsas ex quibus denominantur multiplicatae, supradictam summam concludunt. Et hoc omnino dicit: quota fuerit pars totius multitudinis, sive secunda sive quarta sive alia quaelibet, per ipsam ex qua denominata est multiplicata, totam numeri summam comprehendit.

Dispositiones: in ordine disposita membra sive summae sive termini, quod idem est.

VIIII, 7. Responsura: per multiplicationem, quia octies VIII, LXIIII. Donant: non quod uno nomine participent, sed in ipsa donatione permutant, quia XVI denominatur IIIIa pars a IIIIor, et IIII denominatur XVIa a sedecim; et eodem modo XXXII nominatur a secunda parte, et II denominatur a XXXII, quia est XXXaII<a> pars totius.

VIIII, 8. Hoc ergo: aliud accidens, et spectat ad generationem. Non solum generantur in duplici proportione, sed etiam sic quod una summa iuncta cum sibi proxima facit terciam, uno tamen minus. Verbi gratia: unum et duo faciunt III, cui deest unitas ad completionem IIII. Item eosdem III adiunge cum IIII, et erit octonarius pene uno minus.

VIIII, 11. Coacervata quantitas. Ius, II, III faciunt senarium perfectum. Unus, II, IIII non conficiunt VIII, et ideo est imminutus. Sex et IIII et III et II et I plus conficiunt quam XII, qui est superfluus. Et ubi doctrinam talium faciet, ibi valebit nobis scire de tali coacervatione. Proprio consentire principio: quia ille numerus, qui praecedit, semper est contractior unitate eo quem praecedit.

VIIII, 12. Ut respondentibus: utpote XVI et IIII, et invicem multiplicatis, componitur seu conficitur maior summa eius numeri, id est pariter paris, utpote ex partibus praedictis altrinsecus multiplicatis, conficitur LXIIII. In qua extremus terminus. Tricies et secundo, id est IIIIor vicibus triginta duo CXXVIII sunt.

723 CXXVIII] CXXVIIII ms **728** LXIIII] LXII ms **741** ergo] igitur *ed. Guillaumin*

CAPUT X

X, 1. PARITER IMPAR. Postquam tractavit de una specie paris numeri, scilicet de pariter pari, ingreditur modo tractare de altera specie paris numeri, scilicet de pariter impari, hoc modo: describit quis sit pariter impar, et subdit causam vocabuli, ubi etiam ostendit convenientiam eius cum pariter pari, necnon etiam differentiam; deinde accidentia enumerat, et qualiter procreentur.

X, 8. SI AB UNO. Differentia cuiuslibet imparis ad alium imparem sibi proximum est binarius; et hos impares si per binarium multiplicaveris, nascentur pariter impares.

X, 9. A SUPERIORE, scilicet ternario, quem vocat superiorem vel dignitate, quia vicinior est unitati, vel certe propter ordinem litterae et orationis huius, in qua superius est facta mentio ternarii.

X, 11. IGITUR IN NATURALIS. Quandoquidem impares primo loco pro fundamentis istorum positi binario sese vincebant, et postea per binarium multiplicabantur ad procreandos pariter impares; ideo isti, ex tali multiplicatione surgentes, quaternario se vincunt, et QUINTO LOCO DISTANT A SE invicem. Et facit epilogum eorum quae dixit: QUINTO LOCO DISTANT, ut denarius a senario – VI, VII, VIII, VIIII, X – IIIIOR SE PRAECEDUNT, id est vincunt, tres intersunt – VII, VIII, VIIII; ipsi dico procreati ex imparibus per multiplicationem binarii.

X, 14. IN NUMEROSIORIBUS. Quia videretur hoc non posse fieri nisi in tam parvis numeris et tam parva dispositione, ostendit quod adhuc in ampliori possit fieri, ita tamen ut primo videamus quid conficiant duae medietates, hoc idem conficiunt termini supra se constituti; si enim ab extremitatibus inciperemus, plus impediremur.

758–763 Cf. Euclid ("Adelard II" version), *Elementa* 9.35 (ed. Busard and Folkerts, p. 221): "Numerus cuius medietas impar, est pariter impar. Si enim esset pariter par sive impariter par, qui dicitur secundum eum pariter par impariter, eius medietas esset par. Omnis autem compositus par, id est par omnis preter binarium, aut pariter par est aut pariter impar aut pariter par impariter, qui dicitur secundum Boecium et Nichomacum impariter par." Euclid's definition is here completed with data drawn from Boethius's *De arithmetica*; see above, p. 31.

Caput XI

XI, 3. Sunt et<iam> alii. Non solum hi duo numeri sunt tales, sed etiam sunt alii qui dividuntur in plura membra, sed divisio non pervenit ad unitatem.

XI, 6. Respondent. Quia et par est quantitas et par est denominatio, sicut in numero pariter pari, qui est quasi genus istius, quia impariter par numerus nascitur ex utroque, scilicet ex pariter pari et pariter impari; et recipit accidentia utriusque. Quae denominationes. De solis semel XXIIII ostendit quod par est quantitas et impar denominatio, modo addit de omnibus quae contrarie denominantur.

XI, 7. Ut substantiam naturamque. Naturale esse ostenditur ex his ex quibus procreantur.

XI, 9. Vel si tercius. Istud non est diversum a priori, sed est expositio prioris. Quia obscure dixerat, secundus secundi et tercii, ostendit quia hoc sit tercius et quartus a primo. Superiores multiplice<n>t: sub se positos. In exemplis autem facit quod inferiores multiplicant supra se positos, quia ter IIIIor et quater III idem faciunt.

XI, 10. Duplices: duplicatae simul iunctae et coacervatae. Et hoc fiet sicut cito iam dicet, ad similitudinem pariter imparis, cuius hanc ostendit proprietatem ubi de eo tractavit.

XI, 12. Descriptio quae subposita est in figura illa quae dicitur admirabilis, ibi numerus qui ex duabus medietatibus surgit, tam in longitudine quam in latitudine bis scriptus est. Quod mihi videtur ad ornatum figurae factum esse.

Caput XII

XII, 1. In quo hanc proprietatem. Ut in impari dispositione quod faciunt duae extremitates iunctae, hoc idem facit una medietas per se multiplicata. In pari vero quod faciunt duae extremitates, hoc quoque duae medietates iunctae complent.

Caput XIIII

XIIII, 1. Consequens reperietur. Quia si constiterit eum esse primum et incompositum, consequenter etiam ut sola eum unitas metiatur, et non habeat alias partes se constituentes, neque terciam neque quartam vel aliquam caeterarum.

XIIII, 4. HI AUTEM. Ex his enim procreantur numeri tercii generis, quod in fine capituli de eorum generatione dicet. PRIMAM SUBSTANTIAM, scilicet unitatem quae dat subsistere omnibus numeris, et etiam rebus ad similitudinem numerorum procreatis. Hanc isti numeri solam et principalem sortiuntur, quia sola unitas eos metitur, et ita ipsi sunt fundamenta et initia aliorum de se prodeuntium. VELUT QUAEDAM ELEMENTA: ut ternarius qui est primus et incompositus, si per se multiplicetur facit novenarium, qui est primus in ordine eorum qui sunt secundi et compositi. Similiter V qui est primus et incompositus post ternarium, per ternarium multiplicatus, XV procreat, qui est secundus et compositus. Sic et de reliquis fit in ordine eorum qui sunt secundi et compositi.

CAPUT XV

XV, 1. SUBSTANTIAM PRINCIPALEM, id est non principaliter subsistit ex unitate sicut primus et incompositus, sed etiam ex aliis numeris. ET AB ALIENO VOCAT, id est unam vel duas vel amplius, secundum quod magni sunt ipsi componentes, quia si aliquis ex bis LX surgit, plures habet partes quam si surgeret ex bis X.

XV, 2. PRINCIPALIS INTELLIGENTIAE. Non intelliguntur principaliter constare ex unitate. SED OMNI. Ista est necessitas et natura rerum, ut quicquid sit compositum possit dissolvi, et econverso, nisi solus mundus, qui cum sit compositus et naturaliter possit dissolvi, gratia componentis conservatur in aeternitatem.

CAPUT XVII

XVII, 7. MODUM AUTEM. Modum intermissionis docui, et modo docebo. MODUM, id est quomodo qui numeri metiantur istos, qui sic

834–838 Plato, *Timaeus* 41A–B (ed. Waszink, p. 35): "... opera siquidem vos mea, dissolubilia natura, me tamen ita volente indissolubilia, omne siquidem quod iunctum est natura dissolubile ... quia voluntas mea maior est nexus et vegetatior ad aeternitatis custodiam quam illi nexus vitales ex quibus aeternitas vestra coagmentata atque composita est." See also William of Conches, *Glosae super Platonem* 114 (ed. Jeauneau, p. 205): "Deinde ostendit qualitatem eorum geminam, scilicet naturalem et datam a sua gratia, ut per eam captet illos: DISSOLUBILIA NATURA id est potentia dissolvi. Sed ne timerent ex natura, addit: sed indissolubilia gratia mea."

intermittuntur. Et istud idem est quod etiam superius dixit in clausula: SUSCIPIUNT ERGO. Sed quia difficile fuit, aliis verbis idem repetit. CRIBRUM ideo nominabat istum numerum, quia sicut in cribro multa sunt foramina, ita in dispositione imparium numerorum intermittuntur II vel IIII vel plures numeri, et quasi quaedam relinquunt foramina, ut tales numeri inveniantur.

XVII, 8. NEQUE UT OMNES. Quia ternarius sicut novenarium numerat, non ita numerat quinarius XXV, quia unus, id est novenarius, ter habet III, alter, id est XXV, quinquies quinque habet. VEL QUI METIUNTUR. Hoc deberet dicere, sed non repetit iterum "alios," quia non sunt diversi qui mensi sunt et qui metiuntur, et ideo addit: IPSI. INVENIES: quod nec ab aliquo uno omnes numerantur, nec aliquem unum omnes numerant. AB UNO TANTUM: ut VIIII a tribus.

XVII, 9. VEL ALIENIGENAE. Hoc idem est quod iam dixit, quia denominari ab aliquo et mensuram aliquam recipere idem est. QUI VERO: ut XV, excepta unitate a se denominata, id est quintadecima parte, aliam mensuram habet, id est quinque. Habet etiam vocabulum alienae partis, ut tres; ter enim V, XV. SECUNDOS ATQUE COMPOSITOS. "Secundos" ideo, quia non solum a prima unitate, sed etiam ab alio numero iam composito, componuntur.

CAPUT XVIII

XVIII, 1. QUA VERO. Ostensa generatione eorum qui sunt IIIa species imparis numeri, videretur quod omnes illi numeri, qui vel sola unitate vel altero aliquo impari numero commensurabiles sunt, essent de illa tercia specie imparis numeri. Sed hoc non est. Et ideo subdit quod non solum isti, sed multi alii, sive sint per se secundi et compositi ad alium relati primi et incompositi, sive non, vel unitate vel aliquo impari commensurabiles sint. TALES NUMEROS, id est primos et incompositos, secundos et compositos, et eos qui per se secundi sunt et compositi, ad alterutrum vero comparati, primi sunt et incompositi.

XVIII, 2. NECESSARIO: vel unitas vel aliquis impar numerus impediet. SI, id est si quidem, id est cum hac constantia, quia impares sunt qui proponuntur. Non dicit "si" dubitative sed constanter. SED EUM. Quia videretur

842 suscipiunt] suscipiant *ed. Guillaumin*

---

842 Cf. Papias Vocabulista, *Elementarium doctrinae erudimentum* (Milan, 1476): "Clausula est in oratione brevis sententia."

ille relictus adhuc de alio posse auferri, vel alius ab ipso, ideo dicit eum sibi aequalem remanere.

XVIII, 3. SI IN UNUM scilicet numerum, id est unitatem, illa subtractio incurrerit. CONTRA SE, id est oppositi, a quibus incepit subtractio. Si quaeratur quare XXIIII plus excrescat quam bis XII separatim, responde: quia plures habet partes, scilicet octavam III et vicesimam quartam I, quas non habet XII.

XVIII, 4. HOS ERGO. Qui ad se sunt secundi et compositi; CONTRA SE, id est altrinsecus comparati primi vocantur.

CAPUT XVIIII

XVIIII, 1. INTRODUCTIONIS, id est doctrinae, qua introducitur aliquis ad intima scientiae. SUMMA PARTIUM. “Summa” dicitur quotiens partes totius numeri in unum sunt redactae.

XVIIII, 8. CENTIMANUS GYGAS: a poetis sine nomine profertur, quia quinquaginta capita fertur habuisse. GERIO dicitur fuisse gygas tricorpor, et fuit rex Hyspaniae; qui ideo tricorpor dicitur fuisse, quia tribus insulis imperavit, ad quem Hercules cum aerea olla et bicipiti cane dicitur venisse: cum aerea olla, ob aeratas naves et proelium navale; cum bicipite cane, quia navali et terrestri proelio plurimum valuit. MONSTRUOSUM vel MONSTROSUM est, quod subito extra naturam apparet. CICLOPS gygas fuit monoptalmus, id est unioculus, quem habuit in fronte. Quod ideo fingitur, quia sapientissimus fuit et sedes sapientiae in cerebro est. Eundem autem oculum Ulixes dicitur terebrasse, quoniam ille prudentior fuit. CURTATUS dicitur truncatus. Unde et Iudei curti dicuntur, id est in parte pudenda truncati.

896 pudenda] pudenti MS

883–896 Cf. Caiazzo, “Un commento altomedievale,” 139, ll. 340–352: “INTRODUCTIONIS, id est doctrinae, quia introducitur aliquis ad intima scientiae. CENTIMANUS GIGAS: a poetis sine nomine profertur, qui quinquaginta capita fertur habuisse. GERIO dicitur fuisse gigas tricorpor. MONSTROSUM est quod subito extra naturam apparet; quidam autem dicunt monstruosum, sed potius dicendum est monstrosum. SUBRIPUIT, id est fefellit. GERIO rex fuit Hispaniae, qui ideo tricorpor dicitur fuisse, quoniam tribus insulis imperavit, ad quem Hercules cum aerea olla et bicipiti cane venisse dicitur: cum aerea olla ob aeratas naves et proelium navale; cum bicipiti cane, quia navali et terrestri proelio plurimum volvit. CICLOPS gigas fuit monotalmus, id est unioculus, quem habuit in fronte, quod ideo fingitur, quia prudentissimus fuit, et sedes sapientiae in cerebro est. Eundem autem oculum Ulixes dicitur tenebrasse, quoniam illo prudentior fuit. CURTATUS dicitur truncatus; unde Iudei curti dicuntur, id est in parte pudenda truncati.” Cf. Servius’s commentary at *Aeneid* 7.662.

Caput XX

XX, 1. Alternatim: quia si unus terminatur in VI, alter in VIII, tercius in VI, quartus in VIII. Qualiter et quot modis inordinati nascantur considera. Nota quod aggregare et coacervare diversa sunt hoc loco, quia ex aggregatione alicuius numeri fit coacervatio. Et alter est numerus aggregatus, alter coacervatus, quod in exemplis patet, quia unus est primus aggregatus, II posterior et ultimus aggregatus, unde surgit et coacervatur ternarius, qui est coacervatio prima; et illam coacervationem, scilicet ternarium, duc in secundum aggregatum, scilicet binarium, et inde surget perfectus numerus.

XX, 4. Bis tres faciunt VI, qui habet unam quidem a se denominatam partem. Hoc est: a tota sui quantitate denominat ipse sextam partem, id est unum.

XX, 5. Secundum collectionem. Quia collectis omnibus partibus quas habet aliquis numerus, ex nulla illarum partium tam cito perpenditur quantitas et summa totius numeri, sicut ex binario, scilicet ex secunda parte, quae est medietas. Citius enim possum duplicare numerum quam quadruplicare vel triplicare.

XX, 6. Dudum: quando coacervationem, id est septenarium, duxisti in ultimum aggregatum IIII, unde surrexit perfectus ille XXVIII.

XX, 7. Super illam, id est non solum metitur eum unitas, sed etiam ternarius, pro quo iudicatur esse alterius generis, quia est secundus et compositus.

XX, 8. Igitur prima unitas. Quia de perfectione istorum numerorum mentionem faciens, videbatur derogare perfectioni unitatis, quam in principio huius artis ostenderat esse perfectiorem et principaliorem omnibus, ideo subiungit eius commendationem, ostendens etiam eam simili de causa perfectam esse, sicut isti numeri sunt, sed diverso modo, quia potentia et non actu, quippe cum nullas habeat partes. Et ipsa perfecta est: quia neque superat partes suas, nec superatur ab ipsis.

Caput XXI

XXI, 1. Ad aliquid: quantitatis per se. Prima divisio duplex erat in parem et in imparem, et non solum illius. Vero: pro "sed," sed etiam quantitatis relatae duplex est divisio, quia quicquid alicui comparatur aut aequale aut inaequale est ipsi.

904 prima] illa ms **924** ipsa] ita ms

XXI, 3. Nullus potest dicere: illud est tale, quia tunc non esset ei aequale, sed sic dici potest de aequalibus: istud est tale quale illud, et non alterius modi.

XXI, 5. Inaequalis. Inaequalitatis alia est maior, alia est minor, et utrarumque istarum specierum in V dividitur.

Caput XXIII

XXIII, 1. Custodiunt. Quia omnes species multiplicitatis assignari possunt per eos qui sequentur. Ad primum. Sequens post unitatem aut est duplus unitatis aut triplus, aut decuplus, aut centuplus. Unus subduplus est binarii, subtriplus ternarii, subquadruplus quaternarii.

XXIII, 2. Plus quam semel. Non ideo dictum est quod nullus multiplex alium contineat nisi semel, quia quandoque continet sepius; sed dictum est propter initium, quia omnis multiplicitas incipit a binario. Discriminatus. Discrimen proprie vocatus est acus maior, quo capilli mulierum segregantur; unde et discriminari dicimus segregare. Plus quam semel: quia maior numerus habet minorem plus quam semel, ut binarius bis habet unitatem, ternarius ter, quaternarius quater, quinarius quinquies. Inchoans et determinans: quia ad alium comparatus nec infra relinquitur, nec supravincit eum, ut binarius qui bis continetur a quaternario. Unus subquincuplus est quinarii, subsescuplus senarii, subseptuplus septenarii, suboctuplus octonarii, subnonuplus novenarii, subdecuplus denarii.

XXIII, 4. Cum autem. Quia multiplicare possumus in infinitum vel duplicando vel triplicando vel decuplando, infinitae sunt species multiplicis et submultiplicis, et sunt diversae, ideoque generatio eorum non est una sed diversa. Alio enim modo nascuntur dupli, aliter tripli, sicut videbitur. Si enim positis. Hic ostendit generationem duplorum, et in ipsa generatione quandam eorum proprietatem assignat, quod cum ipsi sint tantum pares, multiplicant tamen sua duplicitate et pares et impares, sicut in ordine naturali se sequuntur.

XXIII, 5. Triplices intermittunt duos, quia post duos tercius qui occurrit multiplicat sua triplicitate tam pares quam impares alternatim, sicut se sequuntur in naturali ordine. Et hoc est quod dicit: ad naturalem numerum comparentur excepto ternario, qui primus triplus unum tantum, scilicet binarium intermittit. Eodem modo posset excipere

944–945 Cf. Caiazzo, "Un commento altomedievale," 139, ll. 359–360: "Discrimen proprie vocatur acus maior, quo capilli mulierum segregantur, unde discriminari dicimus segregare."

in qualibet specie multiplicis primum numerum. Verbi gratia: omnes quadrupli intermittunt III, nisi primus scilicet quaternarius. Omnes quincupli intermittunt IIII, nisi primus scilicet V.

XXIII, 6. SI CRESCAT ADIECTIO. Si crescant species multiplices per adiectionem non quantamcumque, sed per talem quae vincat intermissionem in uno. Verbi gratia: primo duplo adice unum, et excrescit primus triplus, et cum sit proprium triplorum duos intermittere, quanto maior est illa adiectio quam ista intermissio? Uno maior est.

XXIII, 8. ET OMNES QUIDEM. Dupli tantum sunt pares, sed multiplicant et pares et impares. TRIPLI ita fiunt: uno impari termino constituto, altero pari, ut ternarius impar, triplus est unitatis. Senarius par, triplus est binarii. Novenarius impar, triplus est ternarii. XII par, triplus est quaternarii. XV impar, triplus est quinarii. XVIII par, triplus est senarii. Quadrupli natura equipollet duplis, quatuor par, quadruplus est unitatis. VIII par, quadruplus est binarii. XII par, quadruplus ternarii, XVI quaternarii, XX quinarii. CONSTITUUNTUR, id est generantur et incipiunt a quaternario, et semper intermittunt unum parem. Verbi gratia: quaternarius primus quadruplus intermittit binarium, VIIIrius senarium. QUINCUPLI PROPOSITIO, id est constitutio et proprietas habet se ad similitudinem tripli, quia quandoque est par, quandoque impar, et multiplicat tam pares quam impares alternatim. Quinarius enim impar quincuplus est unitati, Xius par quincuplus binarii.

CAPUT XXIIII

XXIIII, 1. SUPERPARTICULARIS. Postquam tractavit de prima specie inaequalitatis, scilicet de multiplici et eius speciebus, modo vult tractare de secunda specie, scilicet de superparticulari, ostendens eius et specierum ipsius generationes.

XXIIII, 3. SECUNDUM MAIORUM. Quia quot fuerint maioritatis species, tot etiam erunt minoritatis.

XXIIII, 5. NAMQUE SESQUALTER. Vere interminabili, quia de sesqualtero progrediemur ad sesquitercium, inde ad sesquiquartum, et caetera. Et primum ostendit generationem sesqualteri. NATURALITER: propter ordinem, quia idem possunt esse et dupli alio respectu.

XXIIII, 6. HEMIOLIA: habens dimidium et totum; "hemis" enim "semis" dicimus, "olon" "totum."

XXIIII, 8. ET IN CUNCTIS: ut si maiores vocentur sesquiquarti, minores vocentur subsesquiquarti.

997–998 Cf. Caiazzo, "Un commento altomedievale," 139, ll. 361–362: "HEMIOLIA habens dimidium et totum; 'hemi' enim 'semis' dicimus, 'holon' 'totum.'"

Caput XXV

XXV, 1. Nam primi. Uno semper minus dimittunt quam ipsi termini sunt. Primi termini nullum dimittunt, secundi unum, tercii duos, quarti III, quinti IIII, sic reliqui.

Caput XXVI

XXVI, 1. PERSTRINGENTES: parum tangentes. Perstringere est parum tangere.

Caput XXVII

XXVII, 1. QUAE FACIUNT ANGULUM. Notandum in figura quod ab uno usque ad X prima unitas vocatur, a X usque <ad> C secunda unitas vocatur, quoniam peracto denario numero rursus quasi ab unitate repetitur, dicendo XI, XII et reliqua. Tercia vero unitas a C usque ad mille dicitur. His tribus unitatibus omnis numerus consistit, ergo ulterius si velit quis numerare, necesse est ut hae unitates semper repetantur, ut II, III, IIII, et caetera.

XXVII, 2. Quod dicit: ITA UT ISTA COMPARATIO PER X LITTERAM FIAT, ideo "per X litteram" dicit, id est propter figuram quae est in modum crucis, id est propter denarium, qui in huius figurae angulis est constitutus, vel etiam propter tetragonos, qui<a> ab angulo ad angulum figurae in modum X litterae sunt digestae, a quibus omnes species numerorum multipliciter digestae ad numerandum inchoantur.

1009 in figura] infra MS

1006–1007 Cf. Caiazzo, "Un commento altomedievale," 139, ll. 363–364: "PERSTRINGENTES, id est parum tangentes. Perstringere est parum tangere."

1009–1014 Cf. Caiazzo, "Un commento altomedievale," 139–140, ll. 364–369: "Notandum in figura quod ab uno usque ad X prima unitas vocatur, a X usque ad C secunda unitas vocatur, quoniam peracto denario numero, rursus quasi ab unitate repetitur, dicendo undecim, duodecim et caetera. Tertia vero unitas a C usque ad mille dicitur; his tribus unitatibus omnis numerus consistit. Ergo ulterius si velit quis numerare, necesse est, ut hae unitates semper repetantur, ut duo milia, tria milia et caetera."

1015–1020 Cf. Caiazzo, "Un commento altomedievale," 140, ll. 369–374: "Nam quod dicit ita ista comparari per x litteram, ideo per decem litteram dicit; vel propter figuram quae est in modum crucis; vel propter denarium qui in binis figurae angulis est constitu[i]tus; vel etiam propter tetragonos quod ab angulo ad angulum figurae in modum x litterae sunt digesti, a quibus omnes species multiplicium per syncopam numerum digestae ad numerandum inchoantur."

XXVII, 8. CUNCTIQUE ALII EIUSDEM MODI SUNT, UT PRIMO ET SECUNDO ORDINE AD ALTERUTRUM MULTIPLICATIS TERMINIS PROCREENTUR, scilicet antelongiores, ad figuram nos dirigit, quia si primum ordinem naturalis numeri per secundum multiplices duplicem, omnes sunt antelongiores.

XXVII, 9. DUO ANGULARIUM TETRAGONORUM ANGULI AEQUUM FACIUNT, QUOD SUB IPSIS CONTINETUR, ILLI QUOD FIT AB UNO ILLORUM QUI EST ALTRINSECUS ANGULORUM, scilicet hoc in figura considerandum est, ubi sunt I et C e regione positi, qui concludunt angulos tetragonorum in se. Hi, si per invicem multiplicentur, tantam summam facient, quanta est si multiplicetur secum unusquisque denarius, qui in reliquis angulis binis est constitutus, agendo decies decem. In utraque enim multiplicatione centenarius est.

XXVII, 1. PER EANDEM SEQUENTIAM, id est in eodem ordine latitudinis sive longitudinis praetereunt se, quilibet sequens praecedentem in latitudine et inferior superiorem in longitudine.

XXVII, 2. PER X LITTERAM. In modum crucis X posita in angulis tetragonorum, uno suo brachio in longitudine innuit quorum illi de IIIo ordine

1037 litteram] litterae MS

1021–1023 This lemma is taken from *De arithmetica* I, XXVII, 8.

1021–1033 Cf. Caiazzo, "Un commento altomedievale," 140, ll. 376–389: "CUNCTIQUE ALII EIUSDEM MODI SUNT UT PRIMO ET SECUNDO ORDINE AD ALTERUTRUM MULTIPLICATIS TERMINIS PROCREENTUR, scilicet antelongiores, ad figuram nos dirigit, quia si primum ordinem naturalis numeri, per secundum duplicem multiplices, omnes sunt antelongiores. PRIMA UNITAS est ab I usque ad X. Secunda autem dicitur unitas denarius numerus, quia sicut per unitates pervenitur ad X, ita rursum quasi per unitates denarii numeri pervenitur ad C. Tertia vero unitas est C, quia quasi per tertiam unitatem centenarii pervenitur ad mille, ac per hoc vicem obtinent unitatis. Notandum est: II ANGULARIUM TETRAGONORUM ANGULI EQUUM FACIANT, QUOD SUB IPSIS CONTINETUR ILLI QUOD FIT AB UNO ILLORUM QUOD EST ALTRINSECUS ANGULORUM, scilicet hoc in figura est considerandum, ubi sunt I et C, e regione positi, qui concludunt angulos tetragonorum in se. Hi, si per invicem multiplicentur, tantam summam faciunt, quanta est si multiplicetur secum unusquisque denarius, qui in reliquis angulis binis est constitutus, augendo decies X; in utraque enim multiplicatione centenarius est."

1026–1028 This lemma is taken from *De arithmetica* I, XXVII, 9.

1037–1043 The following figure, in the lower margin of fol. 15v, explains how one should read the table of *De arithmetica* I, XXVI, 2.

Pi S

I X

X

X C

IIIIa Ter

sint tripli, et alio brachio innuit quorum sint tripli in latitudine dispositionis figurae. Quod dico de prima linea longitudinis, idem serva in prima latitudinis, et de secunda similiter et tercia, quia eadem est natura multiplicitatis et in lato et in longo. SECUNDUM PARITATIS, id est per pares numeros, qui naturaliter dispositi sunt et conexi in secunda linea, a binario incipientes.

XXVII, 3. QUAM REM. Ita integre et plene possumus invenire eorum habitudines, sicut naturaliter se ad invicem habent. EXTRA NATURAM: quia non egemus extrinsecus et aliunde vocatis probationibus, sed consideratio naturae ostendet et inveniet istos numeros se sic habere et sic vincere.

XXVII, 4. PER X LITTERAM. Quia brachia illius litterae innuunt lineas, in quarum summitatibus sunt numeri, quorum isti de quarta linea sint quadrupli, et versus latitudinem et versus longitudinem figurae. PRIMUS. Non in angulari tetragono incipias, sed in principio lineae in qua ipse est tetragonus. PROGRESSIO. Ostendit in quot vincant isti illos de primo ordine. TRIUM. Non similium neque eorumdem, quia quartae lineae primus transgreditur se superiorem in prima linea tribus unitatibus, secundus suum superiorem tribus binariis, tercius suum tribus ternariis, quartus suum tribus quaternariis, et sic deinceps. ADIECTA. Quod potes videre in tercia linea, qualium trium quantitate quilibet quarti ordinis vincat se superiorem.

XXVII, 5. SI QUIS VERO. Non solum in hac figura consideratur species multiplicium, sed etiam superparticularium. EADEM TAMEN. Licet isti de tercio ordine non sint sesqualteri eorum quorum illi de secundo ordine sunt dupli, tamen in hoc conveniunt, id est quota summa supergreditur duplex simplices, ut binarius unitatem uno, eadem sesqualter superat duplices, ut ternarius binarium uno, et ut quaternarius binarium duobus, sic senarius quaternarium duobus superat.

XXVII, 6. ALTERUTRIS COMPARAVERIS: id est ut si quintus ordo comparetur quarto, sesquiquarta comparatio colligitur, ut V ad IIII, vel X ad VIII.

XXVII, 7. PRIMI ORDINIS, quia primae lineae numeri, per se ipsos multiplicati, faciunt omnes tetragonos. Et ideo dicit: AEQUALES, quia unitas unitatem, binarius binarium, et caetera. TETRAGONUS est numerus ille, secundum quem disponitur figura IIIIor angulos rectos et aequales habens, et ipsa figura est tetragona.

XXVII, 8. DUO NASCUNTUR EX UNO ET DUOBUS, propter multiplicationem quae sic fit: semel duo vel bis unus. UNO SUPERANT: quo se vincunt, sicut dictum est, in multiplicatione, cum dicitur bis tres et ter quatuor, et qui inde surrexerint longilateri sunt. AD ALTERUTRUM. Idem tamen sequeretur si unus ordo non multiplicaret alium, sed in uno ordine quilibet

1045 extra naturam] extra machinantibus *ed. Guillaumin* **1048** litteram] litterae *ed. Guillaumin*

praecedens multiplicaret sequentem. Melius tamen est ut primus ordo multiplicet secundum vel econverso, et tunc duc ad proselidem, et videbis longilaterum qui ex illorum multiplicatione processit. UT QUOD NASCITUR. Quandam proprietatem assignat ex longilaterorum generatione surgentem, scilicet quod etiam ipsi de se procreant tetragonos. Verbi gratia: altrinsecus circa secundum tetragonum, scilicet quaternarium, infra sunt VI et supra sunt duo positi, qui, si per aggregationem iunguntur, VIII facient. Cui VIIIrio si tetragonus, qui medius est, scilicet IIIIrius, duplicatus advenerit, facient XVI, qui est tetragonus. Et ad hunc modum si longilateri aggregantur cum tetragono duplicato, surget tetragonus, sive ille contineatur in cribro, sive non. ET RURSUS. Aliam proprietatem insinuat, et hic accipitur "altrinsecus," id est prope se positi, scilicet quorum anguli coniunguntur, hoc modo: iunge unum et IIII, qui sunt primus et secundus tetragoni, et fient V; adde etiam longilaterum qui circa eos est, id est binarium, qui simul iuncti faciunt VIIII, qui tetragonus postmodum invenitur. Sic etiam per caeteros tetragonos faciundum est: iunge per aggregationem IIII et VIIII, et facies XIII; duplica etiam longilaterum, scilicet senarium, et erunt XII; sed XII et XIII simul iuncti faciunt XXV, qui est tetragonus.

XXVII, 9. ET UT ANGULORUM. Non solum in hoc cribro praedicta notari possunt, sed etiam sequentia.

CAPUT XXVIII

XXVIII, 1. IGITUR. Ipse continuat, ostendens quid dixerit et quid dicturus sit.

XXVIII, 2. ET EIUS ALIQUAS. Hoc est: quando numerus continet in se minorem semel et eius aliquas partes, ut quinarius continet in se ternarium et terciam eius partem, id est unitatem, bis. Et octonarius continet in

1078 The term *proselidem* is found in the *De utilitatibus astrolabii* (**J**), published in Gerbert of Aurillac, *Opera mathematica* (ed. Bubnov, p. 126): "Et dum vis per hanc scire signum et gradum solis in quolibet die, considera in utroque latere praesentem mensem et eiusdem mensis praesentem diem et ita binis altrinsecus lineis ingredere usque ad proselidem angularem, et quotum in ibi reperis numerum desuper in eadem linea considerato signo, ipse est gradus solis, quem requiris." It refers to the lines of a table in *De utilitatibus astrolabii* (ed. Bubnov, p. 125). The term *proselidem* is repeated in Hermann of Reichenau's reworking of *De utilitatibus astrolabii*. In the Stuttgart manuscript, Württembergische Landesbibliothek, Cod. math. 4° 33, fol. 81r, a marginal gloss to *De utilitatibus astrolabii* explains the term: "ad proselidem angularem: occurrentem et respondentem illi."

se quinarium et eius quintam partem, id est unitatem, ter. VEL QUOT IPSA TULERIT: secundum quod magni fuerint numeri, qui ad se comparantur.

XXVIII, 3. INCIPIT A DUABUS <PARTIBUS TERTIIS>. Et quare non a duabus medietatibus, cum prior sit in duo dimidia quam in tres partes divisio? Hoc ideo fit, quia si habuerit eum totum et insuper eius duas medietates, non erit ad eum superpartiens sed multiplex, quia duplus; duae enim medietates pares sunt toti. NUMERIS: illis in quibus maior vincit minorem, existentibus IMPARIBUS, id est ab impari denominatis, quia aut sunt duae terciae, aut duae quintae, aut duae nonae, et caetera. Numquam <habebit> duas medietates, quod paulo ante dixit, nec duas quartas, nec sextas, quia tunc non esset superpartiens sed multiplex, quia duplus, aut superparticularis. Si duas medietates habuerit, duplus est; si duas quartas, sesqualter, quia duae quartae medietas una est; si duas sextas, sesquitercius, quia duae sextae IIIa pars est. Senarius habet quaternarium et eius duas quartas, id est medietatem quaternarii, et fit in illis sesqualtera proportio. SI ERGO NUMERUS. Ostendit a generalitate, subdit eius species et earum generationes. Bis VI: XII, bis X: XX; viginti ad XII superbipartiens est, quia habet eum totum et eius duas tercias, id est duos quaternarios, supra; ter enim IIII: XII. SIN VERO TRES, SUPERTRIPARTIENS. Hoc modo: ter XXI: LXIII, ter XII: XXXVI. Horum alter alteri comparatus continet eum in se totum et eius tres partes. Novies IIII XXXVI faciunt, novies VII: LXIII. Continent ergo in se LXIII XXXVI, et insuper eius tercias partes, id est ter novenos, ter VII fiunt XXI, ter IIII: XII; qui simul comparati supertripartientem faciunt. Quater VIIII: XXXVI, quater V: XX; qui comparati superquadripartiens est.

XXVIII, 5. QUI VOCATUR SUPERBIPARTIENS: sive vocetur superbipartiens sive tripartiens, semper superpartiens erit, quia hoc est generale nomen eius.

XXVIII, 6. AT VERO. In superiori figura apparet quis sit primus superbipartiens et primus supertripartiens, et sic de caeteris; sed quomodo quilibet istarum specierum nascatur ab istis, et ita usque in infinitum, HIC MODUS EST.

XXVIII, 8. ET CAETERAE SPECIES: quia debemus superbipartientem duplicare, sequentes triplicare, tercios quadruplicare, quartos quinplicare, et sic de caeteris.

1103 The Guillaumin edition, following Friedlein and two manuscripts from Bamberg, contains *quotquot* instead of *quot*.

Caput XXVIIII

XXVIIII, 1. Igitur relatae: ad aliquid quantitas dicitur cum quantitas maioris numeri comprehenditur ex collatione minoris, sicut in cribro vidimus: prima species inaequalitatis est multiplex, secunda superparticularis, tercia superpartiens. Igitur hae tres sunt primae, respectu duarum sequentium, et sunt simplices respectu earumdem, quia duae sequentes, quaelibet earum videlicet nascitur ex duabus superioribus, sed primarum quaelibet sequitur et descendit ab illa simplici simpliciter.

XXVIIII, 3. Quorum diffinitio. Et primo ostendit de uno, scilicet de multiplici superparticulari.

XXVIIII, 5. Fictum vocabulum ex utriusque nomine compositum, quia dicitur multiplex superparticularis.

XXVIIII, 7. Et hi semper. In hac prima figura disponuntur omnes in ordine, sed postea cuiuslibet istarum generatio ostenditur in singulis subiectis figuris.

Caput XXX

XXX, 1. In priorem descriptionem, scilicet cribrum. Hanc ipse non nominat, sed circumscribit quomodo disposita sit.

Caput XXXI

XXXI, 5. Nec erit alias eius partes, id est triplum superbipartientem, et caetera. Nascitur autem triplus superbipartiens sic: incipiamus a tribus, et omnes illos faciamus sequi qui sese ternario vincunt, et istis comparemus omnes illos qui ab XI incipientes, XI sese praetereunt.

Caput XXXII

XXXII, 1. Restat autem. Postquam relatum numerum divisit in aequalem et in inaequalem, postquam etiam inaequalitatem in suas species

1147 Fictum vocabulum] ficto vocabulo *ed. Guillaumin, and in the apparatus of variants we have* fictum vocabulum ms. *V2 (hand 2 of* ms. *V = Pal. Lat. 1341);* fictum vocabulum *ed. Oosthout-Schilling, after correction of the scribe (?) in* ms. *N = Cologne, DB, 185*

1161–1164 The subjects treated in this chapter are summarized by Thierry of Chartres in the *Tractatus de sex dierum operibus* 43–44 (ed. Häring, pp. 573–574): "Et sicut

divisit, intendit ostendere quomodo inaequalitas omnis procedat ab aequalitate. Sed prius scientiam istam commendat. Dicit enim quiddam difficillimum esse, quo cognito theologiae quidem, mathematicae quoque et physicae perfecta facile possit haberi cognitio. Hoc autem est quod malum omne descendit a bono. Quod quidem cum sit difficillimum et utile, facile cognosci poterit, si diligenter attendatur quomodo inaequalitas procedat ab aequalitate. Commendat itaque scientiam istam, in qua et benivolos nos reddit et dociles et attentos. Malum nihil aliud est quam boni transgressio, quare necesse est prius praecedere bonum, cuius est transgressio; sed substantia boni ex aequalitate consistit, mali vero ex inaequalitate. Cognito igitur quomodo inaequalitas ab aequalitate procedat, cognoscetur nimirum quomodo bonum sit origo et principium mali. Modus autem citra vel ultra quem non est bonum, a quibusdam lex divina vocatur, ab aliis praefinitio aeterna, a nonnullis vero providentia sive mens divina.

RESTAT NOBIS. Docentibus de speciebus inaequalitatis RESTAT – hoc est superest – TRADERE, id est ut tradamus, PROFUNDISSIMAM, id est subtilissimam et valde utilem, QUAE PERTINEAT, cum cognita fuerit, AD OMNEM

ipsa unitas omnes numeros ex se procreat ita ipsa unitatis equalitas omnes proportiones et inequalitates omnium rerum ex se producit. Et in ipsam eadem omnia resolvuntur. Inde etiam rerum mensure et pondera et quicquid omnino est habet existere. Mensura enim omnis vel ab equalitate vel ab inequalitate habet existere. Utrum autem horum fuerit ab equalitate esse necesse est quoniam ipsa inequalitas ab equalitate existit."

1174–1176 Cf. Thierry of Chartres, *Tractatus de sex dierum operibus* 42 (ed. Häring, p. 572): "Istum autem modum sive unitatis equalitatem antiqui philosophi tum mentem divinitatis tum providentiam tum creatoris sapientiam appellaverunt"; and 46 (ed. Häring, pp. 574–575): "Nichil enim aliud est esse Verbum deitatis quam eterna creatoris de omnibus rebus prefinitio: quid quale quantum sit unaqueque earum vel quomodo se habeat in sua dignitate vel tempore vel loco. At huiusmodi prefinitio equalitas existentie rerum est infra (intra *ed. Häring*) quam vel ultra quam nequit aliquid consistere."

1177–1188 Cf. Caiazzo, "Un commento altomedievale," 140–141, ll. 391–402: "RESTAT NOBIS: docentibus de speciebus inaequalitatis. RESTAT: hoc est superest. TRADERE, id est superest, ut tradamus. PROFUNDISSIMAM: obscurissimam et subtilissimam DISCIPLINAM, QUAE PERTINEAT, cum cognita fuerit, AD OMNEM NATURAE VIM, id est ad penetrationem et intellectum phisicae, id est naturalis philosophiae. Et QUAE PERTINEAT AD OMNEM INTEGRITATEM RERUM: quae vestigatur per eam philosophiam quae appellatur phisica, quia nihil in rebus creatis sine numero consistit. Et hac disciplina, quam traditurus est, deprehenditur quomodo manaverint omnes inaequalitates ab aequalitate, quae semper eadem est, et in nullo sibi diversa est, sicut omnis creatura processit ab ea natura, quae est creatrix. Omnis enim natura aut creatrix est aut creata. Creatrix: deus; creata: quaecumque ab eo facta est. In eo similitudo est, quia exordium habent omnes creaturae a deo; non tamen ab eius natura, sed ab eius opificio, sicut istae omnes inaequalitates manant ex aequalitate, 'quae ad omnem vim naturae rerumque integritatem maxima ratione pertineat.'"

NATURAE VIM, id est ad penetrationem et intellectum physicae, id est naturalis phylosophiae. Et QUAE PERTINEAT AD OMNEM INTEGRITATEM RERUM: quia investigatur per eam phylosophia quae appellatur physica, quia nihil rebus creatis sine numero consistit. Omnis autem natura aut creatrix est, aut creatura; creatrix: deus; creatura: quaecumque ab eo facta sunt. In eo est similitudo, quia exordium habent omnes a deo, non tamen ab eius natura, sed ab eius opificio, sicut istae omnes inaequalitates manant ex aequalitate. QUAE AD OMNEM VIM NATURAE RERUMQUE INTEGRITATEM MAXIMA RATIONE PERTINEAT. Omnis vis naturae in monade subsistit, et integritas rerum in monade custoditur, scilicet ut unitatis principium numquam deserat, neque naturae vis, neque rerum integritas. RERUMQUE INTEGRITATEM: id est ad mathematicam et ad theologiam. In mathematica namque docetur veritas circuli, trianguli et caeterarum figurarum, quarum vera quidem confusa est in modo. Docetur itaque in mathematica rerum integritas, id est rerum ipsarum veritas.

Omnis natura quamdiu se in suo statu conservat, integra est et incorrupta, sicut omnis aequalitas. At quae exorbitaverit et declinaverit a suo principio ad vitia deflectens, quasi ab aequalitate ad inaequalitatem labitur, quae fine non clauditur. Natura rerum dominatur rebus, et quasi quadam violentia conservat res in suo statu et in sua proprietate et integritate. Aliter enim excederent naturalem statum et effluerent ipsae res ab ordine suo. Haec autem conservatio rerum habetur a numeris, ad quorum similitudinem factae sunt res. Sunt enim numeri quidam inaequales, sunt quidam aequales. Inaequales numeri oriuntur ab aequalibus, et ipsa aequalitas constringit inaequalitatem, et inde ipsa inaequalitas redit ad aequalitatem. Quae

1196 At quae] atque MS

1183–1184 Augustine, *De Trinitate* 15.1: "Deus est, natura scilicet non creata, sed creatrix." The citation of Augustine is more faithful in the old commentary; see the text cited in the previous note.

1188–1190 Cf. Thierry of Chartres, *Glosa super Boethii librum De Trinitate* 5.18 (ed. Häring, pp. 297–298): "Sed huius rei rationem reddunt sancti huiusmodi secundum mathematicam quod *monas* numerus immutabilitati quia certe coaptatur merito deo applicatur cum sit immutabilitas."

1191–1194 Cf. Thierry of Chartres, *Lectiones in Boethii librum De Trinitate* 2.23 (ed. Häring, p. 162): "Considerat [mathematica *scilicet*] enim formas extra materia in veritate sua sicut verum circulum verum triangulum et cetera in hunc modum."

1195–1198 Cf. Caiazzo, "Un commento altomedievale," 141, ll. 404–406: "Omnis natura, quamdiu se in suo statu conservat, integra est et incorrupta, sicut omnis aequalitas. At postquam exorbitaverit et deviaverit a suo principio ad vitia deflectens, quasi ab aequalitate ad inaequalitatem labitur, quae fine non clauditur."

omnia divina sunt, et ad similitudinem dei fiunt, quia deus, qui vera bonitas est, cum omnia creaverit, malum non creavit, et tamen maliciam nostram sua bonitate coercet et continet, et non patitur eam effluere. Nisi enim contulisset nobis animum et animi rationem, non haberemus, unde stringeremus iram et concupiscentiam; et ipsae in infinitum et in illicitum procederent, nisi a deo temperarentur, et nisi eas stare faceret et finem eis daret. Sicut igitur omnia a deo temperantur, et ad ipsum revertuntur, et finem in eo ponunt, eadem similitudine inaequalitas procedens ab aequalitate stringitur ab ipsa et in ipsam redigitur. Quomodo autem inaequalitas procedat ab aequalitate, hic docet in fine huius libri; quomodo vero eadem resolvatur et redeat in eandem, dicet in principio sequentis libri.

INTEGRITATEM SCIENTIAE. Ignoto enim principio rerum quod est bonum, non integre res poterunt comprehendi. MAXIMA RATIONE, id est propria vi determinata et numquam excedens propriae naturae modum, quia maxima ratio est cum haec aequalitatis et inaequalitatis scientia ad omnia praedicta pertineat. QUOD BONITAS. Saepe dictum est quod unitas gignat aequalitatem, semel enim unum est unum, et sic ab unitate gignitur aequalitas, quae nimirum ipsa est unitas. Quod enim per solum unum gignitur, aequale quidem est unitati, nec aliud inde nisi unitas gignitur. Est autem inter unitatem et unitatis aequalitatem concordia et conexio quaedam, quoniam et ipsam esse dicimus unitatem, videlicet ut prima sit unitas, deinde unitatis aequalitas, quod est conexio. Dicitur autem unitas quasi entitas, quod nihil aliud est quam ipsam esse rerum intelligi. Deus autem rerum est entitas, quare et ipse vere est unitas. Eadem vero unitas entitatis est aequalitas. Haec enim est essendi forma, perfectio et integritas, ultra quam et infra nihil est quod ad esse rei pertineat, quo scilicet nihil plus de rei esse est, vel minus. Quae videlicet aequalitas rerumque perfectio omnem excludit superfluitatem, omnemque perficit et implet indigentiam. Materia namque rei tantum est inchoatio, forma vero ipsius perfectio; quippe quae et aequalitas essendi est. Quamdiu enim demendum aliquid est vel addendum rei, nec est adhuc res ipsa perfecta. Postquam vero nihil demendum est vel addendum, res quoque integre perfecta est, id est formata. Haec est ergo illa essendi forma et entitatis aequalitas, a qua omnes rerum formae

1209 ipsae] ipse MS

1220–1226 Cf. Thierry of Chartres, *Tractatus de sex dierum operibus* 41 (ed. Häring, p. 572); Thierry of Chartres, *Commentum super Boethii librum De Trinitate* 2.30–38 (ed. Häring, pp. 77–80). On *unitas* and *aequalitas*, see above, pp. 50–57.

1236–1240 Thierry of Chartres, *Commentum super Boethii librum De Trinitate* 2.40–41 (ed. Häring, p. 81): "Forma namque rei integritas et perfectio eius est: velut

descendunt, ut humanitas, quae quidem est aequalitas essendi hominem, sicut lapiditas quoque aequalitas est essendi lapidem, ultra quam videlicet nihil est vel infra, quod ad lapidis esse pertineat. Est ergo haec illa entitatis aequalitas quae vere et proprie dicitur unitatis ipsius figura. Quod Apostolus, subtilissime inspiciens, filium dei figuram substantiae patris appellat. Haec est iterum essendi forma, quae nihil aliud putanda est esse quam bonitas. Illam enim bonum dicimus, in quo nihil minus vel plus est quam oporteat esse. Sic enim dicimus istud quidem bonum, id est nihil ultra vel infra in eo est quam esse debeat. Quicquid ergo hanc aequalitatem transgreditur malum est, quoniam quicquid vel ultra vel infra est, vitium est. Iam ergo patet ex his quoniam malum a bono descendit. Malum enim ad bonum dicitur, neque aliquid malum nisi respectu bonitatis vel est vel vocatur. Ponatur enim bonum non esse; nihil erit, ut opinor, ad quod plus minusve dicatur. Sola enim aequalitas inter plus et minus est. Quae quidem aequalitas nihil aliud putanda est esse quam bonitas. Malum vero nihil est, nisi quod plus vel minus bono est, id est quod bonum sive in maiori sive in minori transgreditur, scilicet quod ultra vel infra bonum est; malum quippe nihil, nisi boni transgressio, et ab aequalitate discessio. Malum itaque ad bonum dicitur, ita videlicet quod, bono non existente, nihil est malum;

1238 aequalitas] est *add.* MS   **1240** entitatis] entitas MS

---

humanitas quidem, ut more inusitato loquar, equalitas est essendi hominem. Ut enim aliquid homo sit, nichil in eo ultra est, nichil infra nichil plus nichil minus. Merito ergo ab illa simplici forma divina rerum omnium forme emanare dicuntur quia iuxta formam illam divinam unaqueque res suam habet essendi equalitatem. Sicut enim forma divina equalitas essendi est sicut etiam integritas et perfectio omnium rerum ita quoque, ut ad exemplum descensus fiat, humanitas equalitas quedam et integritas essendi hominem est. Quare humanitas imago quedam forme est. Illa vero divina vere forma est."

1241–1242 Cf. Heb. 1:3: "Qui cum sit splendor gloriae et figura substantiae eius … ." Cf. Thierry of Chartres, *Tractatus de sex dierum operibus* 41 (ed. Häring, p. 572); *Commentum super Boethii librum De Trinitate* 2.32 (ed. Häring, p. 78): "Unde Apostolus filium splendorem et figuram substantie patris appellat. … Bene ergo Apostolus figuram substantie patris filium nominat eo scilicet quod filius essendi, ut diximus, est equalitas." See further above, pp. 58–60.

1246–1247 Cf. Thierry of Chartres, *Commentarius super Rhetoricam ad Herennium* 3.3.5 (ed. Fredborg, pp. 280–281): "SI UNAMQUAMQUE, etc. id est si definite ostenderimus unicuique rei naturalem terminum esse constitutum, iuxta quod dictum est ab Horatio: *est modus in rebus, sunt certi denique fines / quos ultra citraque nequit consistere rectum*. Verbi gratia: si ultra modum comederis vel infra, nocivum est, et sic per cetera." Cf. *Commentarius super libros De inventione* 2.53.159 (ed. Fredborg, p. 209): "Nam modus naturae per vitium exceditur, ad quem modum per virtutem sequendo rationem fit reversio."

quare a bono malum descendat necesse est. Illa ergo quam praediximus bonitas seu forma essendi, seu unitatis figura, vel entitatis aequalitas, regula quaedam est, quam quicquid transgreditur malum est, quoniam quicquid plus vel minus, ultra vel infra est, vitium est. Dico itaque quod aequalitas illa omnium malorum iudex est. Nihil enim malum nisi ad ipsam et ad eius dicitur formam. Illi ergo aequalitati iudicium omne datum est. Haec est virtus quae mala digerit omnia, et ad bonum exitum ducit. Quamvis enim ab aequalitate illa descendat, illa tamen aequalitas ad aequalitatem quandam illud redigit, quoniam illud, ut praedictum est, ad bonum exitum ducit, sicut arithmetica ratio omnem inaequalitatem redigit ad aequalitatem. Patet itaque quod bonitas definita est, quoniam in ea nihil plus, nihil minus, nihil ultra, nihil infra est. Nota autem duas esse bonitates: unam dei et aliam qua nos participamus. Bonitas divina infinita est, et incomprehensibilis est, sicut et ipse deus neque cepit neque destitit, nisi quantum in parte sub scientiam et participationem nostram cadit. Nostra vero bonitas est definita quasi usualis, quia et a divina bonitate procedit, et in ipsam terminatur et revertitur. Et ita de hac bonitate indifferenter agit, et promiscue, nunc illam nunc istam volens intelligi. CADENS SUB SCIENTIAM, id est comprehensa, non tamen secundum quod ipsa est, sed secundum capacitatem intelligentis, ut possit intelligere incompositum esse deum et simplicem. Veracissime itaque dictum est: SUB SCIENTIAM CADENS. Nichil enim agnosci potest, nisi quod finitum et terminatum est. Nec illud nunc dicimus quod quicquid infinitum sit diffinitione demonstrari non possit, sed non ideo infinitum potest cognosci, nec perfecta eius noticia haberi, quippe quod animo circumdari non potest. Unde fit, quoniam malum infinitum est, eo scilicet quod in transgrediendo nullus finis est. Transgressio enim nullum terminum habet; inde, inquam, est, quare malum sub scientiam non cadit, quoniam ab animo circumdari non potest, quippe nullius est naturae ideo, quia malum nihil est. Bonitas vero sub scientiam cadit, quoniam finita est, sed nec istud facilem satis intelligentiam habet, cum bonitas nihil aliud sit quam ipsa divinitas, quae in sua quidem natura incomprehensibilis est. Et de divinitate quidem per se nullus fere haberi potest intellectus, ipsa tamen divinitas in rebus ipsis simpliciter intelligitur. Ut enim Plato sapientissimus testatur, res nulla intelligi potest, nisi deus in ea prius intelligatur. Deus enim omnium rerum est entitas, ut cum humanitas sit entitas hominis, ipsa quoque divinitas entitas

1289–1290 *Locum non inveni.*

1290–1292 Cf. Thierry of Chartres, *Lectiones in Boethii librum De Trinitate* 4.27 (ed. Häring, p. 195): "Ipse enim non est ens sed est ipsa essentia vel entitas omnium rerum"; and 4.42 (ed. Häring, p. 200): "QUONIAM IS scilicet DEUS SIT FORMA simplex scilicet ET UNUM ESSE i.e. una entitas omnium rerum." Cf. *Glosa super Boethii librum De Trinitate* 2.36 (ed. Häring, p. 276): "Hec autem quoniam formis et imaginibus non repugnat in

est humanitatis. Deus itaque in omnibus rebus simpliciter intelligitur, quamvis per se non possit intelligi. Quis enim aliquam rem sine suo esse intelligit? ANIMOQUE SEMPER IMITABILIS. Hoc ideo, quoniam animus omnium rerum forma est, sicut deus essentiae forma est. Animus enim omnibus rebus se conformat, et rerum omnium similitudines in se concipit. Naturale est autem hominis virtutem quam in alio considerat imitari. IMITABILIS: ut nos conformemus ipsi bonitati, vel ideo, quia et nos postquam esse cepimus semper sumus. In essentia habemus communionem, in perpetuitate non habemus, quia ille semper fuit, nos ex tempore cepimus. ET PERCEPTIBILIS, id est quam debet animus imitari et percipere. Nisi enim imitetur illam, non eam percipere poterit. Expositio autem est eius quod perdixerat, quoniam idem est perceptibilem esse, quod semper imitabilem. PRIMA NATURA. Quod ad ipsam refertur bonitatem quae, cum ipsa sit divinitas, nec eam primam esse natura dubium est, quia est fons omnis bonitatis. ET SUAE SUBSTANTIAE DECORE PERPETUA. Ordo: ipsa quae bona per naturam est, et prima PERPETUA est DECORE SUAE, id est aeternae non accidentalis SUBSTANTIAE, quia ei nec accedit aliquid per augmentum, nec decedit per detrimentum. Ita enim deus bonus est, ut sit etiam bonitas. Illud quidem de ipsa aequalitate dicitur. Aequalitas enim vere decora est, quod inde patet, quoniam si homo sine manibus nasceretur, aliquo modo imperfectus turpitudinem sine dubio incurreret, quoniam homine minus haberet; si vero cum tribus oculis nasceretur, sive pluribus, pro turpissimo quoque haberetur, eo quod aequalitatem essendi hominem excederet. Decor enim ex aequalitate. Unde filio dei nihil pulchrius, nihil aeque ei pulchrum est, quippe aequalitas essendi est, quoniam nihil pulchrius, nihil esse aeque pulchrum potest. PERPETUA: non enim potest non esse bona. INFINITUM VERO MALICIAE DEDECUS. INFINITUM, quia videtur compositum ex diversis affectibus, nec capit aliquando finem satietatis executio vitiorum. DEDECUS et depravatio naturae et bonitatis nostrae malicia est, quae in infinitum cresceret, si a divina bonitate non retineretur. NULLIS PROPRIIS PRINCIPIIS NIXUM: quia malicia, quae contraria est bonitati, sine initio est. Si enim

eo quod quodam modo sistunt materiam ut ipsum possibile sit homo secundum hominis formam humanitatem scilicet hac ratione addit ET EX QUA ESSE EST. Nam et imago humanitas quod confert esse hominem et quod humanitati esse convenit ex prima forma provenit que est deus."

1315–1317 Cf. Thierry of Chartres, *Commentum super Boethii librum De Trinitate* 2.32 (ed. Häring, p. 78), cited above in the note to 1251–1252.

1318–1319 Cf. Caiazzo, "Un commento altomedievale," 141, ll. 423–425: "'Infinitum' quia videtur compositum ex diversis affectibus, nec capit aliquando finem satietatis executio vitiorum."

1321–1325 Cf. Caiazzo, "Un commento altomedievale," 141, ll. 425–428: "'Nullis principiis nixum,' quia malitia, quae contraria est bonitati, sine initio est. Si enim malum

malum initium haberet, aliquid esset. Sed quia non habuit initium, nihil est, et nihil aliud est nisi absentia boni. Vel ideo est sine initio, quia deus non est auctor cuiuslibet mali, nec deus facit aliquando malum. Vel ideo, quia transgressio boni aliquod sui generis ut transgressionem vel aliquod huiusmodi principium habere non potest. Detestandus tamen Manichaeus duo rerum principia esse mentitus est, unum boni, aliud vero mali, ut quemadmodum a bono boni principio omnia bona descendunt, ita et a malo mali principio omnia mala descendere putaret; dogma sequens Platonicum qui deum et materiam rerum principia constituit, videlicet ut a deo omnia bona et ordinata descenderent, a materia vero in qua est mutabilitas et confusio, omnia mala defluerent. Quod tantum phylosophum dixisse non puto, cum arithmetica disciplina illud falsum esse convincat. Alteritas enim ab unitate descendit; materia vero est mutabilitas, quae nimirum nihil aliud est quam alteritas. Ab unitate igitur descendat materia necesse est. Deus autem vere est unitas; necesse ergo est, ut a deo descendat materia, itaque deo coaeterna non est, sed ab ipso potius creata. Nullum ergo proprium est mali principium, sed a bono, ut praedictum est, necesse est mala omnia descendant; quae cum a bono defluant principio, ad quandam, ut supra diximus,

initium haberet, et aliquid esset; sed quia non habuit initium, nihil est, et nihil aliud est nisi absentia boni; vel ideo est sine principio, quia deus non est auctor cuiuslibet mali, nec deus fecit aliquando malum."

1330–1331 Calcidius, *Commentarius in Timaeum* 295 (ed. Waszink, p. 297): "… Pythagorae dogmate, cui concinere dicit dogma Platonicum, ait Pythagoram deum quidem singularitatis nomine nominasse, silvam vero duitatis." Cf. Thierry of Chartres, *Commentum super Boethii librum De Trinitate* 2.28 (ed. Häring, p. 77): "Sunt igitur secundum Platonem duo rerum principia: actus scilicet sine possibilitate i.e. deus vel necessitas et – quasi ex adverso posita – materia i.e. possibilitas. … Nemo tamen existimet quod Plato materiam deo coeternam esse voluerit licet deum et materiam rerum principia constituerit. Immo a deo descendere voluit materiam. Ubique enim magistrum suum sequitur Pythagoram qui unitatem et binarium duo rerum principia constituit: unitatem deum appellans, per binarium materiam designans. Quoniam ergo binarius ab unitate descendit constat quoniam materiam a deo descendere voluit: in hoc uno a veritate devians quod nec formam nec materiam creatam esse putavit."

1331–1333 Calcidius, *Commentarius in Timaeum* 296 (ed. Waszink, p. 298): "Deum quippe esse – ut etiam Platoni videtur – initium et causam bonorum, silvam malorum"; and 298 (ed. Waszink, p. 300): "Igitur iuxta Platonem mundo bona sua dei tamquam patris liberalitate collata sunt, mala vero matris silvae vitio cohaeserunt."

1334–1338 Cf. Thierry of Chartres, *Tractatus de sex dierum operibus* 30 (ed. Häring, p. 568); *Lectiones in Boethii librum De Trinitate* 2.25 (ed. Häring, p. 163): "Sed ab inmutabilitate descendit mutabilitas. Materia vero est mutabilitas. Ipsa tamen inter aliquid et nichil est sicut in Platone dicitur. Unde patet error eorum qui dixerunt quod materia coeterna esset deo. Ipsa enim a deo descendit et deus eam creavit i.e. eius causa et principium est. Potest igitur simplicitas eterna que deus est esse sine materia."

rediguntur aequalitatem. Deus enim quoniam summe bonus est, nec etiam mala fieri nisi causa boni permittit. SED NATURA SEMPER ERRANS. Natura errare dicitur, et deviare a suo principio quod est bonum et definitum, cum transgrediendo aequalitatem corrumpitur. Sed quia bonum impressum est ei, sicut OPTIMAE FIGURAE aliquod sigillum cerae vel aeri impressum est, idcirco natura componitur, id est ab illo errore retinetur. Ut enim dictum est, bonitas summa nec etiam mala nisi boni causa fieri permittit. ERRORIS FLUCTU. Bene fluctum errori comparat. Error enim semper fluctuat, quoniam principio nulli sibi proprio innititur. RETINETUR: ne usque ad praecipitium ruinae errando perveniat. NAM NIMIAM CUPIDITATEM. Quasi dicat: "patet in anima quod transgressio boni malum est," et quicquid plus vel minus, quicquid ultra vel infra est. Nam plus cupere quam oportet, vel minus, vitium est; quod tamen animus TEMPERATA BONITATE constringit. Temperatam vero bonitatem appellat aequalitatem, ad quam omnis inaequalitas redigitur. Cum itaque cupiditas videatur esse bonus affectus per desiderium, si excedat modum, fit vitium. IRAEQUE IMMODICAM EFFRENATIONEM, id est intemperatam dissolutionem. Modicum dicimus temperatum, secundum proprietatem latini eloquii.

ASTRINGIT. Sicut, habenis retentis in manibus, eques restringit immoderatum cursum equi quando vult, ita etiam illam effrenationem irae et intemperantiam astringit animus, quasi aliquis rector et aliquis auriga, et intra metas cohibet disciplinae, quia adhuc est in diffinitione boni principii. ET HAS QUODAMMODO INAEQUALITATIS FORMAS, id est quae fiunt diversis affectibus incitatae, TEMPERATA BONITATE constringit. Quia inaequalitas intemperans est, ipsam inaequalitatem redigit rector animus ad aequalitatem, quae bona est.

Per AEQUALITATEM virtus designatur et bonitas, quae deus est. PURA INTELLIGENTIA: quando solum deum contemplatur, et nullis passionibus perturbatur. ET HAS QUODAMMODO. In hoc enim quod inaequalitas pro-

1355–1358 Cf. Caiazzo, "Un commento altomedievale," 142, ll. 441–445: "Cum videatur esse cupiditas bonus affectus per desiderium, si excedat modum, fit vitium. IRAEQUE INMODICAM EFFRENATIONEM, id est intemperatam dissolutionem. Modicum dicimus temperatum, secundum proprietatem latini eloquii."

1359–1366 Cf. Caiazzo, "Un commento altomedievale," 142, ll. 447–453: "ASTRINGIT, id est sicut, habenis retentis in manibus, eques restringit inmoderatum cursum equi quando vult, ita etiam illam effrenationem irae et intemperantiam astringit animus, quasi aliquis rector et aliquis auriga et intra metas continet disciplinae, quia adhuc est in diffinitione boni principii. ET HAS QUODAMMODO INAEQUALITATIS FORMAS, id est quae fiunt diversis affectibus incitatae. TEMPERATA BONITATE CONSTITUIT, quia inaequalitas intemperans est, et ipsam inaequalitatem redigit rector animus in aequalitatem, quae bona est."

cedit ab aequalitate et in ipsam redigitur non est ei similis malicia, quae caret principio, nec terminatur in deo nisi per stricturam, sed in hoc quod, sicut inaequalitas quibusdam regulis artatur ab aequalitate, ita etiam malicia ratione constringitur a bonitate dei ne effluat.

XXXII, 3. SINT ENIM: in quo diligenter intuendum est quod, cum debuisset ostendere quomodo inaequalitas descendat ab aequalitate, ostendit quomodo ab aequalitate descendat proportionalitas, in quo nimirum ab aequalitate omnis aequalitas ostenditur descendere, hoc est aequalitas illa quae ad concordiam redigit omnem naturaliter discordiam. Haec est tetragona<tura> prima, quam ex se gignit unitas, quae multiplicata per se primum efficit tetragonum. Semel enim unum, primus tetragonus est. Haec est aeterna illa tetragoni generatio, de qua hyspana sybilla in suo dicit vaticinio: "Cum pervenitur ad costam tetragoni sedentis aeterni, et ad costas tetragonorum stantium aeternorum, et caetera." Tetragonum quidem sedentem et aeternum appellat dei filium, quem unitatis natura ex sua genuit substantia.

XXXII, 1. RESTAT AUTEM. Tractatis speciebus inaequalitatis, ut de inaequalitate perfecta tradatur scientia, docet scientiam de aequalitate et inaequalitate, ostendens primum quid sit inaequalitatis principium, et quomodo ex ea profluat omnis inaequalitas, et quomodo ad idem redigatur, commendans hanc scientiam docendo eam ad omnes species phylosophiae pertinere, scilicet naturalem, rationalem <et moralem>. Prius ergo doceamus quomodo ad naturalem pertineat. Naturalis est quae vocatur physica, quae causam mundi sensibilis inquirit et compositionem corporum, quae ad vitam animantur. In mundi vero causis demonstrandis ostenditur a philosophis, deum penes quem nulla est vicissitudo, nulla inaequalitas sed identitas, et ex identitate aequalitas, esse mundi et eorum quae in mundo sunt causam efficientem. Sic et in hac parte rerum alteratio et inaequalitas

1388 quid] *lect. dub.* MS, *fortasse* quid *corr. ex* quod

1374–1376 See below *ad Arithmeticam* II, XXXI, 1, and II, XLII, 1. See also above, pp. 67–70.

1378–1385 Cf. Thierry of Chartres, *Commentum super Boethii librum De Trinitate* 2.34 (ed. Häring, pp. 78–79); Thierry of Chartres, *Glosa super Boethii librum De Trinitate* 5.18 (ed. Häring, p. 297); Clarembald of Arras, *Tractatus super librum Boetii De Trinitate* 37–38 (ed. Häring, pp. 121–122). On the prophecy of the *sybilla hyspana*, see above, pp. 57–64.

1394–1402 Cf. Thierry of Chartres, *Tractatus de sex dierum operibus* 2–3 and 45 (ed. Häring, pp. 556–557 and p. 574). On the Trinity and the four causes, see above, pp. 48–49.

ex aequalitate docetur habere principium. Item finalis causa mundi sola dei est benignitas, formalis vero eius sapientia, sed ex eius benignitate et sapientia, in quibus semper est aequalitas, rerum consistit inaequalitas. In materiali docetur esse idem. Constat enim elementa IIII mundi esse materiam, et istorum idem esse principium, scilicet yle. Haec vero elementorum creatio in omni parte consideratur fuisse sibi aequalis. Ex qua tamen partium aequalitate omnis elementorum provenit inaequalitas, cum terra pondus, ignis obtineat levitatem, et aer et aqua multum a se discordent. Sic hanc speciem naturalis philosophiae iuvat aequalitatis et inaequalitatis cognitio. In materiali vero corporum conditione idem est. Quatuor enim humoribus animantium vegetatur natura: sanguine, flegmate, colera nigra, quae melancolia dicitur, et rubea. Qui quidem humores quamdiu adaequantur in corporibus, viget complexio. Cum vero subrepit humoris et caloris inaequalitas, infirmatur animal et plerumque moritur. Ecce quid scientia aequalitatis et inaequalitatis valet ad hanc partem phylosophiae. In rationali idem est, quae continet quadruvium, quia in astronomia, cum scilicet divinamus secundum motum astrorum quando aequaliter in casis suis dicuntur morari planetae, aliquotiens inaequaliter vel aequaliter a casis suis recedunt; et secundum hoc quarumdam rerum futurarum proventus aequales et inaequales praenoscuntur. In geometria quoque secundum quod maior vel minor triangulus ab eadem ypothenusa dependet, consideratur catheti et basis unius trianguli ad cathetum et basim alterius trianguli; et secundum hanc aequalitatem constat certa ratio mensurandi. Si vero fuerint inaequales, erit in mensurando fallacia. Musica etiam ex aequalitate motuum aequalitates sonorum, ex inaequalitate vero motuum sonorum inaequalitates et consonantias metitur. Quomodo autem arithmetica consideret aequalitatem et

1408 flegmate] flecmate MS

1402–1406 Cf. Thierry of Chartres, *Tractatus de sex dierum operibus* 44 (ed. Häring, pp. 573–574): "Inde etiam rerum mensure et pondera et quicquid omnino est habet existere. Mensura enim omnis vel ab equalitate vel ab inequalitate habet existere. Utrum autem horum fuerit ab equalitate esse necesse est quoniam ipsa inequalitas ab equalitate existit. Amplius: modus omnis medium est inter maius et minus. Si enim maius est excedit modum: si autem minus infra modum est. At omne id quod inter maius est et minus equalitas est. Modus igitur omnium rerum equalitas est. At mensura rei omnis ex modo est. Igitur ex equalitate. Idem de ponderibus ratiocinandum est."

1410 Here Thierry uses the term *complexio* as a medical term: see Danielle Jacquart, "De *crasis* à *complexio*: Note sur le vocabulaire du tempérament en latin médiéval," in *Mémoires V: Textes médicaux latins antiques*, ed. Guy Sabbah (Saint-Etienne, 1984), pp. 71–76.

inaequalitatem, post prologum demonstrabit evidenter. In hoc autem prologo, quae ad ethicam pertinent, demonstrabuntur.

XXXII, 28. INGREDIENTIUM ANIMOS. Ideo "animos ingredientium" dicit, scilicet animos legentium, quia iste primus liber totus pene propter introductionem sequentis libri praecessit, quatenus lector, hoc libro quiddam informatus, facilius nexuosas quaestiones mathematicae in sequentibus penetraret.

AB UTILIORIBUS: quia sequentia valent ad musicam et ad phylosophorum lectionem intelligendam.

1426–1430 Cf. Caiazzo, "Un commento altomedievale," 143, ll. 460–463: "Nota: ideo ANIMOS INGREDIENTIUM dicit, scilicet animos legentium, quia iste primus liber pene propter introductionem sequentis libri praecessit, quatinus lector hoc libro quiddam informatus, quatinus facilius nexuosas quaestiones mathematicae in sequentibus penetraret."

LIBER SECUNDUS

CAPUT I

I, 1. In superiori libro demonstravit quomodo inaequalitas ab aequalitate processerit; hic vero demonstraturus est quemadmodum omnis inaequalitas in ipsam, unde processit, aequalitatem revertatur. SUPERIORIS LIBRI. Ipse se continuat, ostendens quid dixerit in praecedenti, et quid dicturus sit in se<quen>ti libro. DISPUTATIONE. Disputatio ad dialecticam, oratio ad rethoricam pertinet. A PRINCIPE. Etenim binarius primus duplus medietatem sui habet monadem, id est unitatem; similiter ternarius terciam. Et hoc est quod dicit SUI GENERIS. Ab ipsa enim generationem accipiunt. PROCESSERIT. Et non solum inaequalitas procedit ab aequalitate, sed etiam ipsa redigitur in aequalitatem. Quod non est mirum, quia haec est natura rerum ut a quibus elementis processerunt, in eadem resolvantur; et hoc per articulatam vocem.

ELEMENTA: dicta quasi elevamenta, vel elimenta a verbo "elimo-elimas," vel etiam, <ut alii> volunt, ilementa, quoniam "ile" "materia." Ipsaque elementa vocat primordia et initia quaedam, UT, id est quemadmodum, ARTICULARIS VOCIS ELEMENTA SUNT LITTERAE. Articularem vocem hic appellat quae scribi potest, quia non resolveretur in litteras, nisi scribi posset. EANDEM VIM, quam tenent litterae in syllabis, eam habet sonus, id est tonus, in musicis; et illud pertinet ad rudimenta grammaticae, istud ad speculationem musicae. Sicut enim in grammatica nihil est minus quam littera, ita in musica nihil est minus quam sonus, id est tonus. Omnis ars suum habet elementum. In grammatica namque littera elementum est, in dialectica proloquium, in rethorica sententia, in arithmetica monas, in geometria centrum, in astronomia illud quod proprio nomine vocatur athomus, in

1 Liber secundus] *add. in marg.* MS

15–20 Cf. Caiazzo, "Un commento altomedievale," 143, ll. 465–469: "ELEMENTA dicta quasi elevamenta, vel elimenta a verbo elimo, id est formo; vel etiam, ut alii volunt, ilementa, quoniam ilen 'materia,' ipsaque elementa vocat primordia et inicia quaedam. UT, id est quemadmodum, QUONIAM ARTICULARIS VOCIS ELEMENTA SUNT LITTERAE. 'Articularem vocem' hic appellatur quae scribi potest, quia non resolveretur in litteras, nisi scribi posset."

20–27 See below *ad Arithmeticam* II, IIII, 4. Cf. the preface to "De astrolabio" preserved in Stuttgart, Württembergische Landesbibliothek, Cod. math. 4° 33, fols. 37va–38ra, published in Caiazzo, "Sur l'astrolabe et le comput au XIIe siècle: nouveaux textes inédits," 37: "Nam et si omnium artium propria sit natura propriumque initium, ut verbi gratia: grammaticae littera, dialecticae pronunciatum, rethoricae sententia, arithmeticae

musica tonus. SONUS in MUSICA vox est aequaliter porrecta, quod minimum est in musica. Et cum iunguntur soni acutus et gravis faciunt tonum. Tonus est vox e duobus sonis consistens.

MARGINE: fine et principio. Ubi enim finis, ibi principium, et ibi finis, ubi principium est. Margo est, ut Plinius dicit, extrema pars latitudinis singulorum planetarum in signifero.

I, 2. PROPORTIONALITER: ut quantum superatur primus, id est VIII, a medio, id est XXXII, tantum superetur medius, id est XXXII, ab extremo, id est CXXVIII. CONSTABIT: cum firma certitudine manebit, solidum et firmum erit.

34 XXXII] XXII MS

---

unitas, geometricae punctus, astronomicae athomus, musicae semitonium, ipsa<e> tamen fieri nequeunt, nisi in aliqua subiecta substantia. Unde videri potest qua<ntu>m digna sit astronomica disciplina, cuius substantia superna est machina." The author copied long passages from *Ad intimas philosophiae*, a text previously edited with the title "Fragmentum libelli de astrolabio, a quodam (an Lupito Barchinonensi) ex Arabico versi," in Gerbert of Aurillac, *Opera mathematica* (ed. Bubnov, pp. 370–375); cf. "Et si omnium artium propria sit natura, propriumque initium, ut verbi gratia grammaticae littera, dialecticae pronuntiatum, rethoricae sententia, arithmeticae unitas, geometricae punctus, astronomicae athomus, musicae semitonium, ipsae tamen fieri nequeunt, nisi in aliqua subiecta substantia – quam digna est tenenda astronomica disciplina, cuius substantia superna est machina" (ed. Bubnov, p. 371). The *Ad intimas philosophiae* is a basic introduction to the study of the astrolabe transmitted through several manuscripts that also contain the *De utilitatibus astrolabii* (J). Another edition of this text was published by José M. Millás Vallicrosa, *Assaig d'història de les idèes fisiques i matemàtiques a la Catalunya medieval* (Barcelona, 1931), pp. 271–275. See also David Juste, *Les 'Alchandreana' primitifs: Etude sur les plus anciens traités astrologiques latins d'origine arabe (Xe siècle)*, p. 247f, and Borrelli, *Aspects of the Astrolabe.*

27–29 Cf. Caiazzo, "Un commento altomedievale," 143, ll. 469–471: "SONUS est IN MUSICA vox aequaliter porrecta, quod minimum est in musica. Et cum iunguntur soni acutus et gravis faciunt tonum. Tonus vero est vox e duobus sonis consistens."

30–31 Cf. Caiazzo, "Un commento altomedievale," 143, ll. 475–476: "MARGINE, id est fine et principio. Ubi enim finis, ibi principium, et ibi finis, ubi principium est."

31–32 On "Plinius," see Christopher McDonough, "Trier, Bibliothek des Priesterseminars Ms. 100 and the Text of Martianus Capella," *Mediaeval Studies* 36 (1974), 56–66, at 58: "(39.20 MARGINE) Margo est – ut Plinius dicit – extrema pars latitudinis singularum planetarum in signifero." According to McDonough, it is a marginal gloss to Martianus Capella, written in a hand from the second half of the ninth century. The Trier manuscript contains books I and II and a part of book III of *On the Marriage of Philology and Mercury*, also written in the second half of the ninth century. Only the first two books are glossed.

I, 3. QUI DE MEDIETATE. Hoc ideo apponit, ne duos tales secundos accipiamus, a quali uno prius abstulimus primum, et cuius remanentiam fecimus secundum terminum, immo illam ipsius remanentiam bis a tercio nos praecipit abstrahere.

I, 7. ELIQUABIMUS, id est resolvemus, et quasi vi exprimemus – quod proprium est huic verbo significare –, liquando et attenuando eam producemus. "Liquitur" defectivum verbum est, id est fluit. Est et aliud verbum defectivum "liquet," id est manifestum est. Est et aliud verbum "liquo-liquas-liquavi," id est fluere facio, sicut cera vel plumbum liquatur ab igne, quod teutonice dicitur "smalzet." Exempli gratia: ostendit Boetius qualiter species multiplicis ad aequalitatem redigantur. Videamus et nos si eadem ratio praeceptorum species superparticularium valeat aequalitati reddere. Videtur autem non sic continue posse fieri, videlicet ut sesquiquartus continue et proximo loco, id est nullo interposito, redeat ad sesquitercium, et eodem modo sesquitercius in sesqualterum, et ille item in aequalitatem. Quia vero sine medio non potest aliquis superparticularis in alium resolvi, quos oporteat medios esse videamus. Cum itaque Boetius a quadruplis inceperit, nos a sesquiquartis incipiamus. Sesquiquarti igitur primo resolvantur in quadruplos, unde nati sunt, hoc ordine: XVI, XX, XXV, tali autem regula. Aufer primum de secundo, et quod remanet sit secundus terminus; quod autem abstulisti de secundo, scilicet primum, ille sit primus terminus; porro de tercio aufer primum et duos secundos, et quod remanet pone tercium terminum. Verbi gratia: XVI aufer de XX, et remanent IIII, eritque primus terminus sedecim et secundus IIII. Deinde de tercio aufer primum, scilicet XVI et duos secundos, id est duos quaternarios, et remanebit tercius terminus unitas, eritque talis dispositio: XVI, IIII, I. Et ecce hi omnes quadrupli sunt, de sesquiquartis nati. Porro converte hunc ordinem et, per eandem regulam supradictam, quadruplos redige in triplos. Conversio itaque talis erit: I, IIII, XVI. Aufer igitur primum de secundo, id est unitatem, et erunt duo termini: I, III. Et de tercio aufer primum et duos secundos, id est ternarios, et erit tercius terminus VIIII, eritque talis dispositio: VIIII, III, I. Et ecce hi omnes sunt tripli, de quadruplis nati. Horum triplorum ordinem si conservaveris, per aliam regulam facies sesquitercios, et tunc erit talis dispositio: VIIII, III, I. Regula autem talis est: pone primum primo aequum, et secundum primo et secundo, tercium primo et duobus secundis et tercio.

41–42 Cf. Caiazzo, "Un commento altomedievale," 143, ll. 479–480: "ELIQUABIMUS: resolvemus, et quasi vi exprimemus, quod proprium est huic verbo significare."

46 The term *smalzet* corresponds to *schmelz* in current German usage (i.e. English *melt*); see *Mittelhochdeutsches Wörterbuch* (Leipzig, 1866), p. 429. It could be a gloss inserted into the commentary by the scribe.

Quod si feceris, erit dispositio talis: VIIII, XII, XVI. Ecce hi omnes sunt sesquitercii, de triplis nati. Porro istos sesquitercios in triplos, unde nati sunt, redige eodem ordine VIIII, XII, XVI, sed per priorem regulam, et surget triplorum dispositio talis: VIIII, III, I; quem ordinem converte per eandem regulam, et istos triplos redige in duplos, quorum dispositio talis est: IIII, II, I. Hoc ordine conservato, per aliam regulam de istis duplis facies sesqualteros, quorum haec erit dispositio: IIII, VI, VIIII. Item hos sesqualteros in duplos, unde nati sunt, rediges, mutata regula mutato ordine, et erunt ipsi sic dispositi: IIII, II, I. Hunc ordinem muta <scilicet I, II, IIII>, et per regulam eandem de duplis incides aequalitatem, et surget talis dispositio: I, I, I. Vocaturque ista reductio superparticularium ad aequalitatem "saltus Gerberti," quia post Boetium omnes arithmetici ignorabant qualiter redirent in aequalitatem superparticulares, per eandem praeceptorum rationem qua et multiplices usque ad tempora Gerberti, qui primus hunc locum, quem ipsi tacendo transilierunt, explanavit.

Caput II

II, 1. Enmusotaton theorema, id est musicalis divinae speculatio. Hinc "amusoteros" dicitur "stultus" et quasi sine musica, unde Symmachus ad Ausonium Mosellam in quadam epistula dicit: "An amusotheros tibi videor." Et ad Platonicam in Thimaeo generationem animae quam Plato in *Thimaeo* ostendit. Thimaeum appellavit quemdam librum Platonis, ubi genituram mundi et animae, sicut ei visum est, composuit. "Thime" enim graece dicitur "anima" a graeco "thimui," id est precioso. Etenim nihil est preciosius anima. <Armonicae>: quasi in armonia animae nostrae omnia nostra sunt posita, aequalia et inaequalia, et intelligibiliter secundum se, et sensibiliter secundum corpus. Et ad intervalla, id est ad

74 VIIII, XII, XVI *add. supra lin.* MS **75** VIIII, III, I *add. in marg.* MS **82** reductio] redactio MS

82–86 This refers to the scholion of Gerbert of Aurillac on *De arithmetica* II, 1, called *saltus Gerberti* from the eleventh century onward; see "Scholium ad Boethii Arithmeticam Institutionem l. II, c. 1," in Gerbert of Aurillac, *Opera mathematica* (ed. Bubnov, pp. 31–35).

89–91 Symmachus, *Epistulae* 14 (ed. Seeck, p. 10): "Aut ἀμουσότερος tibi videbar."

91–92 Plato, *Timaeus* 35A–C (ed. Waszink, p. 27).

92–94 Cf. Caiazzo, "Un commento altomedievale," 143, ll. 482–483: "Timaeum appellavit quendam librum Plato, ubi genituram mundi et animae, sicut ei visum est, exposuit. Time enim graece dicitur anima."

inveniendam armonicam medietatem, quia tres sunt medietates: arithmetica, geometrica, armonica. Intervalla autem symphoniarum hic tangit, quae in ultimis latius disserit. Maius intervallum "diastema," minus "sistema" apud musicos nominatur. IBI ENIM. In musica et in generatione animae et in armonica disciplina, in omnibus his IUBEMUR EXTENDERE tres vel quatuor sesqualteros, et dicamus quis sit primus, quis secundus. Similiter et in sesquiterciis consideratur. Ideo dicit PRODUCERE et EXTENDERE, quia tunc extendimus, cum ab uno sesqualtero pervenimus ad multos sesqualteros. Armonia dicitur omnis consonantia a tercio genere, quod armonicos dicitur, quod est suavius caeteris. Primum enim dicitur cromaticum, quod est dissolutissimum, et constat in cantilenis feminarum. Nam "croma" graece "color," unde cromaticum id est colorabile et varium propter dissolutionem. Secundum vocatur diatonicum, quod est durissimum, et hoc apud Scottos et Brittones est. Tercium vero dicitur armonicum quod est dulcissimum, et ex duobus compositum, ideoque consonantia armonia dicitur. Quomodo autem extendantur et quot, ibi discere poteris facile per istud praescitum.

II, 2. ET INFERACI: quia aut vix aut nullomodo sciri posset quot et quae proportiones in diversis simphoniis musicis continuentur, nisi praemissa hac arte. Hic enim docetur quod quaelibet species multiplicis numeri quoto loco distat ab unitate, tot superparticulares sui generis post se continuat.

II, 4. UT SUPERIUS DISTINCTUM in priori libro ubi dicit: ECCE TIBI ILLA SESQUALTERA QUANTITAS EX TERMINO DUPLICITATIS EXORITUR.

II, 5. AEQUABILIS: numero proportionum se sequentium per locum quo distat ab unitate.

114 Et inferaci] inferaci *ed. Guillaumin ab ed. Friedlein* **118** distinctum] destinatum *ed. Guillaumin,*

100–112 Cf. Caiazzo, "Un commento altomedievale," 144, ll. 484–496: "Maius intervallum 'diastema,' minus intervallum 'sistema' apud musicos nominatur. IBI ENIM, id est in armonica disciplina, IUBEMUR EXTENDERE III VEL QUATTUOR SESQUALTEROS, scilicet ut ponamus ordinatim sesqualteros, et dicamus quis sit primus, quis secundus; similiter et in sesquitertiis consideratur. Ideo dicit: PRODUCERE ET EXTENDERE, quia tunc extendimus, cum ab uno sesqualtero pervenimus usque ad multos sesqualteros. ARMONICAE, id est musicae. Armonia enim dicitur omnis consonantia a tertio genere, quod armonicon dicitur, quod est suavius caeteris. Primum enim dicitur cromaticum, quod est dissolutissimum, et constat in cantilenis feminarum. Nam 'croma' graece 'color,' unde cromaticum id est colorabile et varium propter dissolutionem. Secundum vocatur diatonicum, quod est durissimum, et hoc apud Scottos et Brittones est. Tertium dicitur armonicum, quod est dulcissimum et ex duobus compositum. Ideoque omnis consonantia armonia dicitur."

118–119 Cf. Boethius, *De arithmetica* I, XXXII, 14.

II, 7. Qui sit par loco duplicis supra se positi in superiori linea, quam postea vocabit lineam latitudinis, quia si ultimus videlicet inferior numerus proportionum tercio loco distat a duplici super se posito, duplex etiam cum sua locatione distabit ab unitate tercio loco; et ille non potest dividi in duo aequalia, et ideo non habet sesqualterum ad se. Sicut enim quaternarius tercius est ab unitate, ita in hac proportione novenarius tercius est a quaternario; ergo caret sectione. Novenarius, qui est sesqualter senarii, par est loco duplicis sui, id est quaternarii, quia ipse novenarius a quaternario tercio loco disiungitur, sicut et quaternarius ab unitate. Sciendum etiam quia latitudo harum figurarum ita est videnda: —, longitudo autem ita: |, et sicut primus ordo duplicis secundum longitudinem quidem duplus est, ita in eadem figura omnes sequentes ordines ad longitudinem dupli sunt.

Sic etiam qui tripli sunt in prima longitudine, secundum eandem quoque longitudinem reliqui sunt tripli; eodem etiam modo in reliquis. In prima itaque figura sunt primi duplices et ultimi angulares triplices, et quod habet in latitudine per duplicitatem, hoc habet in longitudine per sesqualteram proportionem. Angularis numerus non habet terciam partem, qua possit aequari praecedenti numero, quia duplex est omnis angularis.

II, 13. Hinc quoque.

II, 14. Erunt duplices. Non ita quod proximus ad proximum versum sit duplex, sed in eodem versu quilibet sequens est duplex praecedentis. Secundum seriem latitudinis. Istud idem dicet melius et evidentius in triplicibus.

122–130 The table proposed by Boethius is helpful in understanding this passage:

1	2	4	8	16	32
	3	6	12	24	48
		9	18	36	72
			27	54	108
				81	162
					243

130–135 Cf. Caiazzo, "Un commento altomedievale," 144, ll. 505–510: "Sciendum autem, quia latitudo harum figurarum ita est videnda: —, longitudo ita: |. Et sicut primus ordo duplicis secundum longitudinem quidem duplus est, ita in eadem figura omnes sequentes ordines ad longitudinem dupli sunt. Sic etiam qui tripli sunt in prima longitudine, secundum eandem quoque longitudinem reliqui sequentes tripli sunt. Eodem etiam modo in reliquis."

Caput III

III, 1. Si igitur duae primae. Quia, sicut dictum est, superparticulares nascuntur a multiplicibus, igitur quamdam habent cum eis affinitatem, hanc scilicet, quod ex sesqualtero et sesquitercio duplex componitur. In omni enim duplo sesqualter et sesquitercius invenitur, sicut in figura apparet: inter duos extremos est dupla proportio, medius vero est sesqualter ad primum, et ultimus sesquitercius est ad medium.

III, 2. Medietas et duplus. Hic medietatem vocat subduplum dupli, quia subduplum est dimidium dupli; et vocatur medietas a continentia. Id vero, quod inter duo extrema ponitur, etiam vocatur medietas a loco sive ab ordine, et ideo cave decipiaris.

III, 4. Continens multiplicis, id est vicina et proximo loco posita. Post duplam enim est tripla, VIIII ad XVIII subduplus, et VI subtriplus. Et cum liber dicat "duplus et triplus," tu converte.

III, 6. Revocatur. Quia cum ex illis constet, etiam istae proportiones inveniuntur in ipso, et hoc idem est quod iam dicet, secundum supradictum modum. Resolutione facta. Si enim resolvamus XVIII in VIIII, dupla erit proportio, si autem VIIII in VI, sesqualtera.

III, 7. Si autem. Verbi gratia: III, IIII, XII. XII ad III quadruplus sed, resolutione facta, XII ad IIII triplus est, IIII ad III sesquitercius. Secundae speciei. Pone sesquitercium, id est XII ad VIIII, qui est secunda species superparticularis, et VIIII ad III, qui est triplus, secunda species multiplicium fit quadrupla proportio.

Caput IIII

IIII, 2. De linearibus numeris: ut II, III, IIII, qui uno adiecto naturali ordine velut lineari ductu progrediuntur. De inaequali. Qui sunt solidi, quia secundum eos disponitur corpus trium dimensionum, sed non aequalium. Et quia numeri quidam sunt solidi et cubici, ideo addit: de solidis etiam et de cubicis, id est de talibus qui, cum sint solidi, sunt etiam cubici. Habent enim omnes dimensiones aequales. Qui sint, et quomodo

173 cubicis] cybis *ed. Guillaumin*

150–151 The sequence proposed by Boethius is the following: 2, 3, 4. Four is the double of two; three is one-and-a-half times two (*sesqualter*); four is one-and-a-third times three (*sesquitercius*).

procreentur omnes isti, postea explicabit. GEOMETRICAE CONSIDERATIONIS. Sed hic ideo de his tractat, ut supra dixit, quia valent ad sequentia, et quia haec ars amat sic alternatim institui. Sed quoniam adhuc diceret aliquis quare hic tractat de illis, propter instantem interrogationem reddit aliam causam, hanc scilicet, quia geometriae principium et fundamentum est arithmetica; ideo convenit, ut dispositiones figurarum geometricalium oriantur a primis numeris arithmeticae. Numerus vero per se consideratus prior est numero ad alium relato. LINEAREM NUMERUM vocat naturalis numeri dispositionem, quia in modum lineae producitur additione unitatis. PLANA DIMENSIO est numerorum, quae fit in longitudine <et latitudine>, ut bis bini; PLANA et aperta, quae verbi gratia in quaternario numero potest contemplari, secundum generalem figurarum rationem. Non habet enim quaternarius nisi duo spacia tantum longitudinis et latitudinis. Et ea est plana figura, quae tali modo gignitur, et profunditate deseritur. Ergo plana figura dimensioque sola superficie continetur, id est longitudine et latitudine. Namque quae est in geometria ratio quadranguli, ea est in arithmetica ratio plani numeri.

CUBUM dicunt Graeci "solidum," unde numerus vocatur cubus, qui in altitudinem, latitudinem longitudinemque extenditur, ut bis bini bis. SPERICI sunt numeri circulares, qui sicut circulus a quo puncto inchoatur in eundem revertitur, ut cum dico: quinquies quinque quinquies, sicut a quinario fit multiplicatio, ita summa multiplicationis in quinarium desinit; sunt enim CXXV. PYRAMIS dicitur graece aedificium infra spaciosum, supra acuminatum, sicut solent esse absidae capsae in quibus sanctorum corpora requiescunt. Unde numeri pyramidae dicuntur, qui super basim latera fundantur, sed in altitudinem exilem porriguntur. LATERCULI sunt qui, in modum laterum, aequalem habent longitudinem et latitudinem, sed in mi-

192–205 Cf. Caiazzo, "Un commento altomedievale," 144–145, ll. 512–526: "CYBUM dicunt Graeci solidum; unde numerus vocatur cybus, qui in altitudinem, latitudinem et longitudinem extenditur, ut bis bini bis. SPHERICI sunt numeri circulares, qui sicut circulus a quo puncto inchoatur in eundem exit, ita a quo numero aequaliter multiplicantur in altitudine, latitudine et longitudine, summa numeri multiplicati in eundem exit, ut cum dico quinquies quini quinquies, sicut a quinario fit multiplicatio, ita summa multiplicationis in quinarium desinit; sunt enim CXXV. PYRAMIS graece dicitur aedificium infra spatiosum, supra acuminatum, sicut solent esse absidae in quibus sanctorum corpora quiescunt; unde numeri pyramidae dicuntur, qui super basim latera fundant, sed in altitudinem exilem porriguntur. LATERCULI sunt qui in modum laterum aequalem habent longitudinem et latitudinem, sed in minimam extenduntur altitudinem, ut sunt ter terni bis. TIGNULI vero sunt qui longitudinem magnam in modum tignorum habent, sed altitudinem et latitudinem minorem, ut quater bini bis. CUNEI autem sunt qui nulla lege constringuntur, sed modo longiores, modo latiores et modo altiores pro libitu formantis existunt."

nimam extenduntur altitudinem, ut ter terni bis. TIGNULI sunt qui longitudinem magnam in modum tignorum habent, sed altitudinem et latitudinem minorem, ut quater bini bis. CUNEI sunt qui nulla lege constringuntur, sed modo longiores modo latiores modo altiores pro libitu formantis existunt.

PLANUM SIQUIDEM. Ostendit quare dicat RADICE AC MATRE, quia sicut in fine secundi prologi dictum est, antiquior est arithmetica reliquis, et verum est quod si ista non esset, reliquae non essent, nec inde sequitur quod, quia ipsa est, et aliae sunt.

IIII, 4. EST IGITUR UNITAS. Quia de numero per se considerato tractandum est, igitur incipiamus ab unitate, quae principium est omnium. Et nota quod apud geometras est punctum, apud arithmeticos est unitas. Ad cuius rei evidentiam pauca prius praemittenda sunt de numero. Numerus ergo alius est cardinalis, ut unum, duo, tria, et similia. Alius est relatus, ut duplum, triplum, quadruplum, et deinceps. Alius est geometricus, ut trigonum et tetragonum, sive mavis dicere triangulum et quadrangulum et huiusmodi, aut bicubitum, tricubitum, vel bipedale, tripedale, et talia. Alius est proportionalis, de quo postea dicemus. Prior autem numerus dicitur cardinalis quasi principalis; "cardo" enim est principium. Est autem cardinalis numerus prior natura omnibus aliis, et ab eo descendunt alii. Ternarius enim causa est trianguli, quaternarius quadranguli. Unde sublato ternario aufertur triangulum, sed non sublato triangulo aufertur ternarius. Possumus autem dicere quod numerus cardinalis nec est relatus nec geometricus, quia ut cardinalis est non refertur, nec ut relatus est, cardinalis est. Possumus tamen dicere quod duplum et duo sunt idem binarius; similiter duplum et bicubitum idem binarius. Illud quoque sciendum quod binarius dicitur linearis numerus, sic et ternarius et huiusmodi. Ideoque dicitur linearis numerus binarius esse, quod est numerus partium lineae, secundum quem ipsa dicitur bicubita vel bipedalis. Eodem modo intelligendum est de caeteris. Unitas vero locum obtinet puncti, et merito, quia sicut unitas est complicatio numeri, et numerus explicatio unitatis, ita quoque punctum est complicatio omnis magnitudinis, et magnitudo est explicatio puncti. De puncto vero pessime senserunt, qui punctum quiddam minimum esse putaverunt et, pro parvitate nimia, indivisibile esse dixerunt, ita scilicet ut punctum puncto

234 indivisibile] invisibile MS

211–212 See above *ad Aritmeticam* II, II, 1.

230–232 On *complicatio* and *explicatio*, see above 00–00.

232–241 See the "Adelard I" version of the *Elements*, the first integral Latin translation of the Arabic version (probably that of al-Hajjaj II), done by Adelard of Bath between 1126 and 1130, edited by Hubert L. L. Busard, *The First Latin Translation of*

adderetur ad constituendam lineam, quemadmodum si unum granum milii adiungatur alii. Dico igitur quod punctus est terminus rei ipsius et perfectio. Haec est ipsa rei totalitas, quae rem totam integre terminat et perfecte, nec est nisi unius rei proprie terminus. Cum enim terminos dicimus, abutimur, quemadmodum cum unitates dicimus. Hanc ergo totalitatem, quam terminum nominamus, explicat ipsius lineae extensio; linea namque est puncti evolutio. Et quoniam magnitudo a puncto incipit, ideo ab ipsa totalitate merito decrescere in infinitum dicitur, sicut pluralitas in infinitum crescere, eo quod ab unitate incipiat. Quoniam enim magnitudo a totalitate inchoat, id est ab ipsa rei perfectione, ultra totalitatem vero rei eiusque perfectionem nihil de ipsa est, eo scilicet quod totalitati et perfectioni rei ipsius nihil addi possit ut ipsa perfecta sit, ideo non crescere sed decrescere in infinitum potest. Perfectio enim prior natura est imperfectione, quia imperfectio ab ipsa descendit, nec aliquid causa perfectionis est. Qui ergo a perfectione inchoat, necesse est in imperfectionem descendat. Ultra perfectionem namque rei nihil est, sed infra. Unde merito dictum est quod magnitudo decrescit in infinitum, eo quod a perfectione et ipsa rei totalitate incipit, ut praemonstratum est, ultra quam eius nihil est, sed infra. Quoniam vero pluralitas ab unitate incipit, infra quam nichil est, merito non decrescere, sed in infinitum dicitur crescere, quia a simplicitate unitatis non fit progressio nisi in multiplicitatem. Merito quoque dictum est quod multitudo crescit in infinitum, magnitudo vero decrescit in infinitum. Illud quoque sciendum quod, quia perfectio prior est omni imperfectione, ideo ipsa perfectio quiddam simplex est. Si enim perfectio compositum quiddam esset, partes eius priores essent. Itaque aliquid prius esset natura perfectione; sed nihil prius est perfectione, quia ab ipsa descendit imperfectio. Quare et perfectio simplex quiddam est, ideoque punctum simplex esse dicitur et indivisibile. Illud quoque sciendum quod, cum dicimus duos terminos lineae esse, abutimur

Euclid's "Elements" Commonly Ascribed to Adelard of Bath (Toronto, 1983). See the various definitions in Busard's edition of the *Elements*, p. 31: "Punctus est illud cui pars non est" (def. 1, p. 31); "Linea est longitudo sine latitudine" (def. 2); "Cuius extremitates quidem duo puncta" (def. 3); "Linea recta est ab uno puncto ad alium extensio, in extremitates suas utrumque eorum recipiens" (def. 4); "Superficies est quod longitudinem et latitudinem tantum habet" (def. 5); "Cuius extremitates quidem linee" (def. 6); "Superficies plana est ab una linea ad aliam extensio, in extremitates suas eas recipiens (def. 7)." Cf. also the so-called "Adelard II" version, the redaction attributed to Robert of Chester (ed. Busard and Folkerts, 1: 113): "Punctus est illud cui pars non est" (def. 1); "Linea est longitudo sine latitudine" (def. 2), "Cuius extremitates quidem duo puncta sunt" (def. 3); "Linea recta est ab uno puncto ad alium extensio in extremitates suas utrumque eorum recipiens" (def. 4); "Superficies est que longitudinem et latitudinem tantum habet" (def. 5), "Cuius termini quidem sunt linee" (def. 6); "Superficies plana est ab una linea ad aliam extensio in extremitates suas eas recipiens (def. 7)."

dicendo terminos, cum unus sit terminus, quia unum punctum solum est, id est una sola totalitas, licet multa sint tota. Nec ob aliud dicimus terminos in linea esse, nisi quia et hic terminatur linea et ibi, id est hucusque est perfectio rei et usque illuc, quod est dicere: "terminatur et hic et ibi." Nec tamen est ubi una rei perfectio, una totalitas quae ipsam magnitudinem includit; ipsa enim est magnitudinis complicatio. Explicare vero totalitatem hoc nihil aliud est quam totum per partes dividere. In unaquaque vero partium est eadem totalitas, quae et in ipso toto, quamvis totum non remaneat idem. Illud quoque sciendum quod, quamvis punctum sit complicatio omnis magnitudinis, tamen soli determinatur ad superficiem, superficies ad lineam, linea vero ad punctum. Sed punctum est prima perfectio, linea vero est imperfectio, non quod non sit perfectio superficiei, sed imperfectio dicitur ad comparationem puncti, quod est prima perfectio, ut praedictum est.

NEC INTERVALLUM. Intervallum vocat spacium quodlibet vel longitudinis, vel latitudinis, vel altitudinis. NEQUE ENIM PUNCTUM. Punctum comparatum ad punctum nec maius eo nec minus est, nec in longo nec in lato nec in spisso. Nullius enim quantitatis est. Quod enim punctum diminutum et parvissimum dicitur, ideo fit, quia indivisibile est. Superponere igitur punctum puncto nihil aliud est nisi punctum puncto comparare. Quod enim antiqui dixerunt, quia punctum puncti non efficit intervallum, fabulosum est. Illud quoque quod alii exposuerunt punctum esse fixuram graphii, cui fixurae si affigatur iterum graphium nulla inde fit longitudo, perridiculosum est. Fabulatur enim quod punctum sic puncto superpositum nullam faceret lineam, sed potius punctum in eadem serie continuatum

268–273 On *complicatio* and *explicatio*, see above 00–00.

277–281 Cf. Peter Abelard, "De quantitate," *Praedicamenta: De praedicamentis Aristotelis* 2 (in *Dialectica*, ed. Lambertus M. De Rijk [Assen, 1970], pp. 59–60): "Afferunt quoque adversus hanc constitutionem linee quod de punctis est, quod in *Arithmetica* Boethius ponit, cum scilicet ait: *Si punctum puncto superponas, nichil efficies, tamquam si nichilum nichilo iungas.* Cuius quidem obiectionis, etsi multas ab arithmeticis solutiones audierim, nullam tamen a me proferendam iudico, quem eius artis ignarum omnino recognosco. Talem autem, memini, rationem Magistri nostri sententia pretendebat, ut ex punctis lineam constare convinceret. 'Cum, inquit, linea ubique possit incidi atque separatis partibus in capite uniuscuiusque puncta appareant, ut dicit Boethius, que prius erant coniuncta, oportet per totam lineam puncta esse; quodsi puncta per totam lineam sint disposita, vel ita quidem sunt de essentia linee vel non; at vero si de essentia linee non sint, non magis partes linee continuare possunt quam albedo superposita vel pars aliqua loci individua.' ... Si puncta interposita, cum de essentia linee non essent, visa sunt Magistro nostro continuationem partium linee dissolvere, cum tamen et ipsa inter quantitates recipiuntur, multo magis ea continuationem destruunt quae omnino sunt a natura quantitatis aliena." On this topic, see also the transcriptions by Yukio Iwakuma of commentaries on Aristotle's *Categories* attributed to William of Champeaux, cited in Irene Caiazzo, *Lectures médiévales de Macrobe* (Paris, 2002), pp. 187–188.

puncto. Quod stare non potest secundum geometriam. Si enim punctum puncto apponas, lineam efficis, sed si ut oculo videre possis punctum stilo tuo feceris, et iterum idem foramen punxeris, nihil creasti nisi punctum. In his itaque verbis Boetius non est obpositus Macrobio, qui dicit punctum puncto additum facere lineam, nec pro istis verbis Boetii oportet nos dicere

289 punxeris] puncxeris *post corr.* MS

289–292 Macrobius, *Commentarii in Somnium Scipionis* 1.5.11. See the explanation of *Glosae Colonienses super Macrobium* in Caiazzo, *Lectures médiévales de Macrobe,* pp. 185–187: "(1.5.11) DUOBUS PUNCTIS. Quidam dicunt puncta intrinsecus tantum terminare, quia illis etiam constituantur lineae. Quidam vero extrinsecus tantum terminare, nec ullam partem esse linearum, exemplo Boetii qui ait: *Si punctum puncto addis, non magis efficies quam si nichilo nichilum addas.* Sed qui dicunt lineam constare de punctis, secundum Macrobium, probant hoc modo. Linea longa duobus tantum punctis terminata <est>: si in plures dividitur lineas, quaelibet ex illis, sine ulla adiectione, duobus punctis terminatur tantum, unde liqueret punctum esse partem lineae constitutivam secundum quosdam; ideoque sic solvitur illud Boetii: *Si punctum puncto addas, non magis efficies quam si nichilum nichilo addas,* id est nichil quod subiaceat visui vel sensui. Quia vero multi, auctore Boetio, dicunt terminum non idem esse cum eo cuius terminus est, sciendum est non perfectam esse similitudinem unitatis et puncti, cum unitas sit pars constitutiva suae lineae, punctum vero non, numquam nisi terminus suae lineae possit esse. Numeralis vero linea quantacumque sit, ex unitatibus constabit. Puncta vero quantacumque conglutines, nil nisi puncta efficies, que nullum integrum componere valent, quia intervallo et omni quantitate carent. Sicut vero superficies lineis constare <non> potest, nam linea est tantum terminus superficiei, sic quoque linea punctis constare non potest, que tantum eius sunt termini. Tollo ergo superficies, lineas et puncta esse terminos, nichil de illis superesse invenies; sed qui hec asserunt terminos, tantum extrinsecus terminantes innuit. Nam duo sunt termini, scilicet intrinsecus et extrinsecus terminans, ut supra dictum est."

William of Conches engaged in a similar discussion in the *versio longior* of his *Glosae super Macrobium.* This might indicate this type of problematic was recurrent among the authors of the first half of the twelfth century. See Munich, Bayerische Staatsbibliothek, MS lat. 14557, fols. 125r–126r: "(*Comm.* 1.6.35) TRIBUS GRADIBUS INCREMENTORUM. Non quia videlicet actualiter crescant vel decrescant sed intellectu. Est enim primum incrementum intelligibile scilicet quod venit ab indivisibili ad divisibile, utpote linea a puncto. Secundum est quod ab divisibili ad magis divisibile venit, ut superficies a linea. Tertium est quod adhuc maiori subiacet divisibilitati, utpote soliditas post superficiem. AUT ENIM. Vere sunt ibi tres gradus incrementi quia isti excrescunt. EX PUNCTO. Non quia ex eo, scilicet constat, sed ab eo incipit esse. Sunt tamen quidam qui lineam ex punctis constare volunt et hoc tali ratione: si enim omnis linea ex minoribus constaret, tunc divisio usque ad infinitum procederet. Certum enim est quod linea que terminat hanc superficiem vel illam in duo media dividi potest, et rursus utrumque illarum in alia duo, et sic inantea; unde nisi dividi possit usque adeo <in> simplicia quibus non sunt simpliciora in quibus dividi habeant. Ratio dividendi in infinitum procedet quod inconveniens ut evitent aiunt quod quedam linea ex duobus constat punctis et ex quo

quod linea constet ex lineis et non ex punctis. VELUT SI NIHIL N<ULLI>, id est Lucretius:

"De nihilo nihil, in nihilum nil posse reverti."

Quemadmodum nihilum nichilo comparatum nec maius nec minus est, ita quoque nec punctum puncto comparatum. EADEM QUIPPE. Quod si aliquos pares numeros per se consideraveris, non multiplicans eos neque per se neque per alium, illi sic considerati nihil creabunt, quia VI, VI, VI iuncti non in aggregatione, sed propria consideratione aequalitatis per se, nullam proportionem, nullam dimensionem facient.

IIII, 6. EXPLICABIT. Extenditur per cunctos numeros. Fit enim quandoque tripunctalis, quandoque centum in ea puncta continuantur et ampliora. QUONIAM PRIMUM INTERVALLUM. Istud iam subdit, quia diceret aliquis: quare dicis longitudinem primam omnium dimensionum? SUPRA QUAE. Has tres dimensiones et his tribus componuntur sex species motus. De illa specie motus quae fit in circulo tacet.

301 Explicabit] explicat *ed. Guillaumin*

---

in ea divisa est, ultra non potest divisio procedere, eo quod simplicius puncto nichil est, et est quasi sine omni quantitate. Nobis vero non placet hec sententia quod linea ex punctis constet. Ut enim ait Boetius quod intervallum intervallum creare non potest, et quod *si punctum puncto addas, nichil plus efficies quam <si> nichilo nichilum addis.* Qui tamen priorem tenent sententiam, ita volunt exponere, quod quemadmodum, scilicet si quis dicat nichil est nichil, hoc nichil similiter puncties nichil est. Alii dicunt quod quemadmodum ad hoc nichil addatur nichilo nichil inde potest effici, ita si quis facto puncto in cera vel aliquo tali, iterum eodem stilo et eiusdem grossitudinis ibidem infigat, nichil efficit, ibi quod iam non esset. Iste quidem expositiones parum vel nichil valent, unde volumus quod omnis linea minores partes habeat et quod nulla ex punctis constet. Puncta vero potest habere ut terminos, non ut partes. Est autem punctum vel principium vel finis ipsius, unde etiam pars eiusdem esse non potest. Si enim aliqua pars ibi esset nondum terminaretur. Quod autem inducunt pro convenienti non solum non fugimus, sed etiam querimus ut divisio, scilicet lineae, procedat in infinitum. Est autem linea magnitudo quedam, ut ait Boetius, *omnis magnitudo decrescit in infinitum veluti et multitudo in infinitum crescit.* Cum ergo in divisione ipsius lineae certum terminum velint ponere, aperte sunt contra Boetium. Unde dicimus quod nulla linea in adeo simplices lineas dividi potest, quin ille adhuc in simpliciores intellectu dividi possint. Eodem modo lineae superficies termini sunt, non vero superficies ex eis constat, immo ex aliis minoribus et per alias minores usque <in> infinitum dividi potest. Similiter dicatur de soliditate, quod superficies scilicet termini ipsius sunt, non partes, immo ex minoribus constat soliditatibus et in minores soliditates usque in infinitum divisionem suscipit. Cum enim superficies omni caret soliditate, qualiter constituet soliditatem hoc non video."

293–294 Cf. Persius, *Saturae* 3.84; Persius here is summarizing Lucretius.

IIII, 7. EST AUTEM. Ostendit ad quid valeant istae dimensiones, et quare earum mentionem fecerit, ideo scilicet, quia omne solidum corpus ex his tribus constat. HAEC ENIM TRIA. Hinc est quod Macrobius dicit: "Quod cogitationi a nobis ad superos prima perfectio incorporalitatis in numeris invenitur." Soliditas namque perfectionem corporalitatis in se continet, quae nulli quidem nisi soli corpori adiacet. Si vero a soliditate ad superficiem veneris, superficies quidem minus habet corporalitatis, sed aliquid corporalitatis habet, quia duas dimensiones. Cum vero ad longitudinem perveneris, et haec quidem minus habet corporalitatis quamvis aliquid, quia unum intervallum. Si vero a linea ad bipunctalem vel tripunctalem animum deduxeris, iam habes numerum in quo nihil est corporalitatis, quia nec numeri est ulla longitudo, nec ullum intervallum, licet numerus et linearis et superficialis et cubicus. Numerus enim dicitur linearis ex eo quod in longum sic numeramus geometrice. Linea haec bicubita est, vel tricubita, et deinceps. Inde vero superficialis dicitur, quia secundum ipsum geometrice iuxta pedes constratos numeramus, vel iuxta palmos vel cubitos vel alias geometricas mensuras, ut cum dicimus "haec superficies est bipedalis vel tripedalis," quod secundum pedes continuatos dicitur. Eodemque modo secundum geometriam iuxta pedes in altum directos numeramus, ut haec soliditas est bipedalis, tripedalis, unde numerus hic vel ille dicitur posse fieri solidus vel cubicus, sed non secundum numerum neque enim hoc nomen

309–311 Macrobius, *Commentarii in Somnium Scipionis* 1.5.4: "Haec est igitur communis numerorum omnium plenitudo, quod cogitationi a nobis ad superos meanti occurrit prima perfectio incorporalitatis in numeris." See the explanation of *Glosae Colonienses super Macrobium* in Caiazzo, *Lectures médiévales de Macrobe*, p. 183: "PLENITUDO ENIM. Macrobius ostensurus plenitudinem numerorum, hunc ordinem probationis oculte innuit: numeri sunt pure incorporei, id est inveniuntur in divinis et supernis rebus; quod autem invenitur in divinis et supernis rebus, illud est plenum, quia non effluit neque influit; et cum deberet hoc dicere - quia non influit neque effluit - dicit, quasi per contrarium, quod omnia corporea et caduca effluant et influant, ubi relinquitur intelligendum quod omnia pure incorporea, id est superna et divina, non effluant neque influant, ergo sunt plena. ... Continuatio. Quandoquidem omnia divina et superna sunt perfecta, igitur numerus est perfectus. COGITATIONI MEANTI A NOBIS id est a corporibus, AD SUPEROS id est accidentia quae dicunt supera et divina, quae sunt pure incorporea. Perfecte incorporei reperiuntur numeri, quasi diceret: quando incipimus cogitare de corpore nostro, istud est corporea natura et istud linquendo progredimur ita usque dum venimus <ad> superficiem terminantem corpus; hec quamvis sit incorporea, non tamen perfecte, quia tantum est in corpore. De superficie progredimur cogitando ad lineas terminantes superficiem, quae nec habent puram perfectionem, quia numquam reperiuntur nisi in corporibus (animae enim nec alicui incorporeae rei accidit linea). Deinde progredimur ad numeros terminantes lineam, qui sunt pure incorporei, quia, sine ullo subiecto, esse possunt tamen in nullo corpore. Nequaquam agitur de substanciis sed de accidentibus, quorum purissimi incorporei sunt numeri."

"cubicum" datum est numero, sed secundum numerum angulorum, ut in sequentibus declarabitur. Unde etiam ternarium primum superficialem dicit Boetius, secundum numerum angulorum, cum alibi dicat quod binarius est superficialis, sed secundum multiplicationem, ut cum dicitur "bis unum." Itaque numerus idem et superficialis esse poterit et linearis et fortasse solidus, quod tamen fieri posse secundum numerum angulorum minime putamus.

INSEPARABILI CONIUNCTIONE: quia nullum corpus est quod non sit longum et latum et altum. ET IN NATURA CORPORUM. Hoc Macrobio docente plane videbis ubi dicit: "Cum totum denique corpus nominas, etiam superficies hoc nomine continetur; de corporibus tamen eam etsi non res sed intellectus sequestrat." Quod ideo dicit, quia licet heae dimensiones non sint corpora, et licet extra corpus sit earum natura, tamen non nisi circa corpus sunt, et in corpore earum natura apparet et perpenditur. Linea et superficies et soliditas non sunt corpus, et ideo non videntur, quia nihil incorporeum videtur. Unde cum dico quia superficiem video, figurativa est locutio, sic exponenda: id est video corpus in superficie.

IIII, 8. QUARE. Infert a toto: quia omne corpus tribus constat dimensionibus, igitur quod una caret, solidum corpus non est. ET HIC: in longitudine et latitudine. Ubi enim hae duae sunt, ibi et superficies.

IIII, 9. PUNCTUM IGITUR ALIO: quam linea vincatur a superficie vel soliditate, quia punctum ab ipsa longitudine vincitur quod est linea.

CAPUT V

V, 1. HIS EXEMPLIS, in quibus hos numeros definit, et figuras subiungit. Binarius est initium linearis numeri, ternarius vero triangulorum; triangulus autem initium omnium figurarum.

336–339 Macrobius, *Commentarii in Somnium Scipionis* 1.5.6: "Cum totum denique corpus nominas, etiam superficies hoc vocabulo continetur; de corporibus eam tamen etsi non res sed intellectus sequestrat." See the explanation of *Glosae Colonienses super Macrobium* in Caiazzo, *Lectures médiévales de Macrobe*, p. 185: "(1.5.6) CUM TOTUM CORPUS. Totum id est omne. Ostendit superficiem affinem esse corpori dicens: CUM TOTUM CORPUS NOMINAS. Sed hoc non videtur verum nisi hoc exponatur littera. Sed CUM TOTUM CORPUS NOMINAS id est omnia significata huius vocis quae est 'corpus'; tunc continetur superficies sub hoc vocabulo secundum mathematicos, qui et lineam et punctum corpora vocaverunt. TOTUM: integrale, omne, universale."

Caput VI

VI, 1. Addita descripta latitudine. Habita enim qualibet, si ex illa aliam facere volueris, dilata angulos secundum ordinem numeri naturalem, et sic quaslibet invenies unam ex alia, scilicet proximam de proxima. Verbi gratia: si hanc triangulam △ velis extendere in quadrangulam, dilata angulos secundum ordinem naturalem. Addita descriptionis latitudine, id est huic descriptae latitudini, quae triangula est, adde quartum angulum, et ipsi extendentur hoc modo: □. Et haec extensio fit secundum naturales numeros se sequentes, quia tribus angulis additur IIIIus. Eadem est in omnibus consideratio. Hi vero. Dat causam quare dixerit a tribus inchoandum esse; ideo scilicet, quia tribus lineis prima superficies continetur. Diceret aliquis: quare non duabus? Ad hoc respondet: quia duae lineae rectae non faciunt superficiem, duae curvae bene faciunt circularem. Sicut nullum est spacium inter unum et binarium, vel hunc et ternarium, sic et in infinitum.

VI, 2. Et omnis. Una est dignitas triangulationis, quod prima superficies triangula est. Ostendit nunc aliud eius accidens singulare, hoc scilicet quod posito puncto in medio cuiuslibet figurae, et ab hoc puncto ductis lineis in quemlibet angulum, dividitur figura in triangulos. Et in quot? In tot certe triangulos quot angulos h<ab>uerit ipsa figura. In modo et mensura, id est secundum hoc quod figura magna fuerit, et plures angulos habens.

VI, 3. At vero. Istud ideo addit, quia videretur ipsam triangulam non dividi. Ita: sicut alias per triangulas.

VI, 4. Neque ab alia latitudine: sicut omnes duxerunt ab ista originem. In sese: quia si alia forma superficiei alicuius esset istius triangulae principium, in illam solveretur.

VI, 5. Idem autem et in numeris: ut triangulus principium sit omnium angularium numerorum, et quaelibet figurae in triangulas resolvantur. In numeris, id est in figuris secundum numeros dispositis. Vel fac ad aliud esse relationem, id est quod sicut figurae resolvantur in id unde principium, ita etiam numeri resolvuntur in id unde ceperunt, scilicet in unitatem.

Caput VII

VII, 1. Secundum superficiei dispositionem: secundum hoc quod disposita est figura secundum triangulationem. Quicumque aequalitatem: id est qui, cum habeant tria latera, habent ea aequalia. De aequalibus

355 Addita descripta latitudine] addita descriptionis latitudine *ed. Guillaumin*
376 duxerunt] *corr. ex* dixerunt MS **385** dispositionem] positionem *ed. Guillaumin*

enim figuris prius aget, et postea de inaequalibus, quia illae sunt istarum principium; ordinatim a tribus semper adduntur sibi numeri, ut ad VI IIII fiunt X, ad X V fiunt XV, et ita in infinitum. Et quot adduntur, tot semper per singula latera cernuntur. Secundum positionem enim triangulorum computabuntur numeri per latera, utpote ternarius, qui est primus actu et opere, secundus autem loci positione, duos habet in latere. Senarius tercius triangulus tres habebit in latere. Denarius, qui est quartus, quatuor habebit in latere.

Caput VIII

VIII, 1. Necesse est: conveniens est, quia probabile est hoc argumentum et non necessarium.

VIII, 2. Crescente unitate, id est addita, quia secundum ordinem naturalis numeri augentur latera figurarum. Crescente naturali, id est addito uno ad ternarium et dualitatem, quia ternarii trianguli dualitas latus est. Vel ideo crescente naturali numero dicit, quia sicut dualitas in illo latus est, ita in isto ternarius.

Caput VIIII

VIIII, 1. Si vero huic: ternario qui ex uno et duobus confectus est, tercium, id est ternarium. Quos per latera disponens. Nota quod in illa dispositione figurarum quidam numeri describuntur in medio, qui cui lateri asscribendi sint ignoro, quia si dividuntur per latera, maior erit quantitas laterum quam ipse dicat. Dicit enim iam quod quot unitatum sit secundus aggregatus, tot unitates habeat quodlibet latus. Porro si contra hanc litteram, has medias unitates lateribus dividendo auxerimus numerum lateralium unitatum, tunc incurrimus inaequalitatem laterum, quod primo videbis in figura cui asscribuntur XXVIII. Ibi enim mediae inveniuntur X unitates, quarum VIIII tribus aequaliter lateribus si parciaris, remanebit unitas; quae cuicumque lateri additur, inaequalitatem faciet, nisi velis vi et eius potestate omnibus eam communiter lateribus attribuere. Quaerere igitur et investigare moneo, an illae mediae unitates ideo ponantur, ut expleant asscriptum numerum, an ideo ut per latera distribuantur. Ultimus: aggregatus, qui non

388 illae] ille MS

---

394 The argument in this chapter is best grasped using the schema found in the *De arithmetica*.

semper erit binarius, sed quandoque alter aliquis numerus secundum ordinem naturalem.

Caput XII

XII, 1. Congregentur. In superioribus triangulis unus duobus iungitur, et fit ternarius triangulus. His iunge ternarium naturalem numerum qui sequitur, et fit senarius triangulus. His, id est VI, iunge quaternarium qui sequitur, et fit denarius triangulus. Si cum superiore. Hoc dicit propter unitatem quae prima est, et ideo ei soli aggregatio fit.

XII, 2. Medium vocat quaternarium, quia relicto binario, ex aggregatione unitatis et ternarii quasi de medio surrexit quaternarius. At vero si his iungam, sic procedendo: unus et III et V et VII faciunt XVI.

XII, 3. Ad coniunctionem propriam: ut ad monadem ternarius additus duo sunt termini. Ergo in secundo quadrato, qui constat ex his duobus terminis, duo sunt in latere.

Caput XIII

XIII, 1. Secundum unitatem, id est ut unitas sit in quolibet angulo.

XIII, 2. Eodem modo: sicut tetragonorum, quia primus pentagonus habebit unum in latere quolibet, secundus II, tercius III. Idem unus, id est unitas eadem, hoc est: unitas erit in quolibet latere.

Caput XIIII

XIIII, 1. Superiori. Hoc dicit propter unitatem, quae prima est potestate pentagona.

XIIII, 2. Namque unitati. Quaternarius vincit unitatem in ternario. Septenarius vincit quaternarium in ternario. Denarius vincit septenarium in ternario.

Caput XV

XV, 1. Secundum modum: quem iam dicturus sum, scilicet ut exagoni tribus intermissis quaternario se vincant. Et ad hoc ostendendum, praemittit et repetit generationes omnium supradictarum formarum. Et erunt radices exagonorum omnes videlicet qui tribus intermissis quaternario se transeant.

CAPUT XVI

XVI, 3. NAM IN TRIANGULO. Nota quod "praecedunt se" et "vincunt se" idem est, necnon "transeunt" et "medii sunt" idem est. Et nota quod alii sunt praecedentes, et alii sunt transeuntes, quia in triangulo ubi uno se praecedunt aggregati, ibi nullus est transiens, sed in tetragono ubi duo sunt praecedentes, unus est medius sive transiens, et in pentagono III praecedunt, duo transeunt, et sic in caeteris. NULLUS MODUS EST. Quia dum crescunt multiangulae formae secundum naturalem ordinem, maiores etiam fiunt numeri se vincentes secundum naturalem ordinem.

CAPUT XVIII

XVIII, 1. QUID IN HAC RE: quam utilitatem consequamur ex ista descriptione.

XVIII, 2. ALTRINSECUS: in praecedenti semper linea triangulorum constituta. NULLA CUNCTATIO: quin omnes pentagoni nascantur a supra se posito tetragono et a triangulo in praecedenti linea constituto.

XVIII, 4. EMINUS: remote, quantum ad altitudinem.

CAPUT XVIIII

XVIIII, 1. SI AD LATITUDINEM, id est in latitudine. Prior enim istorum speculatio fuit in longitudine, modo vero in latitudine. PRO INVENIENDIS DIFFERENTIIS: ad inveniendum in quanto differant.

CAPUT XX

XX, 1. HINC VERO. Ostendit causam quare de superficialibus numeris et figuris egerit, ideo scilicet ut facilior esset doctrina de solidis. QUID OPERETUR: sicut ex triangulis fiunt tetragoni.

Caput XXI

XXI, 2. Ratarum figurarum: quia non omnes figurae quocumque modo dispositae fiunt a numeris, sed certae et determinatae, sicut triangula in planis et in solidis pyramis.

XXI, 3. Pyramis: "pyr" graece, "ignis" latine; inde pyramis figura quae acuitur in altum, ut ignis. A triangula basi: "baseio" graece "fundo," hinc basis dicitur fundamentum. Unde et rex graece dicitur "basileos," quasi "basis laos," hoc est fundamentum populi; "laos" enim graece "populus."

Caput XXII

XXII, 1. Super quam ipsa figura, scilicet pyramis. Noli tamen credere quod basis non sit pars pyramidis.

Caput XXIII

XXIII, 2. Nam quoniam. Epilogat quod dixit, ut pateat qualiter, ex illis praemissis, sequentia proveniant et sciantur. Et est suspensiva constructio illuc usque "nunc vero ad solidorum." Cum distantia. Distat enim tetragonus a triangulo, quod in triangulo sequuntur se numeri naturali dispositione, et se tantum unitate transeunt; in tetragono vero uno intermisso copulantur, seque binario superant, similisque in aliis distantia reperitur.

XXIII, 3. Naturaliter: secundum naturam numerorum triangulorum, tetragonorum.

XXIII, 4. Quaternaria profunditas, id est pyramis habens in sua profunditate IIII unitates, tres in basi et unam in summitate. Denaria. Pyramis continens in sua altitudine decem unitates, VI in basi, III in medio, unam in summitate. His: uni et tribus et VI, quae decem sunt, surget pyramis XX habens in crasso, X in basi, VI deinde, et tres postea, in vertice unitatem.

XXIII, 5. Rata detractio: quia sine ordine fieri non debet, sicut iam dicet. Offendit. Offendere proprie est aliquid insperate invenire. Similis quoque: quod ultimus coniugatus, id est maximus, vicem basis obtineat.

XXIII, 6. Superponam. Quantum ad ipsam figuram pyramidis, unitas superponitur omnibus, sed quantum ad generationem ipsarum et forma-

475 certae] certe MS

477–478 See below note to 855–858.

480 Cf. Papias Vocabulista, *Elementarium doctrinae erudimentum* (Milan, 1476): "Laicus: popularis. Laos λαοσ enim populus graece."

tionem pertinet, unitati superponuntur omnes alii numeri. Verbi gratia: quando fit aggregatio sic: I et IIII et VIIII faciunt pyramidem; ecce quomodo unitati superponuntur reliqui.

Duobus tantum per latera: quia in basi quodlibet latus habebit duas unitates, et una unitas verticem tenebit. Tribus unitatibus concluditur. Verum quidem est, tres numeros esse aggregatos, videlicet I, IIII et VIIII, sed istos tres contineri in quolibet latere non dicit, immo quod iam dixit de superiori pyramide, quae fuit V unitatum, scilicet quod cum ultimus aggregatus videlicet VIIII sit, basis ipsa, quia tetragona est, aequaliter habet in lateribus suis ternarium. Quantae quantitates, id est quot numeri fuerint aggregati, tot unitates asscribantur basi inferius in quolibet latere.

XXIII, 8. Et ad eundem modum: ut sicut triangulae basi omnes triangulos, tetragonae basi tetragonos superaddidisti, sic etiam pentagonae pentagonos et exagonae exagonos superaedifica, ita quod tota aggregatio eiusdem et sui sit generis. Et hoc iam dicit ibi: omnis enim ex figura sui generis: sive illa figura sit triangula vel quaelibet aliarum, quia qualis fuerit basis, talis erit tota sequens aggregatio. Ne te moveat quod dicit: superposita unitati, cum ipsa unitas in summo pyramidis erectae locetur. Hoc enim dictum est secundum aggregationem numerorum, sicut in priori clausula ostendimus, ubi dixit: at si huic tetragono.

XXIII, 9. Et ex hoc equidem. Quod modo dicam, licet idem etiam superius dixerit, repetit pro commendatione triangulae formae. Continetur. Tot enim trianguli per latera cuiuslibet figurae constituuntur, quot angulos ipsa figura perhibetur habere. Quod liquido perpenditur, si a qualibet basi per singula latera lineae usque ad verticem ductae erigantur.

Caput XXIIII

XXIIII, 1. Scire autem. Ostendit formationes quarumlibet pyramidum, sive triangularum sive caeterarum. Sed quia in quolibet genere istae omnes quandoque sunt perfectae quandoque non, idcirco ostendit earum perfectionem et imperfectionem.

XXIIII, 2. Perfecta est, cuius latera sive superficies concurrunt et terminantur in unitate suprema. Recteque vocatur curta, quia unitatem non attingit.

507–508 concluditur] concludatur *ed. Guillaumin* **533** est] enim *ed. Guillaumin*

521–522 Cf. Papias Vocabulista, *Elementarium doctrinae erudimentum* (Milan, 1476): "Clausula est in oratione brevis sententia."

XXIIII, 3. ET HABEBIT. Hoc generaliter dicit de qualibet pyramide quotienscumque curta. SED SUPERFICIEM, id est dispositam secundum quemlibet numerum.

XXIIII, 5. ΚΟΛΟΥΡΟΝ dicitur quasi "colon uron," id est membrum caudae. Curta enim res dicitur, cui aliquid de cauda aufertur. "Colon" enim "membrum" dicitur, "ΟΥΡΑ" "cauda" dicitur graece; "uron" est "bos silvestris," et habent uri curtissimas caudas. Unde Graeci pyramidam dicunt coluron quasi curtam, dicoluron biscurtam, tricoluron tercurtam. Illud quoque notandum quod pyramis est prima species solidarum figurarum; cubus quoque alia species est earum, alia cuneus, alia asseres, alia laterculi. Sed illud hic est considerandum: quod cubus dicitur propter aequalitatem omnium dimensionum, nec corpus cubicum dicitur, nisi omnes aequales dimensiones habeat. Et nota quod cubus a Graecis alio nomine dicitur "thessara." Duo autem genera sunt cubi, quia aliud angulosum est, aliud spericum. Spera ergo species est cubi. Prius autem tractat de cubo anguloso, ad ultimum de sperico, quia sicut pyramis est principium omnium solidarum figurarum, ita et spera est consummatio omnium solidarum. Unde et circulo numerus et figura inscribitur. Inde fit ut mundus, cum sit spericus, omnium omnes rerum formas complectatur. Et haec est figurae spericae virtus. Illud quoque sciendum quod maximam virtutem habeat pyramis in complexionibus rerum. Nulla enim forma complexionum sine pyramide, id est sine virtute pyramidis, potest esse. Unde quia ignis est pyramoides – tendit enim in acumen – et mundus est indissolubilis. Et puta quod pyramis a triangula basi dominatur igni. Terra vero cubica est, et dominatur illi figura quadrata, sive mavis quadratum, quia virtus quadrati in terra potest inveniri. Itaque in quibus melancolia dominatur, in eis dominatur quadrati virtus. In quibus ergo ignis, et in eis pyramica figura dominatur. Sic et in aliis et aliis complexionibus rerum alia et alia figura dominatur. Inde est quod in rithmachia, quae pugna numeri dicitur, ubi summa est victoria

539–543 Cf. Caiazzo, "Un commento altomedievale," 145, ll. 534–537: "ΚΟΛΟΥΡΟΝ dicitur quasi ΚΟΛΟΝ ΟΥΡΟΝ, id est membrum caudae. CURTA enim res dicitur, cui aliquid de cauda aufertur; ΚΟΛΟΝ enim 'membrum,' ΥΡΑ 'cauda' vocatur graece. ΔΥΚΟΛΟΥΡΟΝ: bis curta. ΤΡΙΚΟΛΟΥΡΟΝ: ter curta."

557–558 See below note to 855–858.

559–560 See below note to 855–858.

557–563 Cf. Irene Caiazzo, "La forme et les qualités des éléments: lectures médiévales du *Timée*," in *Il* Timeo*: Esegesi greche, arabe, latine*, ed. Francesco Celia and Angela Ulacco (Pisa, 2012), pp. 307–345.

563–567 See Menso Folkerts, "Die *Rithmachia* des Werinher von Tegernsee," in *'Vestigia mathematica': Studies in Medieval and Early Modern Mathematics in Honour of H.L.L. Busard*, ed. M. Folkerts and J.P. Hogendijk (Amsterdam, 1993), p. 122, where a passage from the *Rithmachia* of Werinher of Tegernsee is cited. See also the *Rhyth-*

maximam armoniam in campis alterius constituere, ex utraque parte ponitur pyramis, quae vim habet maximam ad constituendam maximam armoniam in rithmachia. Itaque virtus numerorum declaratur, ut inde ad rerum complexiones transferatur, et quam vim in rerum complexionibus habeant numeri cognoscamus. Numeri namque in rerum complexionibus operantur, sed diverso modo, quia et cardinaliter et figuraliter et proportionaliter, id est secundum quam proportionem efficiunt. Cum autem X sint proportionalitates, quarum una dicitur arithmetica, alia geometrica, alia armonica, et caeterae, sciendum est quod sic vel sic ab aliqua numerorum virtute nominantur. Ita quoque, quod proportionalitates dicitur disiuncta sive coniuncta, hoc est aliqua numerorum efficit virtus, licet nobis sit ignota. In omnibus itaque rebus virtutes figurarum operantur, ideo quoque numerorum virtutes a quibus figurae descendunt.

Caput XXV

XXV, 2. Et, ut redeamus ad propositum, CUNEUS est contrarius CUBO, quoniam sicut cubus ex omnibus lateribus constat aequalibus, sic et cuneus ex omnibus inaequalibus. Et est nomen latinum per similitudinem sumptum, quia in cuneis nulla est dimensionis aequalitas. Omnes autem aliae figurae, quae praenominatae sunt, inter cubum et cuneum sunt mediae, quia nec omnia latera habent aequalia, nec omnia inaequalia. Omnium autem mediarum commune nomen est PARALLELIPIPEDA, quia epipedon interpretatur superficies, parallelum aequistantia. Inde ergo nominantur sic, quia ex superficiebus constant quarum nulla versus aliam inclinatur, et aequaliter a se distant. Quod non fit in pyramidibus, in quibus una superficies ad aliam inclinatur. Haec autem nomina ab aliis rebus ad numeros mutuata sunt. Nam cubus est saxum, vel quodcumque habet aequalem longitudinem, latitudinem et altitudinem. Unde et cubi dicuntur quod aequaliter cubent in quocumque latere iaciantur. Hoc autem nomen mutuatur in his numeris qui

575 est] hoc MS **586** parallelum] paralellum MS

---

momachy edited by Charles Burnett, "The Instruments which are the Proper Delights of the *Quadrivium*: Rhythmomachy and Chess in the Teaching of Arithmetic in Twelfth-Century England," *Viator* 28 (1997), 175–201, at 195.

567–569 Cf. *Commentary on Boethius's De arithmetica*, Bern, Burgerbibliothek, Cod. 633, fol. 26ra: "Numerus est causa omnium naturalium complexionum. Est enim proportionis causa, proportio vero complexionis, quare et ipse complexionis causa est. Complexiones vero illae sunt quare et numerus est."

habent aequalem longitudinem, latitudinem et altitudinem, sicut bis bini bis, ter terni ter.

ASSERES vero subtegulicia, quae habent quidem longitudinem, latitudinem et altitudinem, sed inaequalem. Namque in his longitudo maior, altitudo brevissima est. Unde et illi numeri, qui ad illorum similitudinem longitudinem, latitudinem et altitudinem inaequalem habent, asseres vocantur, ut quater terni bis.

QUAE SUNT QUOTIENS. Istud quae faciunt quidam spectare ad proximum, scilicet ad parallelipipedas, quidam referunt ad primum, scilicet ad cubum, cuius SUPERFICIES CONTRA SE SUNT: superior contra inferiorem, dextra contra sinistram, anterior contra posteriorem. ET DUCTAE: quia quantumcumque magnas et longas feceris, et quotcumque cuborum superficies feceris, numquam concurrent, id est numquam in unitate convenient, sicut pyramidum superficies.

XXV, 3. DISPOSITIS. Ostendit cuborum formationem. SI PER LATERA, id est per quodcumque unum latus, sive longitudinis sive latitudinis. Pluralem posuit non pro singulari sed pro electione, ut quodcumque illorum laterum velis, accipias. HOS IGITUR DUOS, id est quaternarium qui ex duobus multiplicatis crevit. UNAM LATERIS MULTIPLICATIONEM, id est unius lateris.

XXV, 4. QUARE: id est propter aequalitatem laterum cubicorum et identitatem, etiam unitas erit cubus vi.

XXV, 5. LATERA VERO XII, id est lineas quatuor in basi, quatuor in superiori superficie, quatuor erectas a quatuor inferioribus ad quatuor superiores angulos.

XXV, 4. ET QUONIAM XVI TETRAGONUS. Et ut superius monstratum est, tot unitatum est, quia quot unitates tetragonus habet in latere, tot unitates et cubus, qui ex eo nascitur, per singula habet latera, quorum singulus angulus sub tribus eiusmodi continetur, scilicet tetragonis, quia tres tetragoni continent unum angulum. Tot unitatum est cubus in quolibet suo latere, quot unitatum est basis in quolibet suo latere.

XXV, 5. SUB TRIBUS: superficiebus. Quilibet enim angulus tres superficies in se claudit.

XXV, 8. SCALENOS graece "ascensio," quia "scale" graece "ascendere" dicitur. SPHENISCON graece "cuneus." Supra Ionicum mare Alexander

601 Parallelipipedas] paralellipipedas MS **610** igitur] ergo *ed. Guillaumin*

626–631 Cf. Caiazzo, "Un commento altomedievale," 145, ll. 543–547: "SCALENOS: graece 'ascensio,' quia 'scale' graece 'ascendere' dicitur. SPHENISCON: graece 'cuneus.'

Magnus quandam condidit civitatem, quam ex suo nomine castra Alexandri vocavit; iuxta quam erant quaedam arulae in basi latiores, superius vero acutiores, ad quarum similitudinem BOMISCI vocantur similiter numeri multiplicati. PARALLELIPIPEDAE vocantur numeri, quorum alternatim ponitur latitudo, ut bis terni bis. LATERCULI autem fiunt quotiens aequalibus spaciis in longitudinem et latitudinem porrectis, minor his additur altitudo, ut tres ter bis, quater quatuor ter. ASSERES vero ex aequalibus aequaliter in maius, ut quatuor quater novies. SPHENISCI vel cuneoli, ex inaequalibus inaequaliter ducti, creverunt per inaequalia. CUBI autem ex aequalibus aequaliter per aequalia creverunt.

XXV, 9. IGITUR CUBI. Quia cubus omnia latera habet aequalia, et sibi contrarius gradatus omnia inaequalia; igitur inter hos duos numeros medietates sunt illae formae. QUAE ALTERNATIM, quae sic habet se, ut quandoque longitudo et latitudo conveniant, et differat altitudo, quandoque longitudo et altitudo conveniant, et differat latitudo, quandoque latitudo et altitudo conveniant, differat longitudo.

CAPUT XXVI

XXVI, 1. HUIUSMODI VERO. Postquam egit de pyramidibus et de cubo, et eius contrario gradato, necnon de illorum duum medio, quia fecerat mentionem inaequalitatis laterum gradati, et eius medii, qui utrique sunt solidi, ostendit nunc de quibusdam planis numeris et figuris, quorum latera sunt inaequalia. Ita continua: vidimus qui sunt parallelipipedi, et non solum de his videndum est – VERO pro "sed" – sed etiam videre debemus huiusmodi formas. Ante autem quam tractet de mediis figuris, quas parallelipipedas nominant Graeci, interponit de parte altera longioribus et antelongioribus, et hoc ideo, quoniam hae sunt superficies mediarum figurarum. Quoniam ergo superius non egerat nisi de aequilateris superficiebus, ut de aequilatero triangulo et huiusmodi, merito ante quam tractet de figuris mediis, tractat de parte altera longioribus, id est de inaequilateris superficiebus, ex quibus constant mediae figurae. Illud vero notandum quod parte altera longior

631 Parallelipipedae] paralellipipedae MS **649** parallelipipedi] paralellipipedi MS
651 parallelipipedas] paralellipipedas MS

Supra IONICUM: mare. Alexander magnus quandam condidit civitatem, quam ex suo nomine castra Alexandri vocatur, iuxta quam erant quaedam ARULAE in basi latiores, superius vero acutiores, ad quarum similitudinem BOMISCI vocantur, similiter numeri multiplicati."

630–637 See below *ad Arithmeticam* II, XXVIIII.

numerus dicitur, cum unum latus aliud superat unitate sola, ut duo tribus, tres quatuor. Antelongior autem dicitur, cum unum latus superat aliud in pluribus unitatibus, ut duo quater, tres quinquies. ETHERON Graeci dicunt "alterum," inde ETERONOMIKI dicuntur PARTE ALTERA LONGIORES.

CAPUT XXVII

XXVII, 1. ERGO. Quandoquidem sola unitate latera istorum praecedunt se, ergo ipsi dicuntur p<arte> a<ltera> l<ongiores>, ut unum bis, bis tres, ter quatuor, et reliqui.

XXVII, 2. ALTERUM ENIM. Nota quod dicit: ostendit modo causam vocabuli p<arte> al<tera> lon<gior> hoc modo. Unitas prima est omnium, et ipsa est simplex, et eadem sibi similis. Binarius vero, quia primus multiplex est, ab illa simplicitate et identitate recedit, quia cum de sola unitate dicatur quod ipsa sit sola, de binario posito dici potest quod ipse sit alter ab ea, et ita ipse est principium alteritatis, quia altera unitate praecedit unitatis simplicitatem.

XXVII, 3. Numeri itaque quorum latera sola se unitate praecedunt, illi quoniam sola unitate alterantur a se invicem, ad similitudinem binarii, qui sola unitate alteratur et diversus fit ab unitate, vocantur PARTE ALTERA LONGIORES. Vide quod alteritas incipit a binario, et hic prima est alteritas. Alteritas vero pertinet ad materiam, quia ex alteritate mutabilitas, immo alteritas ipsa est mutabilitas. Unitas vero ad formam pertinet, ideoque ad immutabilitatem. Itaque Pythagoras omnia constare dixit ex unitate et alteritate, id est ex materia et forma, ideoque ex mutabili et immutabili; et sic habent omnia constare ex diverso, materia, et <ex eodem>, forma. Sicut

660–661 Cf. Caiazzo, "Un commento altomedievale," 145, ll. 549–550: "ETEPON: graece 'alterum' dicimus, unde eteromikis dicuntur PARTE ALTERA LONGIORES."

676–679 Cf. Thierry of Chartres, *Commentum super Boethii librum De Trinitate* 2.39–40 (ed. Häring, p. 81): "Ab unitate namque descendit alteritas et ob hoc ab inmutabili mutabilitas. Alteritas namque unitatem semper exigit. Alterum etenim ab uno alterum dicitur. Ut sit ergo alterum preesse necesse est quodam modo unum. Quare et ab unitate descendit alteritas ideoque mutabilitas … . Quare et materia ipsa quidem mutabilitas est. Hec autem ab unitate descendit. Unitas ergo materiam creat ex eo quod unitas est"; and 2.49 (ed. Häring, p. 84). See also Eriugena, *Periphyseon* 1 (ed. Jeauneau, p. 81): "Sive itaque informis materia mutabilitas sit formarum capax secundum Augustinum atque Platonem, sive informitas quaedam speciei, formae ornatus que participatione carens secundum Dionysium, non negabis, ut arbitror, si quodam modo intelligi potest, non nisi solo intellectu percipi"; and 3 (ed. Jeauneau, p. 119): "Materia est mutabilitas rerum mutabilium capax omnium formarum, instabilitas que mutabilis formae, qua ipsa materia specificata formatur."

679–681 See above *ad Arithmeticam* I, XXXII, with notes to 1340–1348.

pyramis vero vim suam habet in rerum complexionibus, ita quoque parte altera longior et quadratum vim suam habent in principiis rerum. Parte altera longior enim vim suam habet in materia propter alteritatem, quadratum vero vim suam habet in forma propter laterum aequalitatem et omnium dimensionum. Forma enim est existendi aequalitas, ex vi vero quadrati aequalitas. Quoniam ergo parte altera longior vim suam habet in materia, quadratum vero vim suam habet in forma propter laterum aequalitatem; ideo omnis res ex materia constat et forma, ita quoque numerus omnis ex quadratis et parte altera longioribus, ut postea declaratur. Illud quoque notandum quod impar ad unitatem attinet, quia per duo media non potest dividi, sicut nec unitas. Unde quoque numerus impar ad immutabilitatem pertinet, eo quod ad unitatem, et inde dictum est:

"Numero deus impare gaudet."

Gaudet enim divinitas immutabilitate. Par vero ad dualitatem pertinet, eo quod per media dividi potest, et sic ad materiam quae gaudet mutabilitate.

XXVII, 6. Et hoc est quod ait: imparitas AB UNITATE FORMATUR, id est formam unitatis retinet, quasi indivisibilitatem. Nec te moveat quod omnia constare diximus ex materia et forma, cum anima ex his constare non habeat. Quamvis enim anima non constet ex materia, habet tamen vim materiei et formae, ut alias demonstrabitur. AMPLIUS. Aliud argumentum quare binarius sit princeps alteritatis, istud scilicet: impar numerus non potest dividi in duo aequalia, sed habet unitatem mediam ex cuius aggregatione conficitur. Tres enim unitates conficiunt ternarium. Et ita medietas imparis numeri est simplex et individua. Parium vero medietas est binarius, quia de omni pari potest dici: "ista est una et ista est altera." Habet enim duas partes aequales,

697 formatur] formetur *ed. Guillaumin*

683–687 See above, pp. 50–57 (on *unitas* and *aequalitas*) and 57–64 (on the prophecy of the *sybilla hyspana*).

694 Virgil, *Eclogae* 8.75. Cf. Thierry of Chartres, *Commentum super Boethii librum De Trinitate* 4.4 (ed. Häring, p. 96): "Inpar vero numerus deo attribuitur. Unde poeta: *Numero deus impare gaudet.* Et merito quia sicut inpar numerus divisionem per equa non suscipit ita deus indivisibilis est nullamque varietatem nullam omnino suscipit mutationem." See the different interpretation of this verse by William of Conches, *Glosae super Platonem* 78 (ed. Jeauneau, p. 137): "Par vero numerus, qui in duo aequa potest dividi, refertur ad dissolubilia, impar vero numerus, qui in duo aequa dividi non potest, ad indissolubilia refertur. Unde Virgilius: *Numero deus impare gaudet.* Ut igitur unam eandemque animam dissolubilia et indissolubilia corpora in longum et in latum posse movere significaret et in spissum, lineares numeros, superficiales et solidos a pari et ab impari in eius compositione posuit." See also *ibid.* 76, p. 134.

et prima divisio paris numeri in aequalia fit per binarium. Et ut breviter dicam, medietas imparium est unitas simplex, parium vero medietas dualitas, alteratur ab unitate, et ideo binarius principium est alteritatis. IMPAR NUMERUS SOLA PERFICI UNITATE MONSTRATUS EST in priori libro, ubi dicit de primo et incomposito, quem sola unitas dimetitur.

XXVII, 5. PAR VERO SOLA DUALITATE PERFICI monstratum est in principio prioris libri, ubi dicit de pariter pari numero quod quaelibet eius pars in duo aequalia dividitur.

CAPUT XXVIII

XXVIII, 1. AT VERO POSITIS. Haec est summa eorum quae dicturus est quod, quia impar pertinet ad unitatem, par vero ad dualitatem ideoque ad alteritatem, quadratus constat ex imparibus, parte altera longior ex paribus constat, unde sit quod quadratum eiusdem atque immutabilis naturae est, parte altera longior alterius. Haec est altissima ad quam tendit phylosophia, ad demonstrandum scilicet quod omnia constent ex eodem et diverso, ut superius demonstratum est. Illud vero diligentissime notandum est quod vis parte altera longioris in materiis rerum est, et vis quadrati in formis, ut superius dictum est. Inde fit quod numerus omnis constat ex quadratis et parte altera longioribus, ut etiam postea demonstrabitur. Sed et illud diligentissime considerandum est, utrum ab alteritate mutabilitas et econverso, vel utrum idem sit alteritas et mutabilitas. Dico autem quoniam illud certum est, quoniam ab alteritate est mutatio, sed utrum sit mutabilitas quaerendum est. Constat autem subtiliter intuenti quod alteritas idem est quod mutabilitas. Sed cum secundum Pythagoram ab alteritate mutabilitas, ita quoque dico quod ab unitate immutabilitas. Ex eo enim quod unus est, immutabilis est; ex eo enim immutabile est quod uno modo se habet; unitas tamen est immutabilitas. QUARE QUONIAM: suspensiva constructio est usque illuc DICENDUM EST, et ubique repete quoniam “haec et haec est natura

719 quadratum] quadrati MS **723** parte altera longioris] quadrati MS quadrati] parte altera longioris MS

710–711 Cf. *De arithmetica* I, XIV.

712–714 Cf. *De arithmetica* I, VIIII.

725–733 See above *ad Arithmeticam* I, XXXII, with notes to 1340–1348. Cf. Thierry of Chartres, *Commentum super Boethii librum De Trinitate* 2.39–40 (ed. Häring, p. 81), cited above in the note to 676–679.

tetragonorum." QUARE DISPONANTUR: in duobus primis primae figurae, et postea, in sequenti figura, videbitur qui ex quibus et quomodo nascantur.

XXVIII, 2. EFFECTRIX. Per coacervationem unitatum nascitur quilibet numerus, praecipue impar qui ab eius simplici identitate non separatur, quia nulla est eius principalis et constitutiva pars, nisi unitas. Par vero praeter unitates aggregatas habet medietatem, terciam partem, et caeteras constitutivas. Pars constitutiva alicuius numeri dicitur quae, vel per se vel per alium multiplicata, totum illum numerum explicat cuius est pars; sed talem partem non habet impar aliquam, nisi solam unitatem.

XXVIII, 3. QUOD CIRCA ALIUM: videlicet ut si seipsum vel alium numerum per se multiplicet, non alium creet quam ipse est, ut bis duo, bis tres.

CAPUT XXVIIII

XXVIIII, 1. QUI FIUNT EX AEQUALIBUS AEQUALITER sed non per omnia. Et ideo addit: IN MINUS, ut ostendat altitudinem superficiei inaequalem esse.

XXVIIII, 2. ASSERES, quos apud nos quidam dicunt esse tignulos, quidam vero tegulitia tecti, quas rustice "lattas" vocamus, in arithmetica numeri sunt QUI EX AEQUALIBUS AEQUALITER DUCUNTUR IN MAIUS, ut bis bini quinquies. E contra LATERCULI SUNT QUI FIUNT EX AEQUALIBUS AEQUALITER IN MINUS, ut ter terni bis. CUNEI sunt SPHENISCI QUI EX INAEQUALIBUS DUCTI INAEQUALITER PER INAEQUALIA CREVERUNT. His vero contrarii sunt CUBI, QUI EX AEQUALIBUS AEQUALITER PER AEQUALIA proferuntur. Inter utrosque hos positi sunt PARALLELIPIPEDI medii, qui non habent cunctas partes aequales neque cunctas inaequales. Medietates namque tales esse debent, quae non per totum sint aequae, nec per totum inaequales, quale est bis tres bis.

750–760 Cf. Caiazzo, "Un commento altomedievale," 146, ll. 560–569: "DOCIDES, id est ASSERES, quos apud nos quidam dicunt esse tignulos, quidam vero tegulitia tecti, id est latas quas rustice vocamus. In arithmetica ASSERES numeri sunt QUI EX AEQUALIBUS AEQUALITER DUCUNTUR IN MAIUS, ut bis bini quinquies. Econtra LATERCULI SUNT QUI FIUNT EX AEQUALIBUS AEQUALITER MINUS, ut ter terni bis. CUNEI sunt SPHENISCI QUI EX INAEQUALIBUS DUCTI INAEQUALITER PER INAEQUALIA CRESCUNT. His vero contrarii sunt CYBI, QUI EX AEQUALIBUS AEQUALITER PER AEQUALIA proferuntur. Inter utrosque hos positi sunt PARALLELIPIPEDI medii, qui non habent cunctas partes aequales, neque cunctas inaequales. Medietates namque tales esse debent, quae non per totum sint aequae, neque per totum inaequales, quale est bis, tres, bis."

751 The word *lattas* is the equivalent of French *lattes* (English *lath, board, slat*). Thierry is reading it in the "old" commentary on the *De arithmetica*.

Circulus est rotunditas aequaliter iacens. Spera est rotunditas in modum globi, tam sursum quam deorsum, ante et retro, aequaliter manens. Item: circulus est circumductio lineae, spera est rotunditas plana cum quadam altitudine, incipiens a diametro et in diametrum desinens. Est autem diametrum linea quae speram dividit, et proprie est diametrum linea intellectualis a polo usque ad polum tendens. Globus est solida rotunditas aequalem habens altitudinem longitudini et latitudini. Dicitur etiam proprie teres vel pila. Et sciendum quia speram hic pro solida ponit rotunditate, sicuti est, sed non tam aequalem altitudine sicuti globus.

XXVIIII, 3. QUOS CUNEOLOS. Repetit et secundo diffinit eos quos supra definiverat. Quod ideo facit, quia ex diffinitione laterculorum et asserum conveniebat cuneolos et cubos facilius et brevius diffiniri quam prius fuerint.

CAPUT XXX

XXX, 1. IPSORUM VERO CUBORUM. Quia non egerat per se de spera, quae cubi species sunt, quia cuborum alii sunt sperici, alii quadrati, agit de sperico numero et circulari. Quia enim circulus superficies sperae est, bene ad ultimum de spera agit, quia spera est omnium figurarum inclusio. QUAE A QUINARIO VEL SENARIO: solis his duobus, ut puto, proficiscuntur. Sed oppones quia aliis numeris idem convenit, sicuti cum dico "quater IIII." Respondeo: in tali multiplicatione extremitas altitudinis, id est extrema altitudo, non desinit in numerum primi lateris unde cepit multiplicatio. Quater enim IIII sunt XVI, sed quater IIII quater faciunt LXIIII, et ibi convenit extremitas cum primo latere. Et tamen ille numerus non est cyclicus, quia prior, scilicet quater IIII, non terminabatur in IIII, et quia in infinitum progredienti multiplicando non eveniet idem. Cyclicus autem, in qualibet multiplicationis progressione, terminat altitudinem in primi lateris numerum.

783 LXIIII] LXIII MS

763–769 Cf. Caiazzo, "Un commento altomedievale," 146–147, ll. 570–576: "CIRCULUS est circumductio lineae. SPHERA, id est rotunditas plana cum quadam altitudine incipiens a diametro et in diametrum desinens. Est autem DIAMETRUM linea quae speram dividit. Et proprie est diametrum linea intellectualis a polo usque ad polum tendens. Globus est rotunditas solida aequalem altitudinem longitudini et latitudini gerens; dicitur etiam proprie et teres vel pila. Et sciendum quod spheram hic pro solida ponit rotunditate, sicuti est, sed non tam aequalis in altitudine sicuti globus."

XXX, 4. Unitas quoque. In parte altera longiorum ordine non praeposuerat unitatem, quia in ea pura identitas est et nulla alteritas. Unde videretur ipsam etiam non convenire cum spera et circulo, quod ipse negat. Vel pro "et": disiunctivam pro copulativa posuit. Quotie<n>s enim: adaptatio similitudinis. Similitudo est de puncto ad unitatem, et convenienter, quia quod apud geometras puncti locum tenet, hoc apud arithmeticos unitas est. Unus redit: ecce longitudo. Si hoc semel: ecce latitudo.

XXX, 5. Igitur. Habita longitudine et latitudine, multiplica longitudinem per latitudinem, vel econverso, et hac una multiplicatione habebis superficiem circularem in plano. Si secunda. Si iterum semel unum feceris, habebis altitudinem. Etenim secunda: quando superficiem multiplicas per unum eius latus.

Caput XXXI

XXXI, 1. Ac de solidis. Quia superius nominaverat naturam eiusdem substantiae et alterius eamque in numeris esse ostenderat, idcirco hinc hoc loco latius disputaturus est. Superius a principio dixerat omnem numerum constare ex pari et impari, sed non docuerat quomodo. Quoniam ergo quadrati ex imparibus descendunt, parte vero altera longiores ex paribus, docere ingreditur quomodo omnis numerus ex quadratis et parte altera longioribus fiat. Docet ergo numerum ex quadratis et parte altera longioribus constare, ad similitudinem naturarum et principalium creaturarum. Naturae namque, id est substantiae, ex materia constant et forma. Cum materia ipsa sit alteritas, necessario ad dualitatem, ideoque ad paritatem, pertinet. Vis enim paritatis est in ipsa materia. Ex vi namque dualitatis, id est paritatis, ipsa est materiae mutabilitas. Quoniam ergo alteritas parte altera longiori figurae est implicita, necessario vis parte altera longioris implicita est materiae; et ex vi parte altera longioris figurae est ipsa mutabilitas. Forma vero aequalitas essendi est, aequalitas vero ex vi unitatis. Et quia formam unitatis retinet imparitas, ideo quoque vis imparitatis est in forma. Ex vi enim unitatis est essendi aequalitas. Quadrato quoque vis unitatis et imparitatis, propter omnium dimensionum aequalitatem, est implicita. Vis ergo quadrati in ipsa forma est. Ex vi namque quadrati forma ipsa quoque essendi est

816 imparitatis] inparitatis MS **817** imparitatis] inparitatis MS **818** implicita] inplicita MS.

792–794 See above *ad Arithmeticam* II, IIII, 4.

aequalitas. Quoniam ergo omnis ratio naturae, id est substantiae, ex materia constat et forma, ideoque ex mutabili et immutabili, quare ex eodem et diverso constant omnia. Anima quoque, licet ex materia non constet et forma, constat tamen ex eodem et diverso, quoniam anima est motus; ubi vero motus, ibi mutatio de uno in aliud, ubi unitas et identitas, ubi alteritas et diversitas. Quare et anima ex eodem constat et diverso. Et ut rem altius exequar, in principalibus causis rerum vires suas habent numeri, et geometricae figurae quae ex ipsis descendunt; immo causae rerum ex vi numerorum et figurarum geometricarum causae ipsarum sunt. Prima enim forma essendi ex vi unitatis est immutabilitas. Ex hoc enim deus immutabilis est. Qui semper unus est nonnisi uno modo habere se potest. Rursus cum forma sit aequalitas essendi, dico quod divinitas ex vi quadrati forma vel causa est. Ex vi namque quadrati aequalitas et forma est existendi aequalitas, ut iam saepe dictum est. Dico itaque quod ex vi quadraturae et unitatis forma habet materiam determinare. Divinitas ergo ex vi quadrati forma vel causa rerum est. Et hoc est quare sybilla hyspana in vaticinio suo filium dei tetragonum aeternum nominat. Materia quoque, id est possibilitas, ex vi alteritatis et parte altera longioris habet ut sit materia vel causa rerum. Ex vi namque alteritatis et illius geometricae figurae habet hoc ut cedat formae et ut sit mutabilitas. Rursus efficiens causa est. Quare et ex vi numeri quia motus ex alteritate, ideoque ex merito dico quod ex vi omnium numerorum, quoniam secundum omnes numeros movet conditor materiam, quod divinitas efficiens causa rerum est, ex eo quod materiam movet, ut ex ea aliquid fiat. Non tamen ideo quis dicat vel quid vel numerum priorem deo esse. Hoc enim aperte falsum est. Dico itaque quoniam vis numerorum et figurarum in principalibus causis est rerum, ex numerorum et figurarum causa habet esse, et efficiens et formalis et materialis, ut praedictum est. Est quoque causa finalis. Habet etiam esse divinitas ex vi numeri et figurae. Finalis enim causa est utilitas ipsa; hoc est conexio, quae alio nomine bonitas, alio spiritus sanctus dicitur, cui inserta est vis proportionalitatis, id est numerorum proportionalium, quibus ista vis est et figurarum. Inde est quod phylosophus

844 quoniam vis] quamvis MS

820–822 See above *ad Arithmeticam* I, XXXII, with notes to 1340–1348, and below *ad Arithmeticam* II, XXVII-XXVIII.

828–838 See above, pp. 57–64; see also the commentary *ad Arithmeticam* I, XXXII, with notes to 1340–1348.

847–850 See below *ad Arithmeticam* II, XLII, 1; see above, pp. 67–70. See especially Thierry of Chartres, *Lectiones in Boethii librum De Trinitate* 7.7 (ed. Häring, p. 225): "Unitas autem numerum facit: equalitas vero proportionem: conexio proportionalitatem."

850–852 Cf. Boethius, *De arithmetica* I, 1, 2.

iste in principio huius tractatus dicit: OMNIA QUAE A PRIMAEVA R<ERUM> NATURA C<ONSTRUCTA> SUNT, NUMERORUM RATIONE VIDENTUR FORMATA, non ad similitudinem numerorum, ut quidam exposuerunt, sed quia numerorum vi omnia formata sunt, quoniam vis numerorum et figurarum est in principalibus rerum causis, ut prius monstratum est. Dico etiam quod terra, ex eo quod cubica est, et ignis, ex eo quod piramoydes, cum extrema sint et duobus mediis ligentur, ex illorum figuris principia plane sunt corporum. Ad hanc ergo praedictam naturarum et creaturarum similitudinem dicturus est phylosophus iste omnem numerum ex quadratis et parte altera longioribus constare, ita scilicet quod parte altera longior loco materiae, quadratus loco formae. Vis enim quadrati loco formae est, vis parte altera longioris in materia, ut saepe iam dictum est. Dico etiam quod ipse quadratus numerus ex quadrato et parte altera longiori constat, et similiter parte altera longior. Unde dictum est a philosopho quod numerus ex seipso videtur compositus. PROPINQUIS RATIONIBUS, id est accedentibus ad ipsam naturam. Nam ratio quasi aliena est, quae non accedit ad rem quae requiritur. Ac si diceret: non quidem audeo dicere quod ipsam naturam potuerint comprehendere, sed "propinquis rationibus" accesserunt illi ut ipsam naturam rerum intelligerent. Propinqua itaque ratio est quae ad veritatem accedit, remota quae non. SUPTILISSIME. Subtilis est qui rei veritatem cito penetrat. PERITISSIMEQUE. Peritus dicitur qui multis experimentis rei veritatem perpendit. IN GEMINA: quia causae ex quibus res constant, aut est materia in qua est mutabilitas, aut forma in qua est immutabilitas, et sic constant omnia ex mutabili et immutabili.

XXXI, 2. DICUNT ENIM illi physici OMNES OMNIUM RERUM SUBSTANTIAS CONSTARE, quia omnis natura aut est creatrix aut est creata. Creatrix

855–858 Calcidius, *Commentarius in Timaeum* 20 (ed. Waszink, p. 71): "… sed inter ignem et terram, quae sunt solida corpora, nullam apparere similitudinem, quando iuxta ipsum Platonem ignis quidem forma et figura pyramoides esse dicatur, id est in modum pyramidis excrescat, terra vero cubus sit, hae porro formae nullam ex se similitudinem mutuentur, quia non sint aequalibus angulis"; and 326 (ed. Waszink, p. 321): "Eodem igitur modo nec ignem, qui est pyramoides, ignem esse respondebimus, sed vel ignitam silvae partem vel igneam qualitatem, nec octahedrum, sed spirabilem silvam, nec icosahedrum nec cubum, sed humectam hanc, terrenam illam silvae soliditatem." Cf. Caiazzo, "La forme et les qualités des éléments: lectures médiévales du *Timée*," in *Il* Timeo: *Esegesi greche, arabe, latine*, ed. Celia and Ulacco, pp. 307–345.

865–867 Cf. Caiazzo, "Un commento altomedievale," 147, ll. 577–579: "VESTIGANTES PROPINQUIS RATIONIBUS, id est accedentibus ad ipsam naturam. Nam ratio quasi aliena est, quia non accedit ad rem quae requiritur."

875–879 Cf. Caiazzo, "Un commento altomedievale," 147, ll. 582–588: "DICUNT ENIM illi, id est phisici, OMNES OMNIUM RERUM SUBSTANTIAS CONSTARE EX EA,

natura deus est, creata natura est quaecumque, creante deo, subsistit. Et deus immutabilis est, et immutabilior loco, tempore omnique re. Creata vero natura mutabilis est et variabilis loco, tempore et diversis affectibus. OMNIUM RERUM SUBSTANTIAS, id est causas rerum praedictas, vel substantias composita<s> ex utraque causa.

XXXI, 3. SECUNDUM UNITATIS SPECIEM: quia sicut unitas non potest dividi, sic nec par per duo altera, id est media.

QUOD NIMIRUM AD UNITATEM PERTINET: quia vis unitatis inserta est formae, vis alteritatis ideoque dualitatis materiae. VEL AB AEQUALITATE FOR<MANTUR> TETRAGONI: quia sicut tetragonus per aequales dimensiones crescit, vel quia ex imparibus nascitur, ut supra monstratum est. ET QUONIAM. Vult ostendere non solam unitatem esse eiusdem substantiae, sed et omnes impares quorum ipsa effectrix est, et omnes tetragonos qui ab imparibus procedunt; deinde vult ostendere non solum binarium diversae esse naturae, sed et omnes pares et omnes parte altera longiores.

XXXI, 4. QUONIAM BINARII FORMAE SUNT: quia secundum speciem binarii, videlicet alteritatis, formati sunt.

XXXI, 5. QUARE. Quia numeri sunt participes eiusdem et alterius substantiae, ideo etiam omnis rerum natura sic se habet, ut participet cum eadem et altera substantia ad similitudinem numerorum ex quibus omnia consistunt. AUT ENIM. Nota quod dicit omne quod est vel esse immutabile, ut deus, anima vel mens – et est istud "vel" subdisiunctivum, sicut "Alexander vel Paris," id est "qui est Paris" – vel mutabile, ut corpus. Quod dicit

892 formae] The scribe added the word *generis* above the line. **898** subdisiunctivum] subdisiunctum MS sicut] non ut MS

---

scilicet natura, QUAE PROPRIAE SUAEQUE SEMPER HABITUDINIS EST NEC ULLO MODO PERMUTATUR, ET EA SCILICET NATURA, QUAE VARIABILIS MOTUS EST SORTITA SUBSTANTIAM: quia omnis natura aut est creatrix aut creata. Creatrix natura deus est; creata natura est quaecumque creante deo subsistit. Et deus INMUTABILIS est et inmotabilis loco et tempore omnique re. Creata vero natura mutabilis est et variabilis loco et tempore et diversis affectibus."

898–899 Thierry of Chartres, *Lectiones in Boethii librum De Trinitate* 1.12 (ed. Häring, p. 136): "Et ita sunt due cause diverse non tamen opposite. VEL subdisiunctivum est non disiunctivum secundum quosdam, sicut est Alexander VEL Paris, i.e. qui et Paris ut sit una causa penitus hoc et illud"; and 1.14, p. 136: "Secundum vero alios eandem causam sub aliis verbis ut tamen sit subdisiunctivum non disiunctivum sicut est illud: Alexander sive Paris." Cf. Priscianus, *Institutiones grammaticae* 16 (*Grammatici latini* 3, p. 98, l. 1): "Subdisiunctivae sunt, quae voce disiunctivarum utrumque tamen esse significant, vel simul, ut copulativae, vel discrete"; and p. 98, l. 9: "Et 'Alexander sive Paris' pro 'Alexander, qui et Paris dicitur.'" See also Petrus Helias, "De specie," *Summa super Priscianum* (ed. Reilly, p. 823).

referendo deum et animam ad formam, propter immutabilitatem, et corpus ad materiam, propter mutabilitatem. Et est sensus quod omnia creata ex materia constant et forma, ut superius dictum est. Ista autem secundum paganos loquitur, qui nescientes deum, animam deum dicebant, asserentes magnum bonum esse maximum bonum. Maximum namque bonum deus est, magnum vero bonum anima est, parvum autem bonum corpus est. Sed illi phylosophi nescientes deum maximum bonum, dicebant animam magnum bonum, deum esse maximum bonum.

Angeli cum sint incorporei, sunt etiam immutabiles, et tamen transeunt de perfecto statu in deteriorem, ut diabolus, vel in perfectiorem, ut illi perfectissimi. Nec ideo dicendum est eos esse mutabiles, quia non agit de hac apertissima mutatione quae est rerum temporalium. Quod in libro *Consolationum* quarto plene insinuat Boetius, ubi dicit: "Quaecumque subsunt fato, subsunt et providentiae, sed non convertitur." Fatum ibi vocat maximam mutationem istarum rerum temporalium, cui non subiacent angeli et caetera caelestia.

XXXI, 6. UT FORMARUM: triangularum, quadrangularum et caetera. Et nota quod numeri, secundum quos disponuntur istae formae, sunt quantitatis relatae. Ipsae autem figurae sunt per se consideratae, nisi quis comparet

902–906 Cf. Caiazzo, "Un commento altomedievale," 147, ll. 591–595: "Ista secundum paganos loquitur, qui, nescientes deum, animam deum dicebant, asserentes magnum bonum esse maximum bonum. Maximum namque bonum deus est, magnum vero bonum anima est; parvum autem bonum corpus est. Sed illi philosophi nescientes dominum maximum bonum, dicebant animam magnum bonum deum esse maximum bonum."

911–913 The quotation is not textual; see Boethius, *Consolatio Philosophiae* 4.pr6: "Quo fit, ut omnia, quae fato subsunt, providentiae quoque subiecta sint, cui ipsum etiam subiacet fatum." Cf. William of Conches, *Glosae super Boetium* 4.pr6 (ed. Nauta, p. 257, ll. 149–150): "Omnia quae subsunt fato subsunt et providentiae; sed non convertitur." Cf. Thierry of Chartres, *Lectiones in Boethii librum De Trinitate* 2.6 (ed. Häring, p. 156): "Que explicatio ab antiquis fatum dicitur. Fatum enim est explicatio divine providentie. Divina vero providentia est ipse deus. Et deus est ipsa divina providentia que precedit fatum. Fatum vero est explicatio divine providentie. Quod Boetius in *Libro de consolatione* testatur. Unde patet divinam providentiam precedere fatum"; and 2.10 (ed. Häring, p. 157): "Absoluta enim necessitas rerum omnium conplicatio est in simplicitate. Necessitas conplexionis earum rerum explicatio in quodam ordine. Qui ordo a phisicis fatum dicitur." Cf. Clarembald of Arras, *Expositio super librum Boetii De Hebdomadibus* 47 (ed. Häring, p. 221): "Boetius quoque in libro *De Consolatione Phylosophiae* divinam providentiam tam sapienter cuncta gubernare non mentitur ut etiam mala quae gerimus ad bonum convertat. Quippe sicut ex falso verum sic ex malo proportionaliter divina ratione ordinante sequitur bonum. Bona igitur i.e. ordinata cuncta sunt quae ab ordinante principio profluxerunt. Neque enim cuncta ordinans recte diceretur principium si non a principio descendentia vere dici possent ordinata."

unam cum alia, et attribuat ei nomen relationis, sic dicendo: ista figura est similis huic et illi dissimilis.

XXXI, 7. ET PRIMO. Videndum est qui sint antelongiores, qui sint parte altera longiores. Et quare de antelongioribus, ideo videndum est, ut pateat eorum differentia ad parte altera longiores. Quod licet superius dixerit, tamen hic repetit, ut apertius exequatur et definitione et exemplo. Similiter de parte altera longioribus licet superius egerit, hic modo repetit propter eandem causam, videlicet ideo ut ostendat eos differre ab antelongioribus et ut hic apertius quid sint definiat. Quod ut facere possit, praemittit aequalitatem laterum cubi ut, visa illa, pateat quae sit inaequalitas laterum parte altera longioris.

XXXI, 10. CUR AUTEM. Et tamen hic dicam, praemittendo naturam quadratorum.

XXXI, 9. UNITATIS DIFFERENTIA. Unitati enim et vi et potestate primo impari si iungatur ternarius actu et opere secundus impar, formabitur quaternarius primus actu et opere tetragonus. Huic quaternario si addatur quinarius tercius impar, procreabitur novenarius secundus tetragonus.

CAPUT XXXII

XXXII, 1. OMNE AUTEM. Dixerat in proximis mundum esse coniunctum ex immutabili et mutabili substantia, hic ostendit numeros ex eisdem constare et ad numerorum similitudinem omnia fieri. Singula singulis reddit: IMMOBILE, quod nulla variatione mutatur; TERMINATUM, quod numquam esse desinit; DEFINITUM, quod numquam esse possit quod non fuit. Haec enim omnia ad laudem unitatis exaggregat. Deo enim non convenit esse terminatum; incomprehensibilis enim est. ET QUAE AB UNITATE FORMANTUR, id est quae formam et similitudinem habent cum unitate in determinatione et in immobilitate. UNITAS quippe et quaeque ex ea FORMANTUR COMPREHENSIBILIS ET DETERMINATAE ET EIUSDEM SUBSTANTIAE ESSE DICUNTUR. Ita et deus semper eiusdem substantiae est et determinatae, non quod de-

943 Et quae ab unitate formantur] et quae unitate formantur *ed. Guillaumin*

945–948 Cf. Caiazzo, "Un commento altomedievale," 147, ll. 598–602: "UNITAS quaeque ex ea FORMANTUR, COMPREHENSIBILIS ET DETERMINATAE ET EIUSDEM SUBSTANTIAE ESSE DICUNTUR. Ita et deus semper eiusdem substantiae et determinatae, non quod comprehendi possit, sed quia idem est semper et inmutabilis, ideoque comprehensibilis et determinabilis quodammodo est, ad huius similitudinem unitas quodammodo comparatur."

prehendi possit, sed quia idem est semper et immutabilis. Ea vero: eiusdem substantiae sunt, et ab unitate procedunt. Formantur: per aggregationem unitatum formantur impares, et ex imparibus coniunctis formantur tetragoni, quia I et III faciunt IIII, qui est tetragonus. At vero binarius, qui est principium alteritatis, et parte altera longiores sunt variabilis substantiae.

XXXII, 2. Constat ergo: quia invariabilis naturae sunt impares et pares sunt variabilis. Ergo omnis numerus constat ex his contrariis. Infinita congeries: quia in imparibus est simplicitas et identitas, in paribus est multiplicitas et mutabilitas. Quae paritas et imparitas, cum sint contraria: ex contrariis sibi invicem concordantibus fit medium, ut apparet in cythara. Si enim acutus sonus gravi sono proportionaliter commensuretur, ita scilicet ut soni aequaliter sibi commisceantur et contemperentur, inde fit consonantia et modulatio quae graece dicitur armonia, quae ex acuti et gravis soni contemperatione fit. Medium quoque ab utroque aequaliter distat. Informatione: per unitatem hoc fit, ex cuius genitura profluunt par et impar.

XXXII, 3. Non ergo. Quandoquidem numerus constat ex contrariis, et omnia ex numeris facta sunt, igitur bene dixerunt phylosophi qui dixerunt omnia constare ex eadem et diversa natura. Hanc divisionem: quod substantia alia eadem, alia diversa.

XXXII, 4. Non sine causa, quia nisi ex contrariis constarent omnia, non iungerentur omnia armonia quadam. Est autem armonia. Tria genera sunt musicae: diatonicum, chromaticum et enarmonicum. Diatonicum genus est perantiquum, durissimum atque difficillimum. Chromaticum est mollissimum et dissolutum. Inter haec duo est medium et elegantissimum genus enarmonicum, ex utroque coniunctum. Inde obtinuit ut omnis consonantia armonia dicatur ab illo tercio genere enarmonico. Armonia est proprie multarum adunatio vocum in unam consonantiam.

Caput XXXIII

XXXIII, 1. Disponantur. Superius dictum est quod omnis numerus fit ex quadratis et parte altera longioribus. Unde quadrati et p<arte> al<tera> lon<giores> priores sunt. Hic iterum dicit quod ex numeris quadrati descendunt et par<te> al<tera> lon<giores>. Unde in natura numeri

970 autem] enim *ed. Guillaumin*

970–976 See above *Ad arithmeticam* II, II, 2.

videntur priores, et inde videtur incurrere contrarietas, sed nulla est. Quadrati enim et p<arte> al<tera> l<ongiores> priores sunt in eo quod numeri fiunt ex eis, ut ex materia et forma, et sic quadrati et p<arte> al<tera> lon<giores> iterum sunt posteriores natura in hoc quod ex numeris descendunt. Nulla ergo dicenda est contrarietas, sed diversis respectibus priores sunt numeris et posteriores. Quod autem dicit quadratum et p<arte> al<tera> l<ongiorem>, sicut et caeteras figuras a numeris descendere, nihil aliud est nisi quod numeri sunt causae figurarum geometricarum. Verbi gratia: ex ternario descendit triangulus, id est ternarius causa est trianguli, quia posito triangulo ponitur ternarius, sed non e converso. Illud quoque diligenter notandum est quod cum deberet ostendere quomodo ex quadratis et pa<rte> al<tera> l<ongioribus> fiat numerus, ostendit quomodo fiunt proportiones et proportionalitates. Ex quadratis et p<arte> al<tera> l<ongioribus>, alternatim dispositis, nascuntur proportiones, quarum differentiae in numero non aliter inveniuntur. Et ob hoc putat se sufficienter docere quomodo omnis numerus ex quadratis constet et par<te> al<tera> l<ongioribus>.

XXXIII, 3. ADIECTIONEM NUMERORUM: istud scilicet quod secundum species superparticularium adicitur unitas priori differentiae. Hoc attinet ad differentiam, id est hoc sunt differentiae, quia unus est differentia dupli, et binarius sesqualteri, et ita differentiae crescunt naturaliter, secundum hoc quod ascendis enumerando species superparticularium.

XXXIII, 4. SUPERIUS: in superiori figura, ubi differentiae ab uno incipiunt.

XXXIII, 5. RURSUS. Differentiae quadratorum sunt impares numeri, differentiae vero p<arte> al<tera> lon<giorum> sunt pares numeri.

XXXIII, 8. SI VERO CONVERTAS. Ista conversio sic deberet fieri, ut cum inter supradictos primus par<te> al<tera> lon<gior> poneretur, hic inter istos medius poneretur primus tetragonus. Liber autem dicit: SI CONVERTAS ET SECUNDUM TETRAGONUM medium PONAS, et ideo determinat secundum ordine, primum autem opere.

CAPUT XXXIIII

XXXIIII, 1. EX UNO. Ex uno tetragono et binario par<te> al<tera> lon<giori> nascitur III triangulus. Ex IIbus et IIII nascitur VI; ex IIII et VI, denarius; ex VI et VIIII, XV; ex VIIII et XII, XXI; ex XII et XVI, XXVIII; ex XVI et XX, XXXVI; ex XX et XXV, XLV; ex XXV et XXX, LV; ex XXX et XXXVI, LXVI; ex XXXVI et XLII, LXXVIII. ALTERNATIM. Quantum ad figuram loquitur divisim, ita scilicet ut inter duos illos versus tetragonorum et p<arte> al<tera> l<ongiorum> locetur medius versus triangulorum, prius tamen coniungi debent illi idem versus tetragonorum et parte al<tera> lon<giorum> ante illam divisionem subiectam. Quae, scilicet figura, par<te> al<tera> lon<gior> vel per additionem vel per subtractionem lateris tetragoni provenit et fit ex magna vi alteritatis, qua participant per binarium qui est alteritatis principium. Cum enim unitas et impares cum tetragonis participent aequalitate et simplicitate, quicquid ab eorum aequalitate discedit in inaequalitem, oportet ut in maius vel in minus prodeat.

Caput XXXV

XXXV, 1. OMNIS TETRAGONUS. De VIIII tolle III, remanebit senarius p<arte> al<tera> lon<gior>. Mitte III, faciet XII p<arte> al<tera> longiorem. De XVI tolle IIII, remanebit XII p<arte> al<tera> lon<gior>. Mitte IIII, erunt XX; sic de caeteris.

Caput XXXVI

XXXVI, 1. CONSTAT IGITUR. Quia tetragonus eiusdem essentiae est per medios impare, impares vero eiusdem propter unitatis participationem, ergo unitas in primo loco est eiusdem substantiae.

Caput XXXVII

XXXVII, 2. DISPONANTUR: alternatim et permixtim in uno versu.

XXXVII, 3. IN SUPERIORI: in supraposito versu tetragonorum et p<arte> al<tera> lon<giorum>. HOC INTUENDUM EST quod supradictum est, scilicet ut ALIAS PRO<PORTIONIBUS> CO<MMUNICENT>. HIC adverbium est, id est in ista numerorum collectione isti numeri DISCREPANT IN PROP<ORTIONIBUS>.

XXXVII, 4. SICUT IN PRIMIS FUIT. Prima enim comparatio easdem habuit proportiones et differentias diversas. Secunda quae ibi est RURSUS SI II AD IIII pares habuit differentias, sed discrepabat proportionibus. Tercia vero IIII ad VI, et VI ad VIIII, convenit cum prima, et quarta cum secunda. Quod iam in sequentibus dicit apertius, ibi videlicet: ET SEMPER ALTERNATIM.

1038 permixtim] permixti MS

Semperque. Ostendit aliam proprietatem istorum tetragonorum et p<arte> al<tera> lon<giorum> altrinsecus positorum, hanc scilicet, quod numeri quibus isti inter se differunt, id est eorum differentiae crescunt secundum ordinem naturalem, quia prima differentia est I, secunda II, tercia III, et sic deinceps.

Quod geminatis. Additio est eiusdem quod modo dixit et determinatio, scilicet quomodo secundum naturalem ordinem crescant differentiae, ita videlicet quod quaelibet differentia geminatur, quia bis agitur per unamquamque secundum naturalis numeri ordinem. Bis enim binarius numerus differentia est: primo inter binarium et quaternarium, secundo inter IIII et VI. Sic et ternarius bis differentia est: primo inter VI et VIIII, deinde inter VIIII et XII. Similiter et quaternarius bis fit differentia: primo inter XII et XVI, deinde inter XVI et XX. Et sic in caeteris bis accipitur quaeque differentia, modo ad praecedentem modo ad sequentem proportionis numerum. Quod – subauditur illa differentiarum geminatio – non debet mirum videri. Nos enim, id est ipsos tetragonos et p<arte> al<tera> lon<giores> geminavimus unumquemque bis comparando et unicuique duas proportiones adaptando, ut binarius habet duas proportiones: unam ad unitatem, aliam ad IIIIrium; et IIIIrius duas: unam ante se, aliam post se. Et hoc est quod dicit: ad primas secundasque prop<ortiones>. Prima proportio est de binario ad I, secunda de binario ad IIII. Binarius enim refertur ad utrumque. Est enim duplus unitatis et subduplus quaternarii. Et ita quemadmodum ipsae summae, id est tetragoni et par<te> al<tera> lon<giores> geminantur, modo ad praecedentem modo ad subsequentem numerum comparatae, sic et summulae, id est differentiae, geminantur secundum ordinem naturalis numeri, sicut supradictum est. Nota quod istud est probatio eius quod geminatis summulis fiat progressio naturalis numeri, ita scilicet ut in eo se superent in quo differunt. Nam IIII binario differunt, et superatur a binario.

XXXVII, 5. Eaedem quoque. Totum esse idem quod aequale sive par in hac littera intellige.

XXXVII, 6. Quae unitatis. Sic intellige: omne totum sui generis totum est. Domus est totum domus, id est talis generis rerum, et unitas unitatis, et binarius binarii. Ergo unitas, unitati comparata, totum est ut illa, sed binario collata, pars est. Et binarius, binario collatus, totum est sicut ille, sed ad quaternarium est pars. Omne enim totum sibi est par. Cum autem dividitur in aliquas quascumque partes, nulla illarum partium est ipsi toto aequalis, sed differt a toto secundum partis quantitatem quam ipsa possidet. Dividitur autem id quod est totum convenienter primo in duas partes, scilicet in duas

1075 ordinem naturalis numeri sicut] *iter.* MS

medietates, deinde consequenter in tres, deinde in IIII, ita videlicet ut, per singulas unitates adiectas, crescat singula species divisionis totius, prima in duo, secunda in tres, et sic amplius. Istae itaque DIFFERENTIAE PROGREDIUNTUR A TOTO PER SEQUENTES PARTES, quia quilibet differentiae numerus totum sui ipsius est, et est aliqua pars alterius. Verbi gratia: unitas est prima differentia, quae sibi aequalis et suum totum est, et binarii medietas. Binarius vero secunda differentia sui ipsius totum est, et medietas quaternarii, et IIIa pars senarii; ternarius media pars senarii, et IIIIa pars duodenarii; quaternarius tercia pars duodenarii, et quarta pars XVI; item quaternarius quarta pars XVI, quinta XX. De caeteris videbis in littera in exemplis. GEMINATIS PARTIBUS: quia quod est medietas unius est pars tercia alterius, et quod est pars tercia unius idem est pars quarta alterius, et sic deinceps. SICUT IPSA. Sicut ipsae summae, id est ipsi tetra<goni> et p<arte> al<tera> l<ongiores>, geminabantur, modo ad praecedentem modo ad subsequentem comparatae. AEQUAS PROGRESSIONES PARTIUM illarum: de secunda in terciam, de tercia in quartam.

CAPUT XXXVIII

XXXVIII, 3. QUONIAM PRIMI SUNT, et primi potestate sunt impares, et ideo in impari loco constituti.

CAPUT XXXVIIII

XXXVIIII, 1. IPSI VERO CUBI. Dixerat tetragonos esse eiusdem substantiae, propter impares unde surgunt. Et idem habent cubi quod tetragoni, cum surgant ab eis. TRIBUS INTERVALLIS: id est tribus dimensionibus; et in hoc imparibus conveniunt. Tamen praeter hoc alia est causa, scilicet aequalis multiplicatio, quia sive numeri efficientes cubum pares sint sive impares, aequaliter multiplicantur, scilicet tripliciter ducuntur, ut bis bini bis, ter terni ter. Si rarius vel sepius multiplicetur, cubus non surget, nisi ter fiat multiplicatio.

CAPUT XL

XL, 1. NUNC RES ADMONET. Hoc loco propriis et signatis verbis est usus: AD MUSICAS SPECULATIONES, in quibus quaeritur quae proportio

1118–1130 Cf. Caiazzo, "Un commento altomedievale," 148, ll. 606–616: "Hoc loco propriis et signatis verbis est usus 'ad musicas speculationes,' in quibus quaeritur quae sit sesqualtera, quae sesquitercia et deinceps, quia duae chordae talem ad se habent convenientiam qualem societatem habent numeri in sesqualtera proportione, vel in sesquitercia, vel in sesquiquarta. VEL AD ASTRONOMICAS UTILITATES, quia ex numeris constant. VEL AD GEOMETRICAE quae talis est, ut in omnibus numeris quos intuemur, sive in maioribus sive in minoribus, aequalis sit proportio. VEL AD VETERUM. Veteres enim in talibus speculationibus versati sunt.

XL, 2. UT AUTEM COMMUNITER DIFFINIAMUS, id est ut non solum de proportionibus dicamus, sed etiam de differentiis proportionum, quia paulo superius de proportionalitate dixit. Hic autem etiam de differentiis dicturus est.

XL, 1. ARITHMETICA<M> INTRODUCTIONEM: quia per arithmeticam introducimur ad praedictorum intelligentiam. COMMUNITER, id est facilius, apertius, communibus et vulgaribus verbis diffiniam.

XL, 2. PROPORTIONALE sive PROPORTIONALITAS idem est quod medietas. QUANTITATIBUS, id est differentiis, quia differentia est quantitas sive numerus per quem unus terminus alterum superabundat.

XL, 4. QUONIAM DUO SUNT TERMINI. Istud non est argumentum quod ideo sit dupla proportio. Potest enim inter duos terminos esse vel tripla vel alia aliqua proportio. Ideo addas QUONIAM isti DUO TERMINI SUNT primi, sola hac proportione invicem se habentes et non alia; ideo duplam obtinet proportionem. UT BINARIUS: hic est exemplo dupla proportionalitate, quae est tamen in tribus terminis. Vel aliter. Binarius ad IIII obtinet duplam proportionem, et non proportionalitatem, quoniam proportionalitas in tribus consistit terminis, sed proportio in duobus.

XL, 6. IGITUR QUOTIENS. Ostenso quid sit proportionalitas, ostenso etiam quod quaedam consisterent tantum in tribus terminis, quaedam in amplioribus, facit eius divisionem, talem scilicet quod alia est coniuncta, alia disiuncta. Et est coniuncta quae tantum in tribus terminis consistit,

1123 utilitates] subtilitates *ed. Guillaumin* **1127** autem] etiam *ed. Guillaumin*

---

proportio sit sesqualtera, quae sesquitertia, et deinceps, quia duae chordae talem ad se habent convenientiam, qualem societatem habent numeri in sesqualtera proportione, vel in sesquitertia, vel in sesquiquarta. 'Vel ad astronomicas subtilitates,' quia ex numeris constant. 'Vel ad geometricae considerationis vim,' quae talis est, ut in omnibus numeris quos intuemur, sive in maioribus sive in minoribus, aequalis sit proportio. Vel ad veterum lectionum intelligentiam. Veteres enim in talibus speculationibus versati sunt. UT AUTEM COMMUNITER DEFINIAMUS haec et caetera, id est ut non solum de proportionalitatibus dicamus, sed etiam de differentiis proportionum, quia paulo superius de proportionalitate dixit; hic autem etiam de differentiis dicturus est."

disiuncta vero quae in amplioribus. Et ita cum iam fecerit eandem divisionem, dicendo quod minima constaret in tribus et alia in amplioribus, hic apertius dividit proportionalitatem in suas species, dando eis nomina et diffiniendo ipsas.

Ut I, II, IIII. Binarius dux est unitatis, quia duplus, comes IIIIrii, quia subduplus. Eodem modo: quia in differentiis est eadem aequalitas proportionis. Differentia unitatis et binarii est unitas, binarii ad quaternarium binarius, quaternarii ad unitatem ternarius.

XL, 7. Si vero alius ad unum, id est primum ut binarius ad unitatem, alius vero ad alium, ut VIII ad IIII, ita scilicet ut unus medius non sit qui communicet utrique extremorum, sicut in coniuncta ostensum est, sed duo vel plures, quorum mediorum nullus habet se aequali proportione ad utrumque extremorum, servata tamen aequalitate proportionis de quolibet proximo ad proximum. Hoc autem est proprie proprium geometricae medietatis exemplum, de qua etiam iam exemplificaverat in coniuncta, ubi dixit: igitur quotiens unus. In sequentibus dabit differentiam inter tres medietates, sed hic exemplificat de geometrica, quia notior est propter proportionum aequalitatem.

Caput XLI

XLI, 1. Quae superius dictis. In priori libro divisit inaequalitatem in quinque species. Illis quinque ostendit alias esse oppositas, scilicet submultiplicem, subsuperparticularem, et caetera. Eodem modo oppositae sunt istis tribus medietatibus aliae tres minores, sicut iam in hac eadem clausula dicet.

XLI, 2. Propter denarii numeri perfectionem. Quia sicut ultra X non progreditur numerus, sic nec proportio ultra X species inaequalitatis, nec proportionalitas ultra numerum X progreditur. Quidam autem hoc de decem digitis manuum vel articulis pedum intelligunt. Alii vero subtilius attendentes in principali corporis membro, id est in capite, hunc denarium accipiunt. Cerebrum enim IIII ossibus in modum crucis compaginatum est. Palatus similiter IIII ossibus constitutis coniunctus est. Et praeterea maxillae duobus ossibus sunt compactae. Qui simul iuncti denarii corporis perficiunt quantitatem.

XLI, 3. Arcitae. Prius quam Aristotiles X praedicamenta scriberet, Archites etiam de decem praedicamentis librum fecerat pro dilectione Pythagorae magistri sui, qui multum dilexit et commendavit denarium

1163–1164 Boethius, *De arithmetica* II, xl, 6.

numerum. QUANDOQUIDEM ET PLATO. Decem sunt medietates propter supradictos auctores, et etiam ideo quod Plato fecit X praedicamenta. Quod autem hic repetit de Archita, hoc ideo facit, quia modo dixerat eum priorem Aristotile. Quod quia quibusdam non constabat, relinquit et ipse dubium dicens: LICET QUIBUSDAM SIT AMBI<GUUM>. Architas et Linus dicuntur fuisse antiquissimi sequaces Pythagorae et, ut quidam volunt, contemporanei Pythagorae fuerunt, sicut in prologo *Rethoricorum* Tullius dicit, ostendens quomodo eos Aristotiles imitatus sit, et ipse Aristotelem.

CAPUT XLII

XLII, 1. NUNC VERO DE PROPORTIONALITATIBUS. Nota quod proportionalitates maximum numerum habent in rerum naturis et complexionibus, ut maxime consideratur in astronomia. Signa namque respiciunt se vel trigona ratione, vel tetragona, vel diametra, vel aliqua alia; et est secundum hoc inter ea aliqua proportionalitas, scilicet quod maximum in rebus habent effectum. Dicunt praeterea astronomici, et inter caeteros Iulius Firmicus, quod planetae quidam habeant respectum et affinitatem ad solem et lunam, sol autem et luna ad nostras naturas, et sic caeteri planetae sole et

1189–1192 Cicero speaks of Aristotle and his predecessors, in particular of Hermagoras, at the beginning of *De inventione*; however, the precise passage to which Thierry of Chartres might be referring is difficult to determine.

1194–1196 Cf. Eriugena, *Periphyseon* 3 (ed. Jeauneau, p. 19): "Ut enim inter numerorum terminos (hoc est inter ipsos numeros sub una ratione constitutos) similes proportiones, ita inter omnes ordines naturales a summo usque deorsum participationes similes sunt, quibus iunguntur. Et quemadmodum in proportionibus numerorum proportionalitates sunt (hoc est proportionum similes rationes), eodem modo in naturalium ordinationum participationibus mirabiles atque ineffabiles armonias constituit creatrix omnium sapientia, quibus omnia in unam quandam concordiam, seu amicitiam, seu pacem, seu amorem, seu quocunque modo rerum omnium adunatio significari possit, conveniunt. Sicut enim numerorum concordia proportionis, proportionum vero collatio proportionalitatis, sic ordinum naturalium distributio participationis nomen, distributionum vero copulatio amoris generalis accepit, qui omnia ineffabili quadam amicitia in unum colligit."

1196–1199 Cf. "De stellis fixis et errantibus," in *Germanici Caesaris Aratea: cum scholiis* (ed. Breysig, p. 223): "Aspiciunt inter se stellae ex quinto signo, quod dicitur trigonum, et habent maximam consensionem. Item a quarto, quod dicitur tetragonum et centron vocatur, et in alterutrum maximum praestant effectum. Item ex contrario, quod est septimum signum et diametron vocatur estque maxime adversum. Cetera dissident vel leviter aspiciunt, ut tertium quodque, quod dicitur hexagonon." See also Firmicus Maternus, *Mathesis* 2.22.

1199–1200 See above, pp. 67–70.

luna mediantibus. Ecce igitur quod secundum proportionalitatem planetae in nostris habent vim maximam naturis. Illud quoque notandum quod proportionalitas non gignit nec gignitur; per nullam enim fit multiplicationem, sed tantum procedit. Sicut enim aequalitas est complexio inaequalitatis, et ab ea inaequalitas omnis descendit, ita quoque conexio complexio est proportionalitatis et proportionis, et a conexione haec et illa descendit. Unitas autem, ut saepe diximus, una est. Est enim unitas, est unitatis aequalitas, est et unitatis conexio et aequalitatis, quae ab unitate et unitatis aequalitate procedit. SECUNDUM QUANTITATIS, id est differentiarum. Arithmetica enim medietas consideratur in differentiis terminorum, non in proportionibus, ibi scilicet ubi aequales et eaedem differentiae, diversae autem proportiones. CONSTITUTORUM: vel trium vel IIII vel quotlibet terminorum, qui aliquo medio et aliqua habitudine coniungendi sunt. IN HIS AUTEM. Determinat quorum sint illae differentiae et quae in quibus haec medietas habet considerari, scilicet in differentiis terminorum et in his quantitatibus, quibus aliquis terminus praecedit alium et differt ab ipso. SUPERIUS: in priori libro, ubi dixit qui essent transeuntes, qui praecedentes.

XLII, 2. IPSA NUMERORUM NATURA: naturaliter positorum, quia inter eos est eadem differentia sed diversa proportio. Primum ideo dicendum est de arithmetica medietate, quia numerorum in naturali ordine positorum eaedem sunt differentiae, et illa aequalitas differentiarum talem potentiam et vim in se habet, ut in eis nulla alia medietas versetur, nisi sola arithmetica. HUIUSMODI PROPORTIONES quae PERTINENT id est versantur in differentiis terminorum, scilicet arithmeticae medietates. DEINDE: alia ratio quare de arithmetica prius tractandum sit medietate.

CAPUT XLIII

XLIII, 1. NEGLECTA: quia non attenditur hic quae sint ipsorum terminorum proportiones, sed qui sint termini et quae eorum differentiae. Debent enim aequales esse differentiae terminorum in arithmetica medietate, quaecumque sint proportiones. Nota quod dicit: TERMINORUM DIFFERENTIARUMQUE; utrumque dicit, quia differentiae non possunt notari

1205–1210 See above *ad Arithmeticam* II, XXXI, 1. Cf. Thierry of Chartres, *Commentum super Boethii librum De Trinitate* 2.38 (ed. Häring, p. 80). See above, pp. 55–56. See especially Thierry of Chartres, *Lectiones in Boethii librum De Trinitate* 7.7 (ed. Häring, p. 225), cited above in the note to 847–850.

1217–1218 Cf. *De arithmetica* I, X, 11: "Igitur in naturalis numeri dispositione pariter impares numeri quinto loco a se distant, solis IIII se praecedunt, III in medio transeuntes, per binarium numerum multiplicatis imparibus procreati."

sine terminorum speculatione qui inter se differunt illis differentiis, sicut etiam in geometrica medietate proportionum aequalitas non potest notari sine terminis, quorum sunt illae proportiones. TANTUM. Excludit proportionum considerationem. SI IGITUR. Ostendit quae sit arithmetica medietas, illa scilicet in qua constituuntur quotcumque numeri, taliter ut aequalis sit eorum ad seipsos differentia. Modo subdividit eam, quod etiam superius fecerat. Sed quia ibi exemplum dederat de geometrica, ideo hic dat de arithmetica, sic quod alia est coniuncta, alia disiuncta. Coniuncta est illa quae in trium terminorum tantum consideratione perpenditur, sive illi termini sint continui sive intermissi, sicut in figura ostenditur. CONSIDERATIO SIT: quia unus medius terminus duorum hinc inde positorum differentia est, ut inter unum et tres binarius differentia est. SIN VERO. Hic diffinit disiunctam; et est idem quod superius dixit de proportionalitatibus, ubi dixit: SI VERO ALIUS AD I, ALIUS AD ALIUM, ubi tamen non de arithmetica sed de geometrica medietate exemplum dederat. HIC. Inter duos primos, utrumque HIC et ILLIC totum de disiuncta intellige, ut in his terminis facile perpendi potest I, II, III, IIII. Namque binarius alias est dux, quia duplus est ad unitatem, alias comes, quippe subsesqualter. Tres vero ad II et IIII ad III UTRIQUE SUNT ALII, tres scilicet ad II sesqualter, IIII vero ad III sesquitercius. Inter quaternarium et unitatem III est differentia, et est quadrupla proportio duobus intermissis, id est binario et ternario. Inter IIII et VII ternarius est differentia, et est supertripartiens quartas, quia continet eum et eius quartam partem ter. In his terminis sunt eaedem differentiae, sed non eaedem proportiones, V ad III superbipartiens tercias, quia continet eum et eius duas tercias. ILLIC: inter sequentes duos. Hic est sensus: medius terminus unus et idem non erit dux et comes, sicut fuit in coniuncta, sed unus erit comes, et alius, diversus ab illo, erit dux.

XLIII, 3. SI IGITUR. Repetit quod dixit, scilicet continuam in tribus tantum terminis constare, disiunctam vero in IIII vel quotlibet aliis. Ideo autem repetit, ut idem quod dixit exemplis declaret.

XLIII, 5. SIN AUTEM. Hic facit aliam subdivisionem utriusque, quia utraque et coniuncta et disiuncta quandoque est intermissa, quandoque non. PERMISCENS. Sic permiscentur, ut ibi fiat praeteritio quandoque unius, quandoque duorum, quandoque plurium, hoc autem modo: SI SINGULOS INTERMITTAS, id est unum tantum, binarius erit differentia, et si

1253 VII] VI MS **1254** supertripartiens quartas] superquadripartiens tercias MS
1256 III] IIII MS

1245–1246 *De arithmetica* II, XL, 7.
1254 Cf. *De arithmetica* I, XXVIII, 4.

duos ternarius, et sic de caeteris. AEQUALES: non dico pares, sed ita ut inter primum et secundum tot intermittas, quot inter secundum et tercium. IN PRIORE: naturalis numeri, quam paulo ante fecit in hac eadem clausula.

XLIII, 9. EST AUTEM PROPRIUM HUIUS, id est arithmeticae medietatis. Inter unum et tres non solum loco medius est binarius, sed et quantitate, quia binarius est medietas quaternarii.

XLIII, 10. EST ILLI. Aliud proprium ostendit quod non multum est magnum, sed puerile dicere. DIFFERENTIAE quidem sunt inter se similes, quia eaedem sunt; sed terminorum quilibet est diversus ab altero: unitas a binario, binarius a ternario. Vult autem sic intelligi: unitas est eadem et similis cum unitate, binarius cum binario, et sic de caeteris, quia aequale aequali est aequale.

XLIII, 11. CONTINUA medietas est in tribus terminis, quando una medietas duas extremitates coniungit. DISIUNCTA vero medietas est in quatuor, vel pluribus terminis, quando non una solum medietas duas extremitates sed duae vel plures coniungunt.

XLIII, 15. QUAE VEL IN MAIORIBUS: istas disiunctivas pro copulativis accipe. Et in maioribus et in minoribus terminis aequalitas proportionis habetur in geometrica medietate, et ideo bene loco medietatis inter duas praedictas ponitur, quia quod aequale est inter duo inaequalia medium ponitur.

CAPUT XLIIII

XLIIII, 1. QUAE SOLA VEL MAXIME, id est praecipue. Quaelibet medietas est proportionalitas, id est habitudo proportionum. Sed geometrica prae aliis melius et iustius dicitur proportionalitas, quia consistit et consideratur in aequalitate proportionum, non in aequalitate differentiarum tantum, ut arithmetica. MULTITUDO NUMERI: qua unus differt ab alio, quia quae illa sit et quam inaequalis non attenditur; in geometrica videlicet.

XLIIII, 4. HOC VERO ETIAM DISIUNCTAE. Non solum haec proprietas convenit disiunctae quae modo diximus, VERO – pro "sed" – etiam istud quod modo dicam convenit disiunctae; et vere LICET. NAM QUEMADMODUM. Noli legere: "etiam disiunctae, sicut coniunctae," quia hoc quod addit non convenit coniunctae, quae tantum in tribus consistit terminis. IDQUE: quod de coniunctis dixi, invenies in coniunctis; quod de disiunctis, invenies in disiunctis.

1293 in aequalitate] inaequalitate MS

XLIIII, 5. IN OMNI DISPOSITIONE, id est in omni geometrica medietate. CONTRA SE SUNT, id est in eadem proportione conveniunt qua et ipsi termini.

XLIIII, 6. NULLI IGITUR: alia consideratio. SUPERIOR ORDO. In figura quae fuit differantiarum reddit subteriores. UNO TERMINO MINUS, quia unitas reddit sibi subpositum binarium uno minus, id est unitate, binarius secunda differentia reddit sibi subpositum IIII uno termino minus, id est binario, et sic de caeteris.

XLIIII, 9. UNA MINUS MULTIPLICATIONE QUAM EST IPSA MINORUM AD MAIORES TERMINOS COMPARATIO, id est una multiplicatione minus quam fuerint ipsi termini maiores ad minores comparati, quia si maior ad minorem fuerit triplus, minor duplicatus erit eorum differentia; si vero quadruplus, minor triplicatus erit differentia, et sic de caeteris. DIFFERENTIAE DUPLI. Subaudit "numeri minores," non quod dupla proportio ibi sit, sed quia minor numerus duplicatur, et facit differentiam. Similiter de triplis.

XLIIII, 10. HAEC AUTEM. Quia dederat exemplum geometricae medietatis in solis tantum multiplicibus, putaret aliquis quod geometrica medietas non inveniretur in aliis speciebus inaequalitatis; ideo ostendit et dicit quod in omnibus eis inveniatur. CONTINUA, id est aequali proportione existente in omnibus. EXEMPLAR, id est documentum. Deberet exemplificare qualiter geometrica medietas inveniatur in omnibus supradictis speciebus. Sed quia hoc longum esset, mittit nos ad finem prioris libri, ubi dat exemplum omnes species inaequalitatis ex aequalitate procedere. In illis enim exemplis omnis illarum specierum geometrica medietas observatur.

XLIIII, 11. VEL IN CAETERIS CONIUNCTIS, id est multiplici superpartic<ulare> et multiplici superpartiente.

XLIIII, 12. VEL IN MAIORIBUS VEL IN MINORIBUS: disiunctivas fac copulativas, quia et in maioribus et in minoribus aequalitas proportionum est, quod non est in arithmetica; ibi enim in minoribus terminis maior proportio et in maioribus minor custoditur. IN BINIS PROPORTIONIBUS: ubi oportet esse tres terminos, id est in coniuncta medietate. Et ideo dicit: "in binis proportionibus," ad ostendendum hoc non posse fieri in disiuncta, quia si alternatim ponimus tetragonos et par<te> al<tera> l<ongiores> et facimus disiunctam unam in IIII terminis, vel in amplioribus, non erit eadem proportionum aequalitas.

CAPUT XLV

XLV, 1. ATQUE IDEO. Istud adhuc dicit pro commendatione geometricae medietatis, et ei optime sedet ista comparatio; duabus vero aliis non bene convenit neque necessario. PARITAS QUAEDAM, id est medietas quaedam in paritate proportionum existens; vel medietas quaedam vicem paritatis et aequalitatis servans inter illas duas, quia aequum ius servat in proportionibus, id est aequales habet proportiones tam in ma<ioribus> quam in minoribus terminis, quod nulli alii medietati convenit.

STATUS dicimus dispositiones et ordines rerum. STATUTA quoque ordinata atque disposita dicimus. Et status populi romani triphariam erant divisi in senatores, milites et plebem. His talibus tres medietates mathematicae comparat. RES PUBLICA dicitur administratio regni romani. OPTIMATES generali nomine dicebantur, qui maiores dignitates post imperatorem administrabant, sicut consules, patricii, aediles, praefecti, agentes in rebus quos in chaldea et palestina lingua satrapas appellamus.

CAPUT XLVI

XLVI, 1. QUIDDAM NIMIS UTILE. Dicit enim Plato quod duo plana una medietate tantum iungantur, et est intelligendum de quadratis superficiebus, quia in caeteris falleret. Dicit etiam quod duo solida duabus medietatibus semper iunguntur, et est intelligendum de cubicis. Medietates vero geometricas debemus intelligere. Unde Plato probat quod ignis et terra duobus mediis necessario sint coniuncta. Facit autem quaestionem Calcidius: quomodo, cum terra sit cubica, ignis vero pyramoides, quomodo inquam duobus mediis coniungantur necessario, cum hoc de cubicis tantum dicatur? Dicit autem postea quod ignis, quamvis sit pyramicus, in mundi tamen compositione rotundus est, et propter rotunditatis perfectionem dicitur cubicus. Unde bene dictum est quod ipse et terra duobus mediis iunguntur, quod in numeris hic ostenditur. Bis bini enim primus quadratus actu est, ter terni secundus, sed bis bini IIII fiunt, ter terni VIIII. Sumatur unum latus quaternarii, id est bis, et aliud latus novenarii, id est ter, et dicatur bis tres, fient VI, qui ad utrumque habet se geometrica medietate. Item bis bini bis

1346–1349 Cf. Caiazzo, "Un commento altomedievale," 148, ll. 632–634: "STATUS dicimus dispositiones et ordines rerum; STATUTA quoque ordinata atque disposita dicimus. Et status populi Romani trifariam: erant divisi in senatores, milites et plebem. His statibus tres medietates mathematicae comparat."

1354–1359 Plato, *Timaeus* 31B–32C (ed. Waszink, pp. 24–25).

1359–1364 Calcidius, *Commentarius in Timaeum* 20 and 326 (ed. Waszink, p. 71 and p. 321, respectively); the texts are cited above in the note to 855–858.

1364–1373 See Caiazzo, "La forme et les qualités des éléments: lectures médiévales du *Timée*," in *Il Timeo: Esegesi greche, arabe, latine*, ed. Celia and Ulacco, pp. 307–345.

primus est cubicus, id est VIII, ter terni ter secundus, id est XXVII. Sumantur ergo duo latera VIII, id est bis bini, et unum latus XXVII, <id est> ter, et dicatur bis bini ter, erunt XII. Sumantur ergo duo latera etiam XXVII, id est ter terni, et unum latus de VIII, id est bis, et dicatur ter terni bis, fient XVIII. Et erit geometrica medietas: XXVII, XVIII, XII, VIII.

Cosmopeia: "cosmos" graece "mundus," "poio" "facio," hinc poeta. Et cosmopeia dicitur mundi factura, quae in libro Platonis qui *Thimaeus* vocatur, continetur. Microcosmus: minor mundus. Marcianus: Calliopeia, id est pulchrifaciens vel pulchrisonans.

Omnes planae figurae. De superficialibus numeris intellige, ut bis bini et ter terni medium habent VIrium. Intervallum hic non accipe pro dimensione, licet tamen et hoc verum sit quod planae figurae duas habent dimensiones, longum et latum, et cubicae tres; tamen ubique intervallum pro distantia accipe. Alia quae iungat: quia nulla alia medietas potest inveniri quae, ad alterum numerum circa se positum, aequalem habeat proportionem.

XLVI, 3. Hoc autem evenit: quod senarius est medietas inter hos duos. Hoc evenit sic accipe: unum latus quaternarii et unum latus VIIIIrii, quod ex illorum multiplicatione surrexerit, hoc erit medietas. Quaternarii latus est IIrius, VIIIIrii IIIrius, XVI habet IIII in latere, XXV habet V in latere. Multiplica haec latera et dic: quater V, faciunt XX. Medietas illorum duorum, XVI videlicet et XXV, XX est, qui ad utrosque sesquiquarta proportione iungitur. Et quotienscumque. Sicut invenimus medietatem IIIIrii et VIIIIrii, ita generaliter quorumlibet tetragonorum medietates inveniendae sunt. Latus tetragoni est numerus ex cuius multiplicatione ipse tetragonus conficitur, ut bis bini, ter terni.

XLVI, 4. Eadem ratio laterum est, quia per multiplicationem laterum et hic invenientur duae medietates.

XLVI, 7. Hoc autem, scilicet de medietatibus et de tali multiplicatione plenius agitur in Platone, in libris de re publica. Phylosophus proprium nomen fuit unius discipulorum Platonis. Illum introducit Plato sub persona

1396 medietates] medietatis MS

1374–1376 Cf. Caiazzo, "Un commento altomedievale," 148, ll. 635–636: "Cosmopeia Timaei, id est factura mundi, quae in libro Platonis, qui Timaeus vocatur, continetur."

1376–1377 Remi of Auxerre, *Commentum in Martianum Capellam* 1.4.3 (ed. Lutz, 1: 70): "Calliopea una est ex novem Musis. Interpretatur autem secundum Fulgentium callion phone, id est pulchra vox; sive Calliopea dicta pulchrifica vel pulchre faciens, quia callos Grece pulcher, poio facio, hinc et poeta dicitur."

Musarum quiddam de Musis dicentem, unde ille LOCUS dicitur NUPTIALIS. Cicero etiam, imitatus Platonem, elegantissimum librum scripsit de re publica, ad cuius similitudinem Martianus duos libros scripsit de nuptiis Phylologiae et Mercurii. Nota EO LOCO QUI DICITUR NUPTIALIS, quia videlicet ibi loquitur de nuptiis Phylologiae et Mercurii ex persona Musarum, quas introducit quasi ad Iovem loquentes.

Caput XLVII

XLVII, 1. ARMONICA MEDIETAS geometricae medietati cognata est, quia sicut geometrica easdem habet proportiones in omnibus, et in maioribus et in minoribus, ita et armonica medietas, quas habet proportiones sub extremis terminis, easdem habet sub differentiis terminorum. Et geometricam medietatem cubum dicunt, quia in omnibus cubis potest considerari. Nam cubus per omnia latera aequaliter multiplicatur, et geometrica medietas per aequales proportiones fit semper, vel in multiplicibus vel in superparticularibus, et caeteris.

XLVII, 3. Ut DICTUM EST: ubi quartum proprium arithmeticae medietatis ostendit. IUSTE IGITUR. Quia tales sunt medietates, arithmetica et armonica, igitur geometrica propter eius proprietates est media inter illas duas.

XLVII, 4. HOC QUOQUE. Alia causa quare iuste media sit geometrica medietas inter illas duas. SED ALIA PARTE. Superat medius minorem, id est IIIus IIum, id est ipsa pars alia est in minore termino, alia in maiore, sicut unitas in binario medietas est, in quaternario quarta pars.

XLVII, 5. RECTE IGITUR. Namque quantum ad ternarium, qui medietas est, eadem pars est semper unitas, id est IIIa. Ipsa tamen unitas in minore, id est binario, medietas est, in maiore vero, id est quaternario, IIIIta pars est.

1412 multiplicatur] multiplicantur MS

1407–1414 Cf. Caiazzo, "Un commento altomedievale," 149, ll. 640–646: "Armonica medietas GEOMETRICAE medietatis COGNATA EST. Quia sicut geometrica easdem habet proportiones in omnibus, et in maioribus et in minoribus, ita et armonica medietas, quas habet proportiones sub extremis terminis, easdem habet sub differentiis terminorum. Et geometricam medietatem CYBUM DICUNT, quia in omnibus cybis potest considerari. Nam cybus per omnia latera aequaliter multiplicatur. Et geometrica medietas per aequales proportiones fit semper, vel in multiplicibus, vel superparticularibus, et caeteris."

Caput XLVIII

XLVIII, 1. PER AEQUAS DIFFERENTIAS. Aequalis enim est terminorum differentia. QUAERIT ENIM, id est exigit ipsa med<ietas>.

XLVIII, 2. IN PRIMI LIBRI, id est in primo libro ubi divisionem omnium rerum fecimus. Hoc autem fuit in secundo prologo, ubi dixit: ILLAM VERO QUAE AD ALIQUID. Ideo autem dicit: IN DIVISIONE PRIMI LIBRI, quia ille secundus prologus est quasi pars libri, quia in eo praelibatur materia. Est enim "ante rem."

PAENE. Quia de semitonio singulariter per se nihil dicit, vel dicamus quod more phylosophorum temperavit. DIATESSERON QUODAMMODO OBTINET VIM ELEMENTI, quia, sicut in fine huius libri dicet, tonus est minimus sonus et communis mensura omnium musicorum sonorum.

XLVIII, 6. NAM IN UTRISQUE. Duplex dispositio est, ubi extremus terminus maior duplus est ad minorem extremum. Triplex autem est illa dispositio, ubi unus extremorum continet alium extremum ter.

XLVIII, 9. ET QUONIAM TRIPLUS. Sensus hic est: ubicumque est triplus, ibi est diapason et diapente. Sed differentiae istarum sic sunt ut una tripla sit alterius. Ergo et ibi in differentiis erit diapason et diapente, sic<ut> in ipsis terminis quorum sunt differentiae.

XLVIII, 11. BIS DIAPASON ex dupla prop<ortione> coll<igitur>. Non inter duos extremos terminos, quorum unus est duplus ad alium, sed inter medium et differentiam sui, ad quam ipse est quadruplus, ac minoris. Et quare non inter duos extremos? Ideo quia ibi consideratur diapason. Et hoc est quod dicit: QUONIAM DIAPASON INTERSERIT SE HUIC IUNCTURAE, id est tali iuncturae ARMO<NICAE> MED<IETATIS>, in qua sit unus extremus duplus ad alium extremum. Diapason nascitur ex duplici, bis diapason ex quadruplo. Et Boetius dicit hic quod nascatur ex duplici, quod etiam verum est, si ad differentiam medii et minoris et ad ipsum medium respexeris, non ad duos extremos inter quos est diapason.

Caput XLVIIII

XLVIIII, 1. ARMONIAE COGNATA, scilicet in similitudine proportionum et differentiarum. Nam quales extremorum terminorum fuerint proportiones, tales et differentiae maiorum ad differentiam minorum terminorum.

1430–1431 Boethius, *De arithmetica* I, 1, 4.

XLVIIII, 3. Diatesseron quidem est. Notandum quod arithmeticae proportiones a maioribus ad minora perveniunt, quia videlicet in minoribus terminis maior proportio repperitur, ut binarius ad unitatem duplus est, ternarius ad binarium non iam duplus sed sesqualter; sicque in infinitum proportiones diminuendo protenduntur, ut sesquitercius, sesquiquartus, sesquiquintus, et rel<iqua>. Musicae autem symphoniae a minoribus inchoant, ut diatesseron id est sesquitercia, diapente <id est> sesqualtera, diapason id est dupla.

Caput L

L, 1. Nos autem praestare debemus: per doctrinam naturam. Dato calamo. Sumit similitudinem ab his qui calamo canunt; sumit etiam ab his qui nervis reddunt sonum, hoc est chordis, quia chordae ex nervis aliquando fiunt. Musicis, cantoribus, est mos, ut per<mutantes> medium foramen. Singulariter dixit, non quin ibi plura sint foramina, sed ideo quia non mutatur nisi unum, quodcumque ad illum vel illum vel illum sonum faciendum. Vel cum duabus. Ibi tres erunt chordae, et duae altrinsecus manent, media trahitur plus, et subtilem et acutum sonum facit, quandoque remittitur et gravem sonum reddit.

Tetragonicum latus inquire. Latus appello numerum illum qui multiplicat. Tetragonicum latus est illud, quando multiplicans, par est multiplicato.

L, 3. Quantum eorum differentiae, id est XV, quia quindecies quindeni CCXXV sunt. Non autem eadem parte minoris. Quia quinarius non ea pars est denarii, quae est in XL, quia denarii medietas est, in XL vero octava pars est.

L, 12. Committe iuxta eum. Non ideo ut connumeres CCC cum L, sed ideo ut multiplicentur L per senarium, et ex illa multiplicatione surgant CCC; et sic appareat senarium esse latus CCCtorum. Committe iuxta eum, id est secundum eum constitue CCC, et vide quem numerum multiplicans faciat CCC. Hoc erit VI, et hic est latitudo CCCtorum, quia multiplicans facit longitudinem.

1469–1471 Cf. Caiazzo, "Un commento altomedievale," 149, ll. 649–651: "Sumit similitudinem ab his, qui calamo canunt; sumit etiam ab his, qui nervis reddunt sonum, hoc est chordis, quia chordae ex nervis aliquando fiunt."

Caput LI

LI, 8. Propter differentiam a minoribus ad maiores conversam, quia in geometrica est proportio differentiarum a maioribus ad minores conversa.

Caput LIIII

LIIII, 1. Restat: in speculatione naturalium quaestionum, quae habentur de elementis, quae elementa ad ipsorum ligaturam duo media exquirunt, et cubicos et solidos numeros habere debent. Perfectius: quia nichil potest adici in corporibus super haec intervalla, super longitudinem, latitudinem et altitudinem.

LIIII, 2. Qui vel ab aequalibus: ut cubi, bis bini bis. Vel ab inaequalibus: ut cunei, bis tres quater. Vel quolibet alio modo: ut bis bini ter, vel ter terni bis.

LIIII, 6. Has igitur omnes: quia tres habent dimensiones, ut VI habet longum semel, latum bis, altum ter, quia semel bis ter faciunt VI. Superius enim habuimus quod omnis numerus per unitatem multiplicatus non plus redderet quam se totum, ut semel duo sunt duo, et rel<iqui>.

Tribus intervallorum dimensionibus, id est tribus intervallis, quae sunt dimensiones quas assignabis in reliquis, sicut in uno bis ter iam diximus.

1491–1492 Propter differentiam a minoribus ad maiores conversam] ms. propter proportionem differentiarum a minoribus ad maiores terminos conversam *ed. Guillaumin*

Appendix

A Handlist of Medieval Commentaries on Boethius's De arithmetica

Over the course of the last few years, I have collected a great deal of material on medieval commentaries on Boethius's *De arithmetica*. I hesitated for a long time before publishing my notes, which are both disparate and of varying significance, in the hopes of someday completing an exhaustive, descriptive catalogue of the commentaries and the manuscripts that contain them. I now realize that I had seriously underestimated the task, and have decided it is perhaps more reasonable to publish my inventory, even in preliminary form, so that it can be of use to other scholars working in the field. Only independent, lemmatic commentaries presented as discrete texts are indexed in the following chronological handlist; neither the abundant interlinear and marginal glosses that survive in manuscript nor summaries and paraphrases have been included.

1. *Commentary of Fleury*

Leiden, Bibliotheek der Rijksuniversiteit, Voss. lat. 8° 88, fols. 42r–v, 45r; tenth century; French hand (Fleury).

2. *Commentary of the Vatican*

Vatican City, Biblioteca Apostolica Vaticana, Reg. lat. 1424, fols. 95r–98v; tenth century; French hand?

3. *Commentary of the School of Johannes Scotus*

† Chartres, Bibliothèque municipale, MS 46, fol. 28v + Leiden, Bibliotheek der Rijksuniversiteit, Voss. lat. Fol. 70 A^{II}, fols. 74r–78v; end of the tenth century; French hand (Fleury?).

Melk, Stiftsbibliothek, Cod. Mell. 740 (228; 1847; G 31), fols. 38v–44r; end of the tenth or beginning of the eleventh century; written by Froumund of Tegernsee (*ca* 960–1012).

Munich, Bayerische Staatsbibliothek, lat. 18764, fols. 2r–v, 5v; end of the tenth or beginning of the eleventh century; written by Froumund of Tegernsee (*ca* 960–1012).

Reims, Bibliothèque municipale, MS 975 (I. 706), fols. 29r–32v; beginning of the twelfth century.

4. *Commentary of Florence*

Florence, Biblioteca Medicea Laurenziana, Plut. 51.14, fols. 78r–86v; end of the eleventh or beginning of the twelfth century; Italian hand.

5. *Commentary of Bern 1*

Bern, Burgerbibliothek, Cod. 633, fols. 19ra–27ra; middle or second half of the twelfth century; French hand.

6. *Commentary of Bern 2*

Bern, Burgerbibliothek, Cod. 633, fols. 27ra–31rb; middle or second half of the twelfth century; French hand.

7. *Commentary of Durham*

Durham, Cathedral Chapter Library, C.IV.7, fols. 50r–66r; first half of the twelfth century.

8. *Commentary of Turin*

Turin, Biblioteca Nazionale Universitaria, D.V.38, fols. 35r–48r; twelfth century.

9. *Commentary of Thierry of Chartres*

Stuttgart, Württembergische Landesbibliothek, Cod. math. 4° 33, fols. 1ra–34ra; middle or second half of the twelfth century; south-German hand (edited in this volume).

10. *Commentary of Paris*

Paris, Bibliothèque nationale de France, lat. 7185, fols. 41r–48v; middle or second half of the twelfth century; French hand.

11. *Commentary of the School of Saint Victor*

Munich, Bayerische Staatsbibliothek, lat. 4643, fols. 100r–107v; end of the twelfth or beginning of the thirteenth century.

12. *Commentary of Simon Bredon*

Oxford, Corpus Christi College, MS 118, fols. 101v–118r; fourteenth century.

13. *Commentary of Oxford*

Oxford, Bodleian Library, Laud. Misc. 644, fols. 132r–139; fourteenth century.

Bibliography

Primary Sources

Achard of Saint Victor. *L'unité de Dieu et la pluralité des créatures ("De unitate <Dei> et pluralitate creaturarum").* Ed. and trans. Emmanuel Martineau. Paris: Authentica, 1987.

Adam of Petit Pont. *Ars disserendi.* In *Twelfth Century Logic: Texts and Studies,* ed. Lorenzo Minio-Paluello, 1: 1–155. 2 vols. Rome: Edizioni di Storia e Letteratura, 1956–1958.

Adelard of Bath. *De eodem et diverso.* In Adelard of Bath, *Conversations with His Nephew: On the Same and the Different, Questions on Natural Science and On Birds,* ed. and trans. Charles Burnett, pp. 1–79. Cambridge: Cambridge University Press, 1998.

—. *The First Latin Translation of Euclid's "Elements" Commonly Ascribed to Adelard of Bath.* Ed. Hubert L.L. Busard. Toronto: Pontifical Institute of Mediaeval Studies, 1983.

—. *Quaestiones naturales.* In Adelard of Bath, *Conversations with His Nephew: On the Same and the Different, Questions on Natural Science and On Birds,* ed. and trans. Charles Burnett, pp. 81–235. Cambridge: Cambridge University Press, 1998.

Alanus, de Insulis. *Alain de Lille: Textes inédits.* Avec une introduction sur sa vie et ses oeuvres par Marie-Thérèse d' Alverny. Paris: J. Vrin, 1965.

Alexander Neckam. *Commentum super Martianum.* Ed. Christopher J. McDonough. Florence: SISMEL–Edizioni del Galluzzo, 2006.

—. *Corrogationes Promethei.* In Paul Meyer, "Notice sur les *Corrogationes Promethei* d'Alexandre Neckam," *Notices et extraits des manuscrits de la Bibliothèque nationale et autres bibliothèques* 35 (1897), 641–682.

—. *Suppletio defectuum.* Ed. Peter Hochgürtel. CCCM 221. Turnhout: Brepols, 2008.

Annales Sancti Iacobi Leodiensis: Chronicon breve Leodiense, ex codice Aureae-vallis, cum proemio et notis L.C. Bethmann, recogn. J. Alexandre. Liège: Abbaye de St. Jacques le mineur, 1874.

Anselm of Havelsberg. "Vita Adalberti II Moguntini." Ed. Philipp Jaffé. *Bibliotheca Rerum Germanicarum,* 3: 565–603. 6 vols. Berlin: Weidmann, 1866.

Aratus. *Germanici Caesaris Aratea: cum scholiis.* Ed. Alfred Breysig. Berlin: Georg Reimer, 1867.

Aristotle. *Categoriae vel praedicamenta*. Ed. Lorenzo Mino-Paluello. Aristoteles latinus 1.1–5. Bruges and Paris: Desclée de Brouwer, 1961.

—. *De sophisticis elenchis (translatio Boethii)*. Ed. Bernard G. Dod. Aristoteles latinus 6.1–3. Leiden: Brill; Brussels: Desclée de Brouwer, 1975.

Arnold of Bonneval. *Commentarius in Psalmum CXXXII*. PL 189: 1569–1590.

Augustine. *De doctrina christiana*. Ed. Joseph Martin. CCL 32. Turnhout: Brepols, 1962.

—. *Retractationes*. Ed. Almut Mutzenbecker. CCL 57. Turnhout: Brepols, 1984.

—. *De Trinitate*. Ed. William J. Mountain. CCL 50–50A. Turnhout: Brepols, 1968.

Avicenna. *Liber de anima seu Sextus de Naturalibus*. Ed. Simone Van Riet. 2 vols. Leuven: Peeters; Leiden: Brill, 1968–1972.

Balderich of Florennes (Baldericus Scholasticus). *Gesta Alberonis archiepiscopi*. Ed. Georg H. Pertz. MGH SS 8. Hannover: Hahn, 1848.

—. *A Warrior Bishop of the Twelfth Century: The Deeds of Albero of Trier by Balderich*. Trans. Brian A. Pavlac. Toronto: Pontifical Institute of Mediaeval Studies, 2008.

Bede. *De natura rerum*. Ed. Charles W. Jones. CCL 123A. Turnhout: Brepols, 1975.

—. *De temporibus*. Ed. Charles W. Jones. CCL 123C. Turnhout: Brepols, 1980.

Bede (pseudo). *De signis caeli*. In *Commentariorum in Aratum reliquiae*, ed. Ernst Maass, pp. 582–594. Berlin: Weidmann, 1898.

Bernard of Utrecht – Conrad of Hirsau. *Dialogus super auctores*. Ed. Robert B. C. Huygens. Leiden: Brill, 1970.

Bernard of Chartres. *Glosae super Platonem*. Ed. Paul E. Dutton. Toronto: Pontifical Institute of Mediaeval Studies, 1991.

Bernard of Clairvaux. *Tractatus de erroribus Abaelardi ad Innocentium II (Epistola 190)*. PL 182: 1053–1072.

Bernard Silvestris. *The Commentary on Martianus Capella's "De nuptiis Philologiae et Mercurii" Attributed to Bernardus Silvestris*. Ed. Haijo J. Westra. Toronto: Pontifical Institute of Mediaeval Studies, 1986.

—. *Cosmographia*. Ed. Peter Dronke. Leiden: Brill, 1978.

—. *Experimentarius*. Ed. Mirella Brini Savorelli. *Rivista critica di storia della filosofia* 14 (1959), 283–342.

—. *Mathematicus*. Ed. Deirdre M. Stone, "Bernardus Silvestris, *Mathematicus:* Edition and Translation." *AHDLMA* 63 (1996), 209–283.

Boethius. *De arithmetica*. Ed. Henri Oosthout and Johannes Schilling. CCL 94A. Turnhout: Brepols, 1999.

—. *De institutione arithmetica*. Ed. Gottfried Friedlein. Leipzig: Teubner, 1867.

—. *Institution arithmétique*. Ed. Jean-Yves Guillaumin. Paris: Les Belles Lettres, 1995.
—. *De consolatione philosophiae*. Ed. Ludwig Bieler. Revised edition. CCL 94. Turnhout: Brepols, 1984.
—. *De differentiis topicis*. PL 64: 1173–1222.
—. *De institutione musica*. Ed. Gottfried Friedlein. Leipzig: Teubner, 1867.
—. *In Categorias Aristotelis*. PL 64: 159–294.
—. *In Isagogen Porphyrii Commenta*. Ed. Georg Schepss and Samuel Brandt. CSEL 48. Vienna: Tempsky, 1906.
—. *In librum Aristotelis Peri hermeneias*. Ed. Karol Meiser. Leipzig: Teubner, 1877–1880.
—. *Opuscula sacra*. In *The Theological Tractates and the Consolation of Philosophy*, ed. H.F. Stewart, E.K. Rand and S.J. Tester. Cambridge, MA: Harvard University Press, 1973.
—. *Opuscula sacra*. Ed. Claudio Moreschini, trans. and comm. Alain Galonnier. Philosophes médiévaux 47, 53. 2 vols. Louvain and Paris: Peeters, 2013.
—. *Opuscula theologica*. Ed. Claudio Moreschini. Munich: Teubner, 2000.
"Boethius" Geometrie II: Ein mathematisches Lehrbuch des Mittelalters. Ed. Menso Folkerts. Wiesbaden: Franz Steiner, 1970.
Byrhtferth (pseudo). *In De temporum ratione*. PL 90: 297–548.
Calcidius. *Timaeus a Calcidio translatus commentarioque instructus*. Ed. Jan H. Waszink. Plato latinus 4. 2nd ed. London: Warburg Institute, 1975.
Cicero. *De l'invention*. Ed. and trans. Guy Achard. Paris: Les Belles Lettres, 2002.
Clarembald of Arras. *The Boethian Commentaries of Clarembald of Arras*. Ed. and trans. David B. George and John Fortin. Notre Dame: University of Notre Dame Press, 2002.
—. *Life and Works of Clarembald of Arras*. Ed. Nikolaus M. Häring. Toronto: Pontifical Institute of Mediaeval Studies, 1965.
Dominicus Gundissalinus. *De divisione philosophiae – Über die Einteilung der Philosophie*. Ed. and trans. Alexander Fidora and Dorothée Werner. Herders Bibliothek der Philosophie des Mittelalters 11. Freiburg: Herder, 2007.
Epitaphium Theodorici. In Wilhelm Wattenbach, "Beschreibung einer Handschrift mittelalterlicher Gedichte (Berl. Cod. theol. oct. 94)," *Sitzungsberichte der königlich preussischen Akademie der Wissenschaften zu Berlin* 8 (1895), 123–156, at 149.
Eriugena, Johannes Scotus. *Periphyseon*. Ed. Edouard Jeauneau. CCCM 161–165. 5 vols. Turnhout: Brepols, 1996–2003.

—. *Annotationes in Marcianum*. Ed. Cora E. Lutz. Cambridge, MA: The Medieval Academy of America, 1939.

Firmicus Maternus. *Mathesis*. Ed. and trans. Pierre Monat. Paris: Les Belles Lettres, 1992–1997.

Garlandus Compotista. *Der "Computus Gerlandi." Edition, Übersetzung und Erläuterungen*. Ed. Alfred Lohr. Stuttgart: Franz Steiner, 2013.

—. *Dialectica*. Ed. Lambertus M. De Rijk. Assen: Van Gorcum, 1959.

Geoffrey of Auxerre. "De sacramentis numerorum a tredenario usque ad vicenarium"; "De creatione perfectorum et sacramento." In Hanne Lange, "Traités du XIIe siècle sur la symbolique des nombres: Geoffroy d'Auxerre et Thibault de Langres," *Cahiers de l'Institut du Moyen Age grec et latin de Copenhague* 29 (1978), 1–22, and 23–28.

Gerbert of Aurillac. *Gerberti postea Silvestri II papae "Opera mathematica."* Ed. Nikolaj Bubnov. Berlin: Friedländer, 1899.

Gilbert of Poitiers. *The Commentaries on Boethius by Gilbert of Poitiers*. Ed. Nikolaus M. Häring. Toronto: Pontifical Institute of Mediaeval Studies, 1966.

Glosae Colonienses super Macrobium. In *Lectures médiévales de Macrobe: Les Glosae Colonienses super Macrobium – Etude et édition*, ed. Irene Caiazzo, pp. 165–275. Paris: J. Vrin, 2002.

Hermann of Carinthia. *De essentiis*. Ed. and trans. Charles Burnett. Leiden: Brill, 1982.

Hermann of Reichenau. *De mensura astrolabii*. In Joseph Drecker, "Hermannus Contractus Über das Astrolab," *Isis* 16 (1931), 200–219, at 203–212.

Hermes Trismegistus. *Liber de sex rerum principiis*. Ed. Mark Delp and Paolo Lucentini. CCCM 142. Turnhout: Brepols, 2006.

Hugh of Saint Victor. *Didascalicon de studio legendi*. Ed. Charles H. Buttimer. Washington, DC: The Catholic University Press, 1939.

Isidore of Seville. *Etymologiae*. Ed. Wallace M. Lindsay. New York: Oxford University Press, 1911.

John of Salisbury. *Metalogicon*. Ed. John B. Hall and Katharine S.B. Keats-Rohan. CCCM 98. Turnhout: Brepols, 1991.

Macrobius. *Commentarii in Somnium Scipionis*. Ed. James Willis. Stuttgart and Leipzig: Teubner, 1963. Repr. 1970, 1994.

Marius Victorinus. *Explanationes in Ciceronis Rhetoricam*. Ed. Antonella Ippolito. CCL 132. Turnhout: Brepols, 2006.

Martianus Capella. *De nuptiis Philologiae et Mercurii*. Ed. James Willis. Leipzig: Teubner, 1983.

Nicholas of Cusa. *De docta ignorantia*. Ed. Paul Wilpert and Hans Gerhard Senger. Philosophisch–Theologische Werke 1. Hamburg: Felix Meiner, 2002.

Odon of Morimond. "Analectica numerorum et rerum in theographyam (I)." In Hanne Lange, "Traités du XIIe siècle sur la symbolique des nombres: Odon de Morimond (1116–1161)," *Cahiers de l'Institut du Moyen Age grec et latin de Copenhague* 40 (1981), 1–191.

—. "Analectica numerorum et rerum in theographyam (II)." In Hanne Lange, "Traités du XIIe siècle sur la symbolique des nombres: Odon de Morimond (1116–1161)," *Cahiers de l'Institut du Moyen Age grec et latin de Copenhague* 58 (1989), 1–322.

Otto of Freising. *Gesta Friderici I Imperatoris*. Ed. Georg Waitz. MGH: Scriptores rerum Germanicarum in usum scholarum. Hannover and Leipzig: Hahn,1912.

Peter Abelard. *Dialectica*. Ed. Lambertus M. De Rijk. 2nd rev. ed. Assen: Van Gorcum, 1970.

—. *Historia calamitatum mearum*. Ed. Jacques Monfrin. Paris: J. Vrin, 1967.

—. *Theologia christiana*. Ed. Eligius M. Buytaert. CCCM 12. Turnhout: Brepols, 1969.

—. *Theologia scholarium (recensio brevis)*. Ed. Eligius M. Buytaert. CCCM 12. Turnhout: Brepols, 1969.

—. *Theologia 'Scholarium.'* Ed. Eligius M. Buytaert and Constant J. Mews. CCCM 13. Turnhout: Brepols, 1987.

—. *Theologia 'Summi boni.'* Ed. Eligius M. Buytaert and Constant J. Mews. CCCM 13. Turnhout: Brepols, 1987.

Petrus Helias. *Summa super Priscianum*. Ed. Leo Reilly. Toronto: Pontifical Institute of Mediaeval Studies, 1993.

Pliny, the Elder. *Histoire naturelle*. Ed. Jean Beaujeu. Paris: Les Belles Lettres, 1950.

Preceptum Canonis Ptolomei. Ed. David Pingree. Louvain-la-Neuve: Academia Bruylant, 1997.

Priscian. *Institutiones grammaticae*. Ed. Martinus Hertz. In *Grammatici latini* 2–3, ed. Heinrich Keil. Leipzig: Teubner, 1855–1859.

Quintilian. *Institutiones oratoriae libri XII*. Ed. Ludwig Radermacher. Leipzig: Teubner, 1965.

Raymond of Marseilles. *Opera omnia*, Vol. 1: *Traité de l'astrolabe; Liber cursuum planetarum*. Ed. and trans. Marie-Thérèse d'Alverny, Charles Burnett, and Emmanuel Poulle. Paris: CNRS Editions, 2009.

Remi of Auxerre. *Commentum in Martianum Capella*. Ed. Cora E. Lutz. Leiden: Brill, 1962–1965.

Robert of Chester. *Robert of Chester's (?) Redaction of Euclid's "Elements." The So-Called Adelard II Version*. Ed. Hubert L.L. Busard and Menso Folkerts. Boston: Birkhäuser, 1992.

Saeculi noni auctoris in Boethii Consolationem philosophiae commentarius. Ed. Edmund T. Silk. Rome: American Academy in Rome, 1935.

Scholica graecarum glossarum. In M.L.W. Laistner, "Notes on Greek from the Lectures of a Ninth-Century Monastery Teacher," *Bulletin of John Rilands Library* 7 (1923), 421–456.

The Berlin Commentary on Martianus Capella's "De nuptiis Philologiae et Mercurii" Book 1. Ed. Haijo J. Westra. Leiden: Brill, 1994.

Thibaut of Langres. "De quatuor modis quibus significationes numerorum aperiuntur." In Hanne Lange, "Traités du XIIe siècle sur la symbolique des nombres: Geoffroy d'Auxerre et Thibault de Langres," *Cahiers de l'Institut du Moyen Age grec et latin de Copenhague* 29 (1978), 29–108.

Thierry of Chartres. *Commentaries on Boethius by Thierry of Chartres and His School*. Ed. Nikolaus M. Häring. Toronto: Pontifical Institute of Mediaeval Studies, 1971.

—. *The Latin Rhetorical Commentaries by Thierry of Chartres*. Ed. Karin M. Fredborg. Toronto: Pontifical Institute of Mediaeval Studies, 1988.

"Tratado sobre el uso del astrolabio planisférico, de Maslama al-Maŷrītī, de Córdoba, en la traducción latina de Johannes Hispanus (perdido el original árabe)." In *Las traducciones orientales en los manuscritos de la Biblioteca Catedral de Toledo*, ed. José M. Millás Vallicrosa, pp. 261–284. Madrid: Consejo Superior de Investigaciones Científicas, 1942.

William of Conches. *Dragmaticon philosophiae*. Ed. Italo Ronca. CCCM 152. Turnhout: Brepols, 1997.

—. *Glosae super Boetium*. Ed. Lodi Nauta. CCCM 158. Turnhout: Brepols, 1999.

—. *Glosae super Macrobium*. In Helen Rodnite, "The Doctrine of the Trinity in Guillaume de Conches' *Glosses on Macrobius*: Texts and Studies." PhD diss. Columbia University, 1972.

—. *Glosae super Platonem*. Ed. Edouard Jeauneau. CCCM 203. Turnhout: Brepols, 2006.

—. *Philosophia mundi*. Ed. Gregor Maurach. Pretoria: University of South Africa, 1980.

William of Lucca. *De arithmetica compendiose tractata: Dal codice 614 (sec. XII) della Biblioteca Capitolare Feliniana di Lucca*. Ed. Gino Arrighi. Pisa: Domus Galilaeana, 1964.

William of Tyre. *Chronicon*. Ed. Robert B.C. Huygens, Hans E. Mayer and Gerhard Rösch. CCCM 63–63A. Turnhout: Brepols, 1986.

Secondary Sources

Albertson, David C. " 'Our Boethius': Traditions of Thierry of Chartres and the Christology of Nicolaus Cusanus in *De docta ignorantia*." PhD diss. University of Chicago, 2008.

—. "A Learned Thief? Nicholas of Cusa and the Anonymous *Fundamentum Naturae*: Reassessing the *Vorlage* Theory." *Recherches de théologie et philosophie médiévales* 77 (2010), 351–390.

—. "A Late Medieval Reaction to Thierry of Chartres's (d. 1157) Philosophy: The Anti-Platonist Argument of the Anonymous *Fundamentum Naturae*." *Vivarium* 50 (2012), 53–84.

—. "Achard of St. Victor (d. 1171) and the Eclipse of the Arithmetic Model of the Trinity." *Traditio* 67 (2012), 101–144.

—. *Mathematical Theologies: Nicholas of Cusa and the Legacy of Thierry of Chartres*. Oxford: Oxford University Press, 2014.

Alessio, Franco. "La filosofia e le *artes mechanicae* nel s. XII." *Studi Medievali*, 3rd ser., 6 (1965), 69–161.

Alverny, Marie-Thérèse d'. "Le cosmos symbolique du XIIe siècle." *AHDLMA* 20 (1953), 31–81.

—. "Astrologues et théologiens au XIIe siècle." In *Mélanges offerts à M.-D. Chenu, maître en théologie*, pp. 31–50. Paris: J. Vrin, 1967.

—. "Abélard et l'astrologie." In *Pierre Abélard-Pierre le Vénérable: les courants philosophiques, littéraires et artistiques en Occident au milieu du XIIe siècle*. Colloque international du Centre national de la recherche scientifique, Abbaye de Cluny, 2 au 9 juillet 1972, pp. 611–630. Paris: CNRS, 1975.

—. *See also* Primary Sources *under* Alanus, de Insulis.

Baeumker, Clemens. "Der Platonismus im Mittelalter." In *Platonismus in der Philosophie des Mittelalters*, ed. Werner Beierwaltes, pp. 3–55. Darmstadt: Wissenschaftliche Buchgesellschaft, 1969.

Bafia, Stanislaw. "Teoria nauki w pismach Teodoryka z Chartres i jego szkoly." *Analecta Cracoviensa* 21–22 (1989–1990), 9–28.

Baldassarri, Guido. "Ancora sulle fonti della *Rettorica*: Brunetto Latini e Teodorico di Chartres." *Studi e problemi di critica testuale* 19 (1979), 41–46.

Bataillon, Louis J. "Bulletin d'histoire des doctrines médiévales, IV: Le douzième siècle; V: Le début du treizième siècle." *Revue des sciences philosophiques et théologiques* 46 (1962), 499–542.

Beaujouan, Guy. "L'enseignement du 'quadrivium.'" In *La scuola nell'Occidente latino dell'alto Medioevo*, Settimane di studio del centro italiano di studi sull'alto Medioevo 19 (Spoleto, 15–21 aprile 1971), pp. 639–667. Spoleto: Centro italiano di studi sull'alto Medioevo, 1972.

—. "The Transformation of the *Quadrivium*." In *Renaissance and Renewal in the Twelfth Century*, ed. Robert L. Benson and Giles Constable, pp. 463–487. Cambridge, MA, 1982. Repr. Toronto: University of Toronto Press, 1991.

—. *Par raison de nombres: L'art du calcul et les savoirs médiévaux*. Aldershot: Ashgate, 1991.

—. "Le *quadrivium* et la Faculté des arts." In *L'enseignement des disciplines à la Faculté des arts (Paris et Oxford XIIIe–XVe siècles)*, ed. Olga Weijers and Louis Holtz, pp. 185–194. Turnhout: Brepols, 1997.

Beierwaltes, Werner. "Unità e Uguaglianza: Una formulazione del problema nel platonismo di Chartres e la sua recezione attraverso Nicola Cusano." In his *Pensare l'Uno: Studi sulla filosofia neoplatonica e sulla storia dei suoi influssi*, pp. 315–328. Milan: Vita e Pensiero, 1991 [Italian translation of Werner Beierwaltes, *Denken des Einen: Studien zum Neuplatonismus und dessen Wirkungsgeschichte* (Frankfurt: Klostermann, 1985)].

Bernhard, Michael. "Glossen zur Arithmetik des Boethius." In *Scire litteras: Forschungen zum mittelalterlichen Geistesleben*, ed. Sigrid Krämer and Michael Bernhard, pp. 23–34. Munich: Bayerische Akademie der Wissenschaften, Philosophisch-Historische Klasse, 1988.

Bertola, Ermenegildo. "Il *De Trinitate* dello pseudo Beda." *Rivista di filosofia neo-scolastica* 48 (1956), 316–333.

Betton, Efrem. "Avicebron è l'unica fonte dell'ilemorfismo universale?" In *Actas del V Congreso internacional de filosofía medieval*, 1: 619–629. 2 vols. Madrid: Editora nacional, 1979.

Bischoff, Bernhard. "Ostertagtexte und Intervalltafeln." In *Mittelalterliche Studien: Ausgewählte Aufsätze zur Schriftkunde und Literaturgeschichte*, 2: 192–226. Stuttgart: Hiersemann, 1967.

Bloch, David. *John of Salisbury on Aristotelian Science*. Turnhout: Brepols, 2012.

Borrelli, Arianna. *Aspects of the Astrolabe: "Architectonica Ratio" in Tenth- and Eleventh-Century Europe*. Stuttgart: Franz Steiner, 2008.

Borst, Arno. "Rithmimachie und Musiktheorie." In *Geschichte der Musiktheorie*, Vol. 3: *Rezeption des antiken Fachs im Mittelalter*, ed. Michael Bernhard, Arno Borst, Detlef Illmer, Albrecht Riethmüller and Klaus-Jürgen Sachs, pp. 253–286. Darmstadt: Wissenschaftliche Buchgesellschaft, 1990.

—. *Die karolingische Kalenderreform*. Hannover: Hahn, 1998.

—. *Der karolingische Reichskalender und seine Überlieferung bis ins 12. Jahrhundert*. 3 vols. Hannover: Hahn, 2001.

Bower, Calvin M. "The Role of Boethius' *De institutione musica* in the Speculative Tradition of Western Musical Thought." In *Boethius and the*

Liberal Arts: A Collection of Essays, ed. Michael Masi, pp. 157–174. Bern: Peter Lang, 1981.

Branner, Robert. *Chartres Cathedral.* London: Thames and Hudson, 1969.

Brunner, Fernard. "*Creatio numerorum, rerum est creatio.*" In *Mélanges offerts à René Crozet*, ed. Pierre Gallais and Yves-Jean Riou, 2: 719–725. 2 vols. Poitiers: Société d'Etudes médiévales, 1966.

—. "*Deus forma essendi.*" In *Entretiens sur la renaissance du 12e siècle*, 1965: Cerisy-la-Salle, ed. Maurice de Gandillac and Edouard Jeauneau, pp. 85–102. Paris: Mouton, 1968.

Burnett, Charles. "The Contents and Affiliation of the Scientific Manuscripts Written at, or Brought to, Chartres in the Time of John of Salisbury." In *The World of John of Salisbury*, ed. Michael Wilks, pp. 127–160. Oxford: Blackwell, 1984.

—. "Adelard, Music and the *Quadrivium*." In *Adelard of Bath: An English Scientist and Arabist of the Early Twelfth Century*, ed. Charles Burnett, pp. 69–86. London: Warburg Institute, 1987.

—. "Scientific Speculations." In *A History of Twelfth-Century Western Philosophy*, ed. Peter Dronke, pp. 151–176. Cambridge: Cambridge University Press, 1988.

—. "A New Source for Dominicus Gundissalinus's Account of the Science of the Stars?" *Annals of Science* 47 (1990), 361–374.

—. Review of: Bernard of Chartres, *Glosae super super Platonem*, ed. Paul E. Dutton, (Toronto, 1991) in *The Journal of Medieval Latin* 3 (1993), 234–235.

—. "Advertising the New Science of the Stars (*circa* 1120–50)." In *Le XIIe siècle: Mutations et renouveau en France dans la première moitié du XIIe siècle*, ed. Françoise Gasparri, pp. 147–157. Paris: Léopard d'or, 1994.

—. "La réception des mathématiques, de l'astronomie et de l'astrologie arabes à Chartres." In *Aristote, l'Ecole de Chartres et la cathédrale*, Actes du colloque européen des 5 et 6 juillet 1997, pp. 101–107. Chartres: AACMEC, 1997.

—. "The Instruments which are the Proper Delights of the *Quadrivium*: Rhythmomachy and Chess in the Teaching of Arithmetic in Twelfth-Century England." *Viator* 28 (1997), 175–201.

—. "King Ptolemy and Alchandreus the Philosopher: The Earliest Texts on the Astrolabe and Arabic Astrology at Fleury, Micy and Chartres." *Annals of Science* 55 (1998), 329–368.

—. "The Blend of Latin and Arabic Sources in the Metaphysics of Adelard of Bath, Hermann of Carinthia, and Gundisalvus." In *Metaphysics in the Twelfth Century: On the Relationship among Philosophy, Science and*

Theology, ed. Matthias Lutz-Bachmann, Alexander Fidora, and Andreas Niederberger, pp. 41–65. Turnhout: Brepols, 2004.

Caiazzo, Irene. "La discussione sull'*Anima mundi* nel secolo XII." *Studi Filosofici* 16 (1993), 27–62.

—. "L'âme du monde: un thème privilégié des auteurs chartrains au XIIe siècle." *Le temps de Fulbert: Actes de l'Université d'été du 8 au 10 juillet 1996*, pp. 79–89. Chartres: AACMEC, 1996.

—. "Un commento altomedievale al *De arithmetica* di Boezio." *Archivum Latinitatis Medii Aevi* 58 (2000), 113–150.

—. "Manegold, *magister modernorum magistrorum*." In *Arts du langage et théologie aux confins des XIe–XIIe siècles: Textes, maîtres, débats*, ed. Irène Rosier-Catach, pp. 317–349. Turnhout: Brepols, 2011.

—. "The Four Elements in the Work of William of Conches." In *Guillaume de Conches: Philosophie et Science au XIIe siècle*, ed. Barbara Obrist and Irene Caiazzo, pp. 3–66. Florence: SISMEL–Edizioni del Galluzzo, 2011.

—. "La forme et les qualités des éléments: Lectures médiévales du *Timée*." In *Il Timeo*: *Esegesi greche, arabe, latine*, ed. Francesco Celia and Angela Ulacco, pp. 307–345. Pisa: PLUS-Pisa University Press, 2012.

—. "Sur l'astrolabe et le comput au XIIe siècle: nouveaux textes inédits." *Sudhoffs Archiv: Zeitschrift für Wissenschaftsgeschichte* 96 (2012), 28–38.

Caiazzo, Irene, and Charles Burnett. "New Manuscripts of *On the astrolabe* by Raymond of Marseilles." *Scriptorium* 65 (2011), 338–349.

Castan, Auguste. "La bibliothèque de l'abbaye de Saint-Claude du Jura. Esquisse de son histoire." *Bibliothèque de l'Ecole des Chartes* 50 (1889), 301–354.

Catalogue général des manuscrits des bibliothèques publiques de France: Départements, Vol. 11: *Chartres*. Paris: Plon-Nourrit et Cie, 1890.

Catalogue général des manuscrits des bibliothèques publiques de France. Départements, Vol. 39: *Reims*. Paris: Plon-Nourrit et Cie, 1904–1906.

Chance, Jane. *Medieval Mythography: From Roman North Africa to the School of Chartres, A.D. 433–1177*. Gainesville: University Press of Florida, 1994.

Châtillon, Jean. "Les écoles de Chartres et de Saint-Victor." In *La scuola nell'Occidente latino dell'alto Medioevo*, Settimane di studio del centro italiano di studi sull'alto Medioevo 19 (Spoleto, 15–21 aprile 1971), pp. 795–839. Spoleto: Centro italiano di studi sull'alto Medioevo, 1972.

Chenu, Marie-Dominique. "Platon à Citeaux." *AHDLMA* 29 (1954), 99–106.

—. "*Involucrum*: Le mythe selon les théologiens médiévaux." *AHDLMA* 30 (1955), 75–79.

—. "Une définition pythagoricienne de la vérité au Moyen Age." *AHDLMA* 36 (1961), 7–13.

—. *La théologie au douzième siècle*. Paris: J. Vrin, 1957.

Clerval, Alexandre. *L'enseignement des arts libéraux à Chartres et à Paris dans la première moitié du XIIe siècle d'après l'Heptateuchon de Thierry de Chartres.* Mémoire présenté au Congrès scientifique international des Catholiques tenu à Paris en 1888. Paris: Les Annales de philosophie chrétienne, 1889.

—. *Les écoles de Chartres au Moyen Age, du Ve au XVIe siècle.* Chartres: Librairie R. Selleret, 1895.

Connochie-Bourgne, Chantal. "Leçon de chose: Thierry de Chartres physicien et poète." In *Par les mots et les textes ... : Mélanges de langue, de littérature et d'histoire des sciences médiévales offerts à Claude Thomasset,* ed. Danielle Jacquart, Danièle James-Raoul and Olivier Soutet, pp. 177–190. Paris: PUPS, 2005.

Copeland, Rita. "Thierry of Chartres and the Causes of Rhetoric: From the *Heptateuchon* to Teaching the *Ars Rhetorica.*" In *The Classics in the Medieval and Renaissance Classroom: The Role of Ancient Texts in the Arts Curriculum as Revealed by Surviving Manuscripts and Early Printed Books,* ed. Juanita Feros Ruys, John O. Ward, and Melanie Heyworth, pp. 81–102. Turnhout: Brepols, 2013.

Counet, Jean-Michel. "Les mathématiques au service d'une théologie de la matière chez Thierry de Chartres et Clarembaud d'Arras." In *Vie spéculative, vie méditative et travail manuel à Chartres au XIIe siècle (autour de Thierry de Chartres et des introducteurs de l'étude des arts mécaniques auprès du 'quadrivium'),* Actes du colloque international des 4 et 5 juillet 1998, pp. 103–114. Chartres: AACMEC, 1998.

—. *Mathématiques et dialectique chez Nicolas de Cuse.* Paris: J. Vrin, 2000.

Dales, Richard C. "The Use of Thierry of Chartres' *Hexaemeron* by Anonymi *De elementis* and Robert Grosseteste." *Journal of the Rocky Mountain Medieval and Renaissance Association* 1 (1980), 11–20.

Dickey, Mary. "Some Commentaries on the *De inventione* and *Ad Herennium* of the Eleventh and Early Twelfth Centuries." *Mediaeval and Renaissance Studies* 6 (1968), 1–41.

Doucet, Dominique. "*De ideis*: Eclipse ou dissémination? Les *Lectiones in Boethium De Trinitate* attribuées à Thierry de Chartres (II, 35–37)." *Revue thomiste* 103 (2003), 363–384.

Dreyer, Mechthild. *Razionalità scientifica e teologia nei secoli XI e XII.* Milan: Jaca Book, 2001.

Dronke, Peter. "New Approaches to the School of Chartres." *Anuario de estudios medievales* 6 (1969), 117–140.

—. "Thierry of Chartres." In *A History of Twelfth-Century Western Philosophy,* ed. Peter Dronke, pp. 358–385. Cambridge: Cambridge University Press, 1988.

—. "*Hermes* and the *Sibyls*: Continuations and Creations." In *Intellectuals and Poets in Medieval Europe*, pp. 219–244. Rome: Edizioni di Storia e Letteratura, 1992.

—. *The Spell of Calcidius: Platonic Concepts and Images in the Medieval West.* Florence: SISMEL–Edizioni del Galluzzo, 2008.

Duhem, Pierre. "Thierry de Chartres et Nicolas de Cuse." *Revue des sciences philosophiques et théologiques* 2 (1909), 525–531.

Dutton, Paul E. "*Filiolitas:* The Short History of One of Eriugena's Inventions." *American Catholic Philosophical Quarterly* 79 (2005), 549–566.

Evans, Gillian R. "The Influence of *Quadrivium* Studies in the Eleventh- and Twelfth-Century Schools." *Journal of Medieval History* 1 (1975), 151–164.

—. "*Alteritas:* Sources of the Notion of 'Otherness' in Twelfth-Century Commentaries on Boethius' *Opuscula Sacra.*" *Archivum Latinitatis Medii Aevi* 40 (1975–1976), 103–113.

—. "A Commentary on Boethius's *Arithmetica* of the Twelfth or Thirteenth Century." *Annals of Science* 35 (1978), 131–141.

—. "Introductions to Boethius's *Arithmetica* of the Tenth to the Fourteenth Century." *History of Science* 16 (1978), 22–41.

—. "The *Saltus Gerberti:* The Problem of the 'Leap'." *Janus* 67 (1980), 261–268.

—. "Boethian and Euclidean Axiomatic Method in the Theology of the Later Twelfth Century." *Archives internationales d'histoire des sciences* 30 (1980), 36–52.

—. "The Uncompleted *Heptateuchon* of Thierry of Chartres." *History of Universities* 3 (1983), 1–13.

—. "John of Salisbury and Boethius on Arithmetic." In *The World of John of Salisbury*, ed. Michael Wilks, pp. 161–167. Oxford: Blackwell, 1984.

—. *Philosophy and Theology in the Middle Ages.* London and New York: Routledge, 1993.

Fidora, Alexander. "La metodología de las ciencias según Boecio: Su recepción en las obras y traducciones de Domingo Gundisalvo." *Revista española de filosofía medieval* 7 (2000), 127–136.

—. *Die Wissenschaftstheorie des Dominicus Gundissalinus: Voraussetzungen und Konsequenzen des zweiten Anfangs der aristotelischen Philosophie im 12. Jahrhundert.* Berlin: Akademie Verlag 2003.

—. "Zum Verhältnis von Metaphysik und Theologie bei Dominicus Gundissalinus." In *Metaphysics in the Twelfth Century: On the Relationship among Philosophy, Science and Theology*, ed. Matthias Lutz-Bachmann, Alexander Fidora, and Andreas Niederberger, pp. 67–84. Turnhout: Brepols, 2004.

—. "Le débat sur la création: Guillaume de Conches, maître de Dominique Gundisalvi?" In *Guillaume de Conches: Philosophie et science au*

XIIe siècle, ed. Barbara Obrist and Irene Caiazzo, pp. 271–288. Florence: SISMEL–Edizioni del Galluzzo, 2011.

Fingernagel, Andreas. "Der Romanische Bucheinband." In *Romanik*, ed. Andreas Fingernagel, 1: 355–408. 2 vols. Graz: Akademische Druck- u. Verlagsanstalt, 2007.

Folkerts, Menso. "Die *Rithmachia* des Werinher von Tegernsee." In *Vestigia mathematica: Studies in Medieval and Early Modern Mathematics in Honour of H.L.L. Busard*, ed. Menso Folkerts and Jan P. Hogendijk, pp. 107–142. Amsterdam: Rodopi, 1993.

—. "La rithmomachie et le manuscrit Avranches 235." In *Science antique, science médiévale (Autour d'Avranches 235)*, Actes du Colloque International (Mont-Saint-Michel, 4–7 septembre 1998), ed. Louis Callebat and Olivier Desbordes, pp. 347–357. Hildesheim and New York: Olms–Weidemann, 2000.

—. "Remarks on Mathematical Terminology in Medieval Latin: Greek and Arabic Influences." *Archivum Latinitatis Medii Aevi* 63 (2005), 149–160.

Fredborg, Karin M. "The Commentary of Thierry of Chartres on Cicero's *De inventione*." *Cahiers de l'Institut du Moyen Age grec et latin de Copenhague* 7 (1971), 1–36.

—. "The Dependence of Petrus Helias' *Summa super Priscianum* on William of Conches' *Glose super Priscianum*." *Cahiers de l'Institut du Moyen Age grec et latin de Copenhague* 11 (1973), 1–57.

—. "*Tractatus Glosarum Prisciani* in MS Vat. lat. 1486." *Cahiers de l'Institut du Moyen Age grec et latin de Copenhague* 21 (1977), 27–44.

—. "Some Notes on the Grammar of William of Conches." *Cahiers de l'Institut du Moyen Age grec et latin de Copenhague* 37 (1981), 21–41.

—. "Petrus Helias's *Summa* on Cicero's *De inventione*." *Traditio* 64 (2009), 139–182.

—. "The Grammar and Rhetoric Offered to John of Salisbury." In *The Classics in the Medieval and Renaissance Classroom: The Role of Ancient Texts in the Arts Curriculum as Revealed by Surviving Manuscripts and Early Printed Books*, ed. Juanita Feros Ruys, John O. Ward, and Melanie Heyworth, pp. 103–130. Turnhout: Brepols, 2013.

Fuehrer, Markus L. "*Cusanus Platonicus:* References to the Term 'Platonici' in Nicholas of Cues." In *The Platonic Tradition in the Middle Ages: A Doxographic Approach*, ed. Stephen Gersh and Maarten J.F.M. Hoenen, pp. 345–371. Berlin: Walter de Gruyter, 2002.

Garfagnini, Gian Carlo. "*Ratio disserendi* e *ratiocinandi via*: il *Metalogicon* di Giovanni di Salisbury," *Studi medievali*, 3rd ser., 12 (1971), 915–954.

Garin, Eugenio. *Studi sul platonismo medievale*. Florence: Le Monnier, 1958.

Gersh, Stephen. "Platonism, Neoplatonism, Aristotelianism: A Twelfth Century Metaphysical System and Its Sources." In *Renaissance and Renewal in the Twelfth* Century, ed. Robert L. Benson and Giles Constable, pp. 512–534. Cambridge, MA, 1982. Repr. Toronto: University of Toronto Press, 1991.

—. *Middle Platonism and Neoplatonism: The Latin Tradition.* 2 vols. Notre Dame: University of Notre Dame Press, 1986.

—. "Calcidius' Theory of First Principles." *Studia Patristica* 18 (1989), 85–92.

—. *Concord in Discourse: Harmonics and Semiotics in Late Classical and Early Medieval Platonism.* Berlin and New York: Walter de Gruyter, 1996.

—. *Reading Plato, Tracing Plato: From Ancient Commentary to Medieval Reception.* Aldershot: Ashgate, 2005.

Giacone, Roberto. "Le arti liberali nella tradizione letteraria del secolo XII." *Atti della Accademia delle Scienze di Torino II: Classe di Scienze Morali, Storiche e Filologiche* 108 (1973–1974), 435–476.

—. "Arti liberali e classificazione delle scienze: L'esempio di Boezio e Cassiodoro." *Aevum* 48 (1974), 58–72.

—. "Masters, Books and Library at Chartres according to the Cartularies of Notre-Dame and Saint-Père." *Vivarium* 12 (1974), 30–51.

Gibson, Margaret. "The *Artes* in the Eleventh Century." In *Arts libéraux et philosophie au Moyen Age,* Actes du quatrième Congrès international de philosophie médiévale (Montréal 27 août–2 septembre 1967), pp. 121–126. Paris: J. Vrin; Montreal: Institut d'études médiévales, 1969.

—. "The Study of the *Timaeus* in the Eleventh and Twelfth Centuries." *Pensamiento* 25 (1969), 183–194.

Giesebrecht, Wilhelm von. *Geschichte der deutschen Kaiserzeit.* 6 vols. in 8. Braunschweig: C.A. Schwetschke und Sohn, 1873–1895.

Gorman, Michael. "The Glosses on Bede's *De temporum ratione* Attributed to Byrhtferth of Ramsey." *Anglo-Saxon England* 25 (1996), 209–232.

Gracia, Jorge J.E. "Thierry of Chartres and the Theory of Individuation." *New Scholasticism* 58 (1984), 1–23.

Gregory, Tullio. *Anima mundi: La filosofia di Guglielmo di Conches e la Scuola di Chartres.* Florence: Sansoni, 1955.

—. *Platonismo medievale: Studi e ricerche.* Rome: Istituto italiano storico per il Medio Evo, 1958.

—. "L'idea di Natura nella filosofia medievale prima dell'ingresso della fisica di Aristotele." In *La Filosofia della Natura nel Medio Evo,* Atti del III Congresso internazionale di Filosofia Medievale (Passo della Mendola, 31 agosto–5 settembre 1964), pp. 27–65. Milan: Vita e Pensiero, 1966.

—. "La nouvelle idée de Nature et de savoir scientifique au XIIe siècle." In *The Cultural Context of Medieval Learning,* Proceedings of the First international Colloquium in Philosophy, Science and Theology in the Middle Ages (September 1973), ed. John E. Murdoch and Edith D. Sylla, pp. 193–218. Dordrecht-Boston: Reidel, 1975.

—. "Le platonisme du XIIe siècle." *Revue des sciences philosophiques et théologiques* 71 (1987), 243–259.

—. "The Platonic Inheritance." In *A History of Twelfth-Century Western Philosophy*, ed. Peter Dronke, pp. 54–80. Cambridge: Cambridge University Press, 1988.

—. "Forme di conoscenza e ideali di sapere nella cultura medievale." In *Mundana Sapientia: Forme di conoscenza nella cultura medievale*, pp. 1–59. Rome: Edizioni di Storia e Letteratura, 1992.

Gross, Charlotte. "Twelfth-Century Concepts of Time: Three Reinterpretations of Augustine's Doctrine of Creation *Simul*." *Journal of History of Philosophy* 23 (1985), 325–338.

Guillaumin, Jean-Yves. "Boethius's *De institutione arithmetica* and Its Influence on Posterity." In *A Companion to Boethius in the Middle Ages*, ed. Noel H. Kaylor, Jr. and Philip E. Phillips, pp. 135–161. Leiden: Brill, 2012.

Hadot, Pierre. "*Forma essendi*: Interprétation philologique et interprétation philosophique d'une formule de Boèce." *Les études classiques* 38 (1970), 143–156.

Hamesse, Jacqueline. "Un nouveau glossaire des néologismes du latin philosophique." In *Aux origines du lexique philosophique européen: L'influence de la "latinitas,"* ed. Jacqueline Hamesse, pp. 237–254. Louvain-la-Neuve: FIDEM, 1997.

Häring, Nikolaus M. "The Creation and Creator of the World according to Thierry of Chartres and Clarembaldus of Arras." *AHDLMA* 30 (1955), 137–216.

—. "A Short Treatise on the Trinity from the School of Thierry of Chartres." *Mediaeval Studies* 18 (1956), 125–134.

—. "A Third Manuscript of Peter Abelard's *Theologia 'Summi boni'* (MS Oxford, Bodl. Lyell 49, fols. 101–128)." *Mediaeval Studies* 18 (1956), 215–224.

—. "Thierry of Chartres and Gundissalinus." *Mediaeval Studies* 26 (1964), 271–286.

—. "Chartres and Paris Revisited." In *Essays in honour of A.C. Pegis*, ed. J. Reginald O' Donnell, pp. 268–329. Toronto: Pontifical Institute of Mediaeval Studies, 1974.

Hauréau, Barthélemy. *Histoire de la philosophie scolastique*. 2 vols. Paris: Durand et Pedone-Lauriel, 1872.

Hoenen, Maarten J. " 'Ista prius inaudita': Eine neuentdeckte Vorlage der *Docta ignorantia* und ihre Bedeutung für die frühe Philosophie des

Nikolaus von Kues." *Medioevo: Rivista di storia della filosofia* 21 (1995), 375–476.

Hunt, Richard W. "The Introductions to the *Artes* in the Twelfth Century." In *Studia mediaevalia in honorem Raymundi Josephi Martin*, pp. 85–112. Bruges: De Tempel, 1948.

—. *The Schools and the Cloister: The Life and Writings of Alexander Nequam (1157–1217)*, ed. and rev. Margaret Gibson. Oxford: Clarendon Press, 1984.

Ilkhani, Mohammad. *La philosophie de la création chez Achard de Saint-Victor*. Bruxelles: Editions Ousia, 1999.

Irtenkauf, Wolfgang. "Der *computus ecclesiasticus* in der Einstimmigkeit des Mittelalters." *Archiv für Musikwissenschaft* 14 (1957), 1–15.

Jacquart, Danielle. "De *crasis* à *complexio*: Note sur le vocabulaire du tempérament en latin médiéval." In *Mémoires V: Textes médicaux latins antiques*, ed. Guy Sabbah, pp. 71–76. Saint-Etienne: Centre Jean Palerne, 1984.

—. "*Theorica* et *practica* dans l'enseignement de la médecine à Salerne au XIIe siècle." In *Vocabulaire des écoles et des méthodes d'enseignement au Moyen Age*, Actes du colloque (Rome, 21–22 octobre 1989), ed. Olga Weijers, pp. 102–110. Turnhout: Brepols, 1992.

—. *La science médicale occidentale entre deux renaissances (XIIe s.–XVe s.)*. Aldershot: Ashgate, 1997.

Jeauneau, Edouard. "Le *Prologus in Eptatheucon* de Thierry de Chartres." *Mediaeval Studies* 16 (1954), 171–175.

—. "Simples notes sur la cosmogonie de Thierry de Chartres." *Sophia* 23 (1955), 172–183.

—. "L'usage de la notion d'*integumentum* à travers les Gloses de Guillaume de Conches." *AHDLMA* 24 (1957), 35–100. Repr. in his *Lectio philosophorum*, pp. 127–192.

—. "Deux rédactions des gloses de Guillaume de Conches sur Priscien." *Recherches de théologie ancienne et médiévale* 27 (1960), 212–247. Repr. in his *Lectio philosophorum*, pp. 335–370.

—. "Gloses de Guillaume de Conches sur Macrobe: Note sur les manuscrits." *AHDLMA* 27 (1960), 17–28.

—. "Macrobe, source du platonisme chartrain." *Studi Medievali*, 3rd ser., 1 (1960), 3–24. Repr. in his *Lectio philosophorum*, pp. 279–300.

—. "Mathématiques et Trinité chez Thierry de Chartres." In *Die Metaphysik im Mittelalter, ihr Ursprung und ihre Bedeutung*, ed. Paul Wilpert, Miscellanea Mediaevalia 2, pp. 289–295. Berlin: Walter de Gruyter, 1963.

—. "Note sur l'Ecole de Chartres." *Studi Medievali*, 3rd ser., 5 (1964), 821–865.

—. "*Nani gigantum humeris insidentes*: Essai d'interprétation de Bernard de Chartres." *Vivarium* 5 (1967), 79–99. Repr. in his *Lectio philosophorum*, pp. 53–73.

—. "Un représentant du platonisme du XIIe siècle: Maître Thierry de Chartres." In his *Lectio philosophorum*, pp. 77–86.

—. *Lectio philosophorum: Recherches sur l'Ecole de Chartres.* Amsterdam: Hakkert, 1973.

—. *L'âge d'or des écoles de Chartres.* Chartres: Houvet, 1995.

—. *Rethinking the School of Chartres.* Trans. Claude P. Desmarais. Toronto: University of Toronto Press, 2009.

Jeudy, Colette. "L'œuvre de Remi d'Auxerre: Etat de la question." In *L'Ecole carolingienne d'Auxerre: De Murethach à Remi 830–908*, ed. Dominique Iogna-Prat, Colette Jeudy, and Guy Lobrichon, pp. 373–397. Paris: Beauchesne, 1991.

—. "*Remigii Autissiodorensis opera (Clavis)*." In *L'Ecole carolingienne d'Auxerre: de Murethach à Remi 830–908*, ed. Dominique Iogna-Prat, Colette Jeudy and Guy Lobrichon, pp. 457–500. Paris: Beauchesne, 1991.

Jeudy, Colette, and Yves-François Riou. *Les manuscrits classiques latins des bibliothèques publiques de France*, Vol. 1: *Agen-Evreux*. Paris: CNRS Editions, 1989.

Jolivet, Jean. "Quelques cas de platonisme grammatical du VIIe au XIIe siècle." In *Mélanges offerts à René Crozet*, ed. Pierre Gallais and Yves-Jean Riou, 1: 93–99. 2 vols. Poitiers: Société d'Etudes médiévales, 1966.

—. "Rapports entre la grammaire et l'ontologie au Moyen Age." In *Sprache und Erkenntnis im Mittelalter*, Akten des VI internationalen Kongresses für Mittelalterliche Philosophie der Société internationale pour l'étude de la philosophie médiévale (Bonn, 29 August – 3 September 1977), ed. Jan P. Beckmann and Wolfgang Kluxen, pp. 135–164. Berlin and New York: Walter de Gruyter, 1981.

—. "Rhétorique et théologie dans une page de Gilbert de Poitiers." In *Gilbert de Poitiers et ses contemporains: Aux origines de la "logica modernorum,"* Actes du septième symposium européen d'histoire de la logique et de la sémantique médiévales (Centre d'études supérieures de civilisation médiévale de Poitiers, Poitiers, 17–22 juin 1985), ed. Jean Jolivet and Alain de Libera, pp. 183–197. Naples: Bibliopolis, 1987.

—. "Platonisme et sémantique, de Bernard de Chartres aux Porretains." In *Vestigia, Imagines, Verba: Semiotics and Logic in Medieval Theological Texts (XIIth- XIVth century)*, Acts of the XIth Symposium on Medieval Logic and Semantics (San Marino, 24–28 May 1994), ed. Costantino Marmo, pp. 9–17. Turnhout: Brepols, 1997.

Juste, David. *Les Alchandreana primitifs: Etude sur les plus anciens traités astrologiques latins d'origine arabe (Xe siècle)*. Leiden: Brill, 2007.

Katzenellenbogen, Adolf. *The Sculptural Programs of Chartres Cathedral: Christ, Mary, Ecclesia*. Baltimore: Johns Hopkins University Press, 1959.

Keats-Rohan, Katharine S. B. "John of Salisbury and Education in Twelfth Century Paris from the Account of his *Metalogicon*." *History of Universities* 6 (1986–1987), 1–45.

—. "The Chronology of John of Salisbury's Studies in France: A Reading of *Metalogicon* II.10." *Studi Medievali*, 3rd ser., 28 (1987), 193–203.

Kibre, Pearl. "The Boethian *De institutione Arithmetica* and the *Quadrivium* in the Thirteenth Century University Milieu at Paris." In *Boethius and the Liberal Arts: A Collection of Essays*, ed. Michael Masi, pp. 67–80. Bern: Peter Lang, 1981.

Kijewska, Agnieszka. "Mathematics as a Preparation for Theology: Boethius, Eriugena, Thierry of Chartres." In *Boèce, ou La chaîne des savoirs*, Actes du Colloque international de la Fondation Singer-Polignac (Paris, 8–12 juin 1999), ed. Alain Galonnier, pp. 625–647. Louvain: Peeters, 2003.

Klibansky, Raymond. *The Continuity of the Platonic Tradition during the Middle Ages*; with a new preface and four supplementary chapters ... . London: Warburg Institute, 1981.

—. "The School of Chartres." In *Twelfth-Century Europe and the Foundations of Modern Society*, Proceedings of Humanities of the University of Wisconsin and Wisconsin Institute for Medieval and Renaissance Studies (12–14 November 1957), ed. Marshall Clagett, Gaines Post, and Robert Reynolds, pp. 3–14. Madison: University of Wisconsin Press, 1961.

—. "Peter Abailard and Bernard of Clairvaux: A Letter by Abailard." *Mediaeval and Renaissance Studies* 5 (1961), 1–27.

Kneepkens, Corneille H. "Grammar and Semantics in the Twelfth Century: Petrus Helias and Gilbert de la Porrée on the Substantive Verb." In *The Winged Chariot: Collected Essays on Plato and Platonism in Honour of L.M. De Rijk*, ed. Maria Kardaun and Joke Spruyt, pp. 237–275. Leiden: Brill, 2000.

—. "Clarembald of Arras and the *Notionistae*." In *La tradition médiévale des Catégories (XIIe–XVe siècles)*, Actes du XIIIe Symposium européen de logique et sémantique médiévales (Avignon, 6–10 juin 2000), ed. Joel Biard and Irène Rosier-Catach, pp. 105–126. Leuven: Peeters, 2003.

Lampe, Kurt. "A Twelfth-Century Text on the Number Nine and Divine Creation: A New Interpretation of Boethian Cosmology?" *Mediaeval Studies* 67 (2005), 1–26.

Lange, Hanne. "Les données mathématiques des traités du XIIe siècle sur la symbolique des nombres." *Cahiers de l'Institut du Moyen Age grec et latin de Copenhague* 32 (1979), 1–128.

Leclercq, Jean. "Textes sur saint Bernard et Gilbert de la Porrée." *Mediaeval Studies* 14 (1952), 107–128.

—. "L'encyclique de saint Bernard en faveur de la croisade." In *Recueil d'études sur saint Bernard et ses écrits*, 4: 227–246. 5 vols. Rome: Edizioni di Storia e Letteratura, 1962–1992.

—. "Pour l'histoire de l'encyclique de saint Bernard sur la croisade." In *Recueil d'études sur saint Bernard et ses écrits*, 4: 247–263. 5 vols. Rome: Edizioni di Storia e Letteratura, 1962–1992.

Lejbowicz, Max. "Les disciplines du *quadrivium*." In *L'enseignement des disciplines à la Faculté des arts (Paris et Oxford XIIIe–XVe siècles)*, ed. Olga Weijers and Louis Holtz, pp. 185–194. Turnhout: Brepols, 1997.

—. "Remarques sur l'anthropologie de Thierry de Chartres: A propos du *Tractatus de sex dierum operibus*, sect. 18–29." In *Vie spéculative, vie méditative et travail manuel à Chartres au XIIe siècle (autour de Thierry de Chartres et des introducteurs de l'étude des arts mécaniques auprès du 'quadrivium')*, Actes du colloque international des 4 et 5 juillet 1998, pp. 51–62. Chartres: AACMEC, 1998.

—. "Le premier témoin scolaire des *Eléments* arabo-latins d'Euclide. Thierry de Chartres et l'*Heptateuchon*." *Revue d'histoire des sciences* 56 (2003), 347–368.

Lendinara, Patrizia. "The *Scholica graecarum glossarum* and Martianus Capella." In *Carolingian Scholarship and Martianus Capella: Ninth-Century Commentary Traditions on 'De nuptiis' in Context*. Ed. Mariken Teeuwen and Sinéad O'Sullivan, pp. 301–361. Turnhout: Brepols, 2011.

Lépinois, Eugène de, and Lucien Merlet. *Cartulaire de Notre-Dame de Chartres*. 3 vols. Chartres: Société archéologique d'Eure-et-Loir, 1862–1865.

Lesieur, Thierry. *Sciences des nombres et spiritualité entre Danube et Meuse (IXe–XIIe siècles)*. Paris: Editions Classiques Garnier, 2009.

Libera, Alain de. "L'onto-théo-logique de Boèce: doctrine des catégories et théorie de la prédication dans le *De Trinitate*." In *Les Catégories et leur histoire*, ed. Otto Bruun and Lorenzo Corti, pp. 175–222. Paris: J. Vrin, 2005.

Lucentini, Paolo. "L'*Asclepius* ermetico nel secolo XII." In *From Athens to Chartres: Neoplatonism and Medieval Thought – Studies in Honour of Edouard Jeauneau*, ed. Haijo J. Westra, pp. 397–420. Leiden: Brill, 1992.

Luscombe, David. "La création du monde chez Thierry de Chartres." In *Vie spéculative, vie méditative et travail manuel à Chartres au XIIe siècle (autour de Thierry de Chartres et des introducteurs de l'étude des arts mécaniques auprès du 'quadrivium')*, Actes du colloque international des 4 et 5 juillet 1998, pp. 79–91. Chartres: AACMEC, 1998.

Maccagnolo, Enzo. *Rerum universitas: Saggio sulla filosofia di Teodorico di Chartres*. Florence: Le Monnier, 1976.

—. *Il divino e il megacosmo: Testi filosofici e scientifici della scuola di Chartres*. Milan: Rusconi, 1980.

—. "Il platonismo nel XII secolo: Teodorico di Chartres." *Rivista di filosofia neo-scolastica* 73 (1981), 283–299.

Mainoldi, Ernesto S. "Le fonti del *De praedestinatione liber* di Giovanni Scoto Eriugena." *Studi Medievali*, 3rd ser., 45 (2004), 651–697.

Maioli, Bruno. *Gilberto Porretano: Dalla grammatica speculativa alla metafisica del concreto*. Rome: Bulzoni, 1979.

Marenbon, John. "Vocalism, Nominalism and the Commentaries on the *Categories* from the Earlier Twelfth Century." *Vivarium* 30 (1992), 51–61.

—. *Boethius*, New York: Oxford University Press, 2003.

—. *Le temps, l'éternité et la prescience de Boèce à Thomas d'Aquin*. Paris: J. Vrin, 2005.

—. "Humanism, Scholasticism and the School of Chartres." *International Journal of the Classical Tradition* 6 (2000), 569–577.

Martello, Concetto. *Fisica della creazione: La cosmologia di Clarembaldo di Arras: Tractatulus super librum Genesis – Testo, traduzione e commento*. Catania: CUECM, 1998.

Masi, Michael. "Boethius' *De institutione arithmetica* in the Context of Medieval Mathematics." In *Atti del congresso internazionale di studi boeziani* (Pavia, 5–8 ottobre 1980), ed. Luca Obertello, pp. 263–272. Rome: Herder, 1981.

—. "The Influence of Boethius' *De Arithmetica* on Late Medieval Mathematics." In *Boethius and the Liberal Arts: A Collection of Essays*, ed. Michael Masi, pp. 81–95. Bern: Peter Lang, 1981.

McGinn, Bernard. "Does the Trinity Add Up? Transcendental Mathematics and Trinitarian Speculation in the Twelfth and Thirteenth Centuries." In *Praise No Less Than Charity: Studies in Honor of M. Chrysogonus Waddell Monk of Gethsemani Abbey*, ed. E. Rozanne Elder, pp. 237–264. Kalamazoo: Cistercian Publications, 2002.

McKeon, Richard. "Rhetoric in the Middle Ages." *Speculum* 17 (1942), 1–32.

—. "Medicine and Philosophy in the Eleventh and Twelfth Centuries: The Problem of Elements." *The Thomist* 24 (1961), 211–56.

Merlet, Lucien, and René Merlet. *Dignitaires de l'Eglise Notre-Dame de Chartres: Listes chronologiques*. Paris: Picard, 1900.

Meyer, Paul. "Notice sur les *Corrogationes Promethei* d'Alexandre Neckam." *Notices et extraits des manuscrits de la Bibliothèque nationale et autres bibliothèques* 35 (1897), 641–682.

Mews, Constant. "In Search of a Name and Its Significance: A Twelfth-Century Anecdote about Thierry and Peter Abaelard." *Traditio* 44 (1988), 171–200.

—. "The Council of Sens (1141): Abelard, Bernard, and the Fear of Social Upheaval." *Speculum* 77 (2002), 342–382.

Meyier, Karel A. de. *Codices Vossiani Latini. Pars I: Codices in folio*. Leiden: Universitaire Pers, 1973.

Michels, G. "Komburg (S.-Nicholas)." In *Dictionnaire d'histoire et de géographie ecclésiastiques,* ed. André Baudrillart et al. (Paris, 1912–), 29: 533–534.

Millás Vallicrosa, José M. *Assaig d'història de les idèes fisiques i matemàtiques a la Catalunya medieval.* Barcelona: Institució Patxot, 1931.

Minio-Paluello, Lorenzo. *Opuscula: The Latin Aristotle.* Amsterdam: Hakkert, 1972.

—. "Note sull'Aristotele Medievale Latino VIII: I *Primi Analitici*: La redazione carnutense usata da Abelardo e la 'Vulgata' con scolii tradotti dal greco." In his *Opuscula,* pp. 229–241.

—. "Note sull'Aristotele Medievale Latino XI: Le due redazioni boeziane (carnutense e fiorentina) e la contaminazione vulgata (ambrosiana) dei *Primi Analitici.*" In his *Opuscula,* pp. 377–383.

—. "The Text of Aristotle's *Topics* and *Elenchi*: The Latin Tradition." In his *Opuscula,* pp. 299–309.

Moreau, Joseph. "*Opifex, id est Creator:* Remarques sur le platonisme de Chartres." *Archiv für Geschichte der Philosophie* 56 (1974), 33–49.

Nauta, Lodi. "The *glosa* as Instrument for the Development of Natural Philosophy: William of Conches' *Commentary on Boethius.*" In *Boethius in the Middle Ages: Latin and Vernacular Traditions of the "Consolatio Philosophiae,"* ed. Marteen J. F.M. Hoenen and Lodi Nauta, pp. 3–39. Leiden: Brill, 1997.

Obertello, Luca. "Le scienze del quadrivio e la cultura medievale." In *Boezio e dintorni: Ricerche sulla cultura altomedievale,* pp. 157–178. Florence: Nardini, 1989.

Onofrio, Giulio d'. *Storia della teologia nel Medioevo*. Casale Monferrato: Piemme, 1996.

Otten, Willemien. "Nature and Scripture: Demise of a Medieval Analogy." *Harvard Theological Review* 88 (1995), 257–284.

Paré, Gérard, Adrien Brunet, and Pierre Tremblay. *La renaissance du XIIe siècle: Les écoles et l'enseignement.* Paris: J. Vrin; Ottawa: Institut d'études médiévales, 1933.

Parent, Jean-Marie. *La doctrine de la Création dans l'Ecole de Chartres*. Paris: J. Vrin; Ottawa: Institut d'études médiévales, 1938.

Payen, Jean-Charles. "L'utopie chez les Chartrains." *Le Moyen Age* 90 (1984), 383–400.

Pingree, David. "The *Preceptum Canonis Ptolomei*." In *Rencontres de cultures dans la philosophie médiévale: Traductions et traducteurs de l'antiquité tardive au XIVe siècle*, Actes du colloque international de Cassino (15–17 juin 1989), ed. Jacqueline Hamesse and Marta Fattori, pp. 355–375. Louvain-la-Neuve and Cassino: Publications de l'Institut d'études médiévales, 1990.

Poole, Reginald L. *Illustrations of the History of Medieval Thought and Learning*. 2nd ed. London: Society for Promoting Christian Knowledge, 1920.

—. "The Masters of the Schools at Paris and Chartres in John of Salisbury's Time." *English Historical Rewiew* 35 (1920), 321–342.

Quain, Edward A. "The Mediaeval *accessus ad auctores*." *Traditio* 3 (1945), 215–264.

Reinhardt, Elisabeth. "La causalidad en la creación según Thierry de Chartres." *Anuario Filosófico* 44 (2011), 53–74.

Riccati, Carlo. *"Processio" et "Explicatio": La doctrine de la création chez Jean Scot et Nicolas de Cues*. Naples: Bibliopolis, 1983.

Ricklin, Thomas. "Plato im zwölften Jahrhundert: Einige Hinweise zu seinem Verschwinden." In *The Platonic Tradition in the Middle Ages: A Doxographic Approach*, ed. Stephen Gersh and Maarten J.F.M. Hoenen, pp. 139–163. Berlin: Walter de Gruyter, 2002.

Riesenhuber, Klaus. "Arithmetic and the Metaphysics of Unity in Thierry of Chartres: On the Philosophy of Nature and Theology in the Twelfth Century." In *Nature in Medieval Thought: Some Approaches East and West*, ed. Chumaru Koyama, pp. 43–73. Leiden: Brill, 2000.

Rodrigues, Vera. "La conception de la philosophie chez Thierry de Chartres." *Mediaevalia: Textos e Estudos* 11–12 (1997), 119–137.

—. "Thierry de Chartres: Profil philosophique d'un maître chartrain." In *Roma, magistra mundi: Itineraria culturae medievalis – Mélanges offerts au Père L.E. Boyle à l'occasion de son 75e anniversaire*, ed. Jacqueline Hamesse, pp. 317–327. Louvain-la-Neuve: FIDEM, 1998.

—. "Thierry de Chartres, lecteur du *De Trinitate* de Boèce." In *Boèce, ou La chaîne des savoirs*, Actes du Colloque international de la Fondation Singer-Polignac (Paris, 8–12 juin 1999), ed. Alain Galonnier, pp. 649–663. Louvain: Peeters, 2003.

—. "*Creatio numerorum*: Nature et rationalité chez Thierry de Chartres." PhD diss. Ecole Pratique des Hautes Etudes, IVème section, 2006.

—. "Pluralité et particularisme ontologique chez Thierry de Chartres." In *Arts du langage et théologie aux confins des XIe–XIIe siècles: Textes, maîtres, débats*, ed. Irène Rosier-Catach, pp. 509–536. Turnhout: Brepols, 2011.

Ronsin, Albert, Raymond Etaix, and André Vernet. *La Bibliothèque Bouhier: Histoire d'une collection formée du XVIe au XVIIIe siècle par une famille de magistrats bourguignons.* Dijon: Académie des Sciences, Arts et Belles Lettres, 1971.

Sauerländer, Willibald. *Das Königsportal in Chartres: Heilsgeschichte und Lebenswirklichkeit.* Frankfurt: Fischer Taschenbuch, 1984.

Silvestre, Hubert. "Le schéma 'moderne' des *accessus*." *Latomus* 16 (1957), 684–689.

Southern, Richard. *Medieval Humanism and Other Essays.* New York: Harper and Row, 1970.

—. *Platonism, Scholastic Method, and the School of Chartres.* The Stenton Lecture 1978. Reading: University of Reading, 1979.

—. "The Schools of Paris and the School of Chartres." In *Renaissance and Renewal in the Twelfth Century*, ed. Robert L. Benson and Giles Constable, pp. 113–137. Cambridge, MA, 1982. Repr. Toronto: University of Toronto Press, 1991.

—. *Scholastic Humanism and the Unification of Europe.* 2 vols. Oxford: Blackwell, 1995–2001.

Spallone, Maddalena. "I percorsi medievali del testo: *accessus*, commentari, florilegi." In *Lo spazio letterario di Roma antica*, Vol. 3: *La ricezione del testo*, ed. Guglielmo Cavallo, Paolo Fedeli, and Andrea Giardina, pp. 387–471. Rome: Salerno, 1990.

Speer, Andreas. "Reception – Mediation – Innovation." In *Bilan et perspectives des études médiévales en Europe*, Actes du premier Congrès européen d'études médiévales (Spoleto, 27–29 mai 1993), ed. Jacqueline Hamesse, pp. 129–149. Louvain-la-Neuve: FIDEM, 1995.

—. *Die entdeckte Natur: Untersuchungen zu Begründungsversuchen einer "scientia naturalis" im 12. Jahrhundert.* Leiden: Brill, 1995.

—. "The Discovery of Nature: The Contribution of the Chartrians to Twelfth-Century Attempts to found a *scientia naturalis*." *Traditio* 52 (1997), 135–151.

—. "*Studium sapientiae*: La philosophie et l'unité du savoir théorique chez Thierry de Chartres." In *Vie spéculative, vie méditative et travail manuel à Chartres au XIIe siècle (autour de Thierry de Chartres et des introducteurs de l'étude des arts mécaniques auprès du 'quadrivium')*, Actes du colloque international des 4 et 5 juillet 1998, pp. 93–102. Chartres: AACMEC, 1998.

—. "The Hidden Heritage: Boethian Metaphysics and Its Medieval Tradition." *Quaestio: Annuario di storia della metafisica* 5 (2005), 163–181.

Stirnemann, Patricia. "Où ont été fabriqués les livres de la Glose ordinaire dans la première moitié du XIIe siècle?" In *Le XIIe siècle: Mutations et renouveau en France dans la première moitié du XIIe siècle*, ed. Françoise Gasparri, pp. 257–301. Paris: Léopard d'or, 1994.

Stock, Brian. *Myth and Science in the Twelfth Century: A Study of Bernard Silvester*. Princeton: Princeton University Press, 1972.

Stollenwerk, Anneliese. *Der Genesiskommentar Thierrys von Chartres und die Thierry von Chartres zugeschriebenen Kommentare zu Boethius "De Trinitate."* Cologne: F. Hansen, 1971.

Toneatto, Lucio. *Codices artis mensoriae: I manoscritti degli antichi opuscoli latini d'agrimensura (V–XIX sec.)*. 3 vols. Spoleto: Centro italiano di studi sull'alto Medioevo, 1994–1995.

Troncarelli, Fabio. "Per una ricerca sui commenti altomedievali al *De Consolatione* di Boezio." In *Miscellanea in Memoria di Giorgio Cencetti*, pp. 363–380. Turin: Bottega d'Erasmo, 1973.

—. *Tradizioni perdute: La "Consolatio philosophiae" nell'alto medioevo*. Padua: Antenore, 1981.

Trottmann, Christian. "*Unitas, aequalitas, conexio*: Alain de Lille dans la tradition des analogies trinitaires arithmétiques." In *Alain de Lille, le docteur universel: Philosophie, théologie et littérature au XIIe siècle*, ed. Jean-Luc Solère, Anca Vasiliu, and Alain Galonnier, pp. 401–427. Turnhout: Brepols, 2005.

Valente, Luisa. "*Illa quae transcendunt generalissima*: Elementi per una storia latina dei termini trascendentali (XII secolo)." *Quaestio: Annuario di storia della metafisica* 5 (2005), 217–239.

—. "*Ens, Unum, Bonum*: Elementi per una storia dei trascendentali in Boezio e nella tradizione boeziana del XII secolo." In *Ad Ingenii Acuitionem: Studies in Honour of Alfonso Maierù*, ed. Stefano Caroti, Ruedi Imbach, Zénon Kaluza, Giorgio Stabile, and Loris Sturlese, pp. 483–545. Louvain-la-Neuve: FIDEM, 2006.

—. "Names That Can Be Said of Everything: Porphyrian Tradition and 'Transcendental' Terms in Twelfth-Century Logic." *Vivarium* 45 (2007), 298–310.

—. *Logique et Théologie: Les écoles parisiennes entre 1150–1220*. Paris: J. Vrin, 2008.

Van de Vyver, André. "Les plus anciennes traductions latines médiévales (Xe–XIe siècles) de traités d'astronomie et d'astrologie." *Osiris* 1 (1936), 658–691.

Verbist, Pieter. *Duelling with the Past: Medieval Authors and the Problem of the Christian Era (c. 990–1135)*. Turnhout: Brepols, 2010.

Verger, Jacques. *La renaissance du XIIe siècle*. Paris: Editions du Cerf, 1996.
Vernet, André. "Une épitaphe inédite de Thierry de Chartres." In *Recueil de travaux offert à M. Clovis Brunel par ses amis, collègues et élèves*, 2: 660–670. 2 vols. Paris: Société de l'Ecole des Chartes, 1955.
—. "Thierry de Chartres, moine cistercien?" In *Vie spéculative, vie méditative et travail manuel à Chartres au XIIe siècle (autour de Thierry de Chartres et des introducteurs de l'étude des arts mécaniques auprès du 'quadrivium')*, Actes du colloque international des 4 et 5 juillet 1998, pp. 11–17. Chartres: AACMEC, 1998.
Ward, John O. "The Date of the Commentary on Cicero's *De inventione* by Thierry of Chartres (*ca.* 1095–1160?) and the Cornifician Attack on the Liberal Arts." *Viator* 3 (1972), 219–273.
—. "From Marginal Gloss to *catena* Commentary: The Eleventh-Century Origins of a Rhetorical Teaching Tradition in the Medieval West." *Parergon* 13 (1996), 109–120.
Webb, Charles C. J. "Note on Books Bequeathed by John of Salisbury to the Cathedral Library of Chartres." *Mediaeval and Renaissance Studies* 1 (1941–1943), 128–129.
Weijers, Olga. "The Chronology of John of Salisbury's Studies in France (*Metalogicon*, II.10)." In *The World of John of Salisbury*, ed. Michael Wilks, pp. 109–116. Oxford: Blackwell, 1984.
Wetherbee, Winthrop. *Platonism and Poetry in the Twelfth Century: The Literary Influence of the School of Chartres*. Princeton: Princeton University Press, 1972.
—. "Philosophy, Cosmology and the Twelfth-Century Renaissance." In *A History of Twelfth-Century Western Philosophy*, ed. Peter Dronke, pp. 21–53. Cambridge: Cambridge University Press, 1988.
White Peden, Alison M. "Boethius in the Medieval *quadrivium*." In *Boethius: His Life, Thought and Influence*, ed. Margaret Gibson, pp. 162–205. Oxford: Blackwell, 1981.
—. "*De semitonio:* Some Medieval Exercises in Arithmetic." *Studi Medievali*, 3rd ser., 35 (1994), 367–403.
Wickersheimer, Ernest. *Les manuscrits latins de médecine du Haut Moyen Age dans les bibliothèques de France*. Paris: Centre national de la recherche scientifique, 1966.
Widmer, Berthe. "Thierry von Chartres, ein Gelehrtenschicksal des 12. Jahrhunderts." *Historische Zeitschrift* 200 (1965), 552–571.
Wittig, Joseph S. "King Alfred's Boethius and Its Latin Sources: A Reconsideration." *Anglo-Saxon England* 11 (1983), 157–198.
—. "The 'Remigian' Glosses on Boethius's *Consolatio Philosophiae* in Context." In *Sources of Wisdom: Old English and Early Medieval Latin Studies*

in Honour of Thomas D. Hill, ed. Charles D. Wright, Frederik M. Biggs, and Thomas N. Hall, pp. 168–200. Toronto: University of Toronto Press, 2007.

Zerbi, Pietro. *"Philosophi" e "Logici": Un ventennio di incontri e scontri – Soissons, Sens, Cluny (1121–1141)*. Milan: Vita e Pensiero, 2002.

Zimmermann, Albert. "Die Kosmogonie des Thierry von Chartres." In *Architectura poetica: Festschrift für Johannes Rathofer zum 65. Geburtstag*, ed. Ulrich Ernst and Bernhard Sowinski, pp. 107–118. Cologne-Vienna: Böhlau, 1990.

Zinner, Ernst. *Verzeichnis der astronomischen Handschriften des deutschen Kulturgebietes*. Munich: Beck, 1925.

Ziomkowski, Robert M. "Science, Theology, and Myth in Medieval Creationism: Cosmogony in the Twelfth Century." PhD diss. Cornell University, 2000.

Index of Manuscripts

Index verborum textus

General Index